Practical
Algorithms
in C++

 CORIOLIS GROUP BOOK

Practical
Algorithms
in C++

Bryan Flamig

John Wiley & Sons, Inc.

New York • Chichester • Brisbane • Toronto • Singapore

Publisher: Katherine Schowalter
Editor: Paul Farrell
Managing Editor: Robert Aronds
Composition: Rob Mauhar, Coriolis Group

Trademarks
Many words in this publication in which the Author and Publisher believe trademark or other proprietary rights may exist have been designated as such by use of Initial Capital Letters. However, in so designating or failing to designate such words, neither the Author nor the Publisher intends to express any judgment on the validity or legal status of any proprietary right that may be claimed in the words.

Library of Congress Cataloging-in-Publication Data

Flamig, Bryan.
  Practical algorithms in C++ / Bryan Flamig.
   p. cm.
  "Coriolis Group book."
  Includes bibliographical references and index.
  ISBN 0-471-00955-5 (paper with disk)
   1. C++ (Computer program language) 2. Computer algorithms.
  I. Title.
  QA76.73.C153F518 1995
  005.13'3--dc20
                       94-33105
                         CIP

Printed in the United States of America

10 9 8 7 6 5 4 3 2 1

Contents

Preface   **xiii**

    How This Book Is Different   xiii
    Who Should Read This Book   xiv
    Why C++?   xiv
    How This Book Is Organized   xv
    What You Need   xvi
    Acknowledgments   xvii
    Contacting the Author   xvii

Chapter 1 Basic Concepts   **1**

    Algorithms: Recipes for Action   1
    From Algorithms to C++: Selection Sorting   2
       An Alternate Selection Sort Algorithm   3
       From English to Pseudo Code   4
       From Pseudo Code to C++   5
       Can We Do Better?   6
    Complexity Functions   6
       Defining the Steps in an Algorithm   8
       Finding Time Complexities   9
    Comparing Algorithms: The Big-Oh Notation   11
       Combining Upper Bounds   14
    Hitting a Brick Wall   15
    Practical Considerations in Algorithm Analysis   18
       Considering the Lost Terms   18
       Obtaining the "Mileage" Estimate of an Algorithm   19
    Practical Design Issues   21
    Algorithm Analysis in This Book   23

Chapter 2 Algorithm Construction in C++   **25**

    Iterative Styles   25
       Exits from the Middle   26
       Gotos Not Always Harmful   30
       Use Gotos with Caution   30
    Recursion   32
       Well-defined Recursion   33
       Another Divide-and-Conquer Example   34
       Implicit Recursion Stacks   36

Reducing Parameter Lists 37
Recursion Removal 38
    Tail Recursion Optimization 38
    Stack-based Iteration 39
    Conversion Example: The Exact-Sum KnapSack Problem 43
    Going Toward More Structure 46
    Is Iteration Worth It? 49
    When to Leave Things Recursive 50
    When Not to Leave Things Recursive 51
Queue-based Iteration 52
Indirect Recursion 52
Template-based Algorithms 53

Chapter 3 Algorithmic Generators **55**

Turning Algorithms Inside Out 55
    Creating Generators 57
    A KnapSack Generator 60
    A Canonical Form for Generators 63
    Multifunction Algorithms 65
Number Generators 65
Generators as Debugging Aids 67
Instruction Generators 70
Generator-Iterator Duality 77
Generators and Data Flow 77
Algorithm Inheritance 78
    Breaking Up Algorithms Laterally 80
    Breaking Up Algorithms Serially 81
Performance Issues 82
More Examples of Generators 82

Chapter 4 Primes, Factors, and Permutations **85**

A Prime Number Generator 85
    Testing for Primes 90
Factor Generators 90
    Testing for Primes Revisited 94
    Factoring Performance 95
Permutation Generators 96
    Using Permutation Generators 100
Combinations 101
    A Combination Generator 102

Chapter 5 Linear Congruential Generators 105

Linear Congruential Sequences   105
   Full-Cycle Generators   106
   A Full-Cycle Generator Class   108
Random Number Sequences   112
The Minimal Standard Generator   114
   Scaled Random Numbers   116
   Local vs Global Randomness   117
Uniform Deviates   118
   Non-Uniform Deviates   119
The Spectral Test   119
Combined Random Number Generators   120
Random Bit Sequences   123
Caveat Emptor   125

Chapter 6 Random Combinations and Permutations 127

Random Combinations   127
   Shuffle Sampling   127
   Selection Sampling   129
   Reservoir Sampling   132
Random Permutations   136
   Piles: Contiguous Random Priority Queues   138
   A Random Permutation Generator   141
   Seeding the Permutation Generator   144
Image Dissolving   145

Chapter 7 Hashing Algorithms 149

Hashing—Random Sampling in Reverse   149
Hash Functions   151
   Numerical Hashing   151
   Hashing Character Strings   153
Collision Resolution Strategies   155
Separate Chaining   155
Open Addressing   161
   Linear Probing   161
   Quadratic Probing   163
   Uniform Probing   164
   Double Hash Probing   164
   Random Probing   166
The OpenHashTable Class   167
   Deletions with Open Addressing   173

Coalesced Chaining	174
File-based Hashing	178
Rehashing	179
Comparison of Techniques	180

Chapter 8 Heaps — **183**

Priority Queues	183
Heaps—Complete Binary Trees	185
The Heap Class	186
Heap Operations	188
Heap Insertion—Bubbling Up	189
Heap Extraction—Trickling Down	190
Replacing Items	192
Deleting Items	193
Building a Heap Bottom Up	194
Using a Heap to Sort	195
Deaps—Double Ended Heaps	197
A Deap Class	199
Deap Insertions	200
Extracting the Minimum Item from a Deap	203
Extracting the Maximum Item from a Deap	205
Replacing and Deleting Items in a Deap	207
Deap Performance	208
Delayed Construction of Heap Elements	208
Indirect Heaps	209

Chapter 9 Basic Sorting Algorithms — **213**

Selection Sort	213
Insertion Sort	215
Shell Sort	216
Heap Sort	220
Quick Sort	222
Partitioning	223
Choosing a Pivot	225
Optimizing Quick Sort	227
Tread Carefully over the Quick Sort Code	230
Using the Built-in Quick Sort Routine	230
The Final Optimized Version of Quick Sort	231
Comparison of the Sorting Algorithms	232
Indirect Sorting	234
Map Sorting	236
Specialty Sorting Algorithms	238

Chapter 10 More Sorting Algorithms — 241

The Merge Sort Algorithm — 241
Simple Merge Sorting for Linked Lists — 246
Balanced 2-Way Merge Sorting — 248
Natural Merge Sorting — 251
External Merge Sorting — 254
    Replacement Selection — 258
Variations on a Theme — 260
    Multiway Merging — 260
    Separate Disks for Temporary Files — 261
    Tuning the I/O Buffer Sizes — 261
Sorting Variable-Length Records — 261
External Random-Access Sorting — 262
    External Quick Sorting — 263
    External Bubble Sorting — 267
Comparison of the External Sorting Algorithms — 271

Chapter 11 Text Searching — 273

Straightforward Searching (SFS) — 274
    Optimizing SFS — 276
    Using Sentinels — 277
Creating a Search Class — 280
Boyer-Moore Searching — 282
    Scan Loop Skipping — 283
    Skipping after the Match Loop — 285
    A Boyer-Moore Search Function — 289
Variations on Boyer-Moore — 291
    The Tuned-Boyer-Moore Algorithm — 292
    The Boyer-Moore-Horspool Algorithm — 292
    The Quick-Search Algorithm — 293
    Least-Frequent Character Optimization — 295
The Knuth-Morris-Pratt Algorithm — 296
Text Searching in Files — 296
Summary — 297

Chapter 12 Shift-OR Text Searching — 299

The Search State Vector — 300
Exact Matching — 300
Optimizing the Shift-OR Algorithm — 304
Character Class Matching — 305
Wild Card Matching — 312
Summary — 317

Chapter 13 Graph Data Structures   **319**

Graph Terminology   320
Representing Graphs in C++   323
Adjacency Matrices   323
   Triangular Adjacency Matrices   328
Adjacency Lists   332
Packed Graphs   336
Toward an Adjacent Edge Iterator Class Hierarchy   338

Chapter 14 Basic Graph Traversals   **341**

Depth-First Traversal   341
   A Depth-First Visitor Class   343
   A Depth-First Iterator   347
Graph Connectivity   351
Connectivity in Digraphs   354
Biconnectivity   355
Strongly Connected Components Revisited   361
Transitive Closure   363
Breadth-First Traversal   365
Topological Sorting   369
Toward More General Traversals   372

Chapter 15 Graph Minimization Algorithms   **373**

Priority-First Traversal   373
A Priority-First Visitor Class for List-Based Graphs   375
Minimum Spanning Trees   379
Single Source Shortest Paths   381
A Priority-First Visitor Class for Array-Based Graphs   383
Prim's Algorithm and Dijkstra's Algorithm   385
Floyd's Algorithm   386
State-Space Searching   389
   Heuristic Searching   392
   Optimal Searching with the A* Algorithm   392
   Solving the Roadway Problem   397
Intractable Minimization Problems   402
Approximate Algorithms   403

Chapter 16 Finite State Machines   **405**

State Transition Diagrams   406
Hard-wired Finite Automata   408

State Transition Matrices 410
List-Based Transition Matrices 415
Hash-Based Transition Matrices 416
Deterministic vs. Non-deterministic Finite Automata 419
The Aho-Corasick Machine 422
   The ACS Class 424
   Building a Deterministic Aho-Corasick Machine 429
   Applications of the Aho-Corasick Machine 432

Bibliography **433**

Index **437**

Preface

This book is about designing practical algorithms in C++. The key word here is *practical*. The practical orientation of this book is due, in no small part, to the fact that I was first trained as a mechanical engineer, and so my tendency is to design things from a engineering point of view, always balancing the ideals of theory with the reality of practice. The engineer in me has always been eager to understand the trade-offs involved in using one approach over another.

This book is part of a two-volume set. The companion volume, *Practical Data Structures in C++* [Flamig 93], is the first of the series, laying a groundwork of tools and techniques to be used in this book. The companion volume is written from the same practical perspective.

In writing these two books, I hope to impart to you some of the interest and knowledge I have gained over the years on the subjects of data structures and algorithms. I hope to show you how elegantly and efficiently these constructs can be implemented in C++.

How This Book Is Different

A quick trip down to your local bookstore will show you that surprisingly few books cover the topic of algorithms from a practical sense, let alone show you how to code the algorithms in C++. This is surprising because algorithms are a fundamental part of the programming process, and C++ is becoming such a popular language.

In the books that do exist on algorithms, the subject matter is covered more from a theoretical viewpoint, with sparse and incomplete coding examples. Many of these books leave a lot of the more interesting and complicated techniques as exercises left to the reader, leaving readers frustrated. This book, in contrast, was designed to cover algorithms with as many complete examples as possible. *There are no explicit exercises left to the reader.* (It is hoped that your imagination will be stimulated to think beyond the book.)

This is not a textbook, but rather a cookbook. You won't have to spend days or weeks trying to figure out how to program the major algorithms described in the book. If you're like me, you want to put the algorithms to use immediately, to see if they are promising for your own applications. Only then do you want to dissect these algorithms to see what makes them tick.

Another unique feature of this book is that I spent a great deal of time looking for practical algorithms that not only stand up to theory but also work well in practice. And I didn't just code up the algorithms the first way that came to mind. Instead, many different designs were implemented and studied to find those that seemed to work the best. The resulting code is near-production quality. It's not production quality only in that I didn't always fill in all of the blanks, and I didn't always cover every error condition that could arise. These were sacrificed in the name of clarity and expediency. There is enough meat here, however, for you to develop production-quality algorithm libraries easily.

The code in this book was crafted specifically with C++ in mind. You won't find warmed-over Pascal or C code in this book, with a few programs changed to use classes. With each algorithm, I spent months trying to determine how C++ could best be put to use. I also explored ways of how the object-oriented nature of C++ could be put to use in the *design* of algorithms themselves, not just their implementations.

Since it's not feasible to show *all* of the code in the manuscript itself (due to space limitations), only the salient features of each algorithm is given. The disk included with this book has everything typed in, ready to run. Many of the programs on disk are testing harnesses that I used while developing the code.

WHO SHOULD READ THIS BOOK

This book is written primarily for intermediate-level C++ programmers. It is assumed that you know the basics of C++. The primary focus is on applying C++ to the design algorithms, not on learning C++ itself. However, the book has been designed to be useful to both those programmers with less C++ experience as well as those at the advanced level. This is accomplished by approaching each subject from the basic level, giving simple techniques, and then progressing to more exotic and advanced techniques.

This trend also occurs at a more global level as well. Each chapter gets progressively more advanced. By the time you reach the last chapters of the book, the focus is less on the code itself and more on the algorithms.

WHY C++?

In my early years of programming, I was somewhat frustrated by the lack of a programming language that was both expressive and efficient enough for every-day use. The languages at hand were either high level but unfortunately too slow, or fast and regrettably too low level. Then came C++, and I found the language I was looking for. C++ is both expressive enough to code algorithms in an elegant way, and yet low level enough to allow efficient programs to be constructed.

Traditionally, most books covering algorithms have used Pascal or some other Algol-like language. Pascal has been heavily favored, due to its reputation as a language for teaching. The problem is that many programmers in industry do not use Pascal, but rather C and C++. That's because standard Pascal lacks many features needed for writing real-world applications. Many dialects of Pascal have been invented to make up for these deficiencies, but these dialects are not standard, and you are tied to specific compiler vendors for their implementations.

C, on the other hand, is powerful enough to be used for a wide range of applications, without needing wildly divergent dialects. But because of C's relative terseness and low-level characteristics, many authors have rejected using it as a language for teaching data structures. However, C++ overcomes many of these difficulties, due to its superior type checking and encapsulation facilities. It does this without losing the efficiency of C. Thus, I feel C++ makes an excellent choice for showcasing the most practical algorithms in current use.

How This Book Is Organized

This book is made up of sixteen chapters, progressing from the basics to more advanced material. The chapters are:

1. *Basic Concepts* looks at the design of algorithms from a general perspective. Central to this discussion is the *big-Oh* notation, which is often used as a yardstick in comparing the performance of algorithms.

2. *Algorithm Construction in C++* discusses the two fundamental methods of constructing algorithms: *iteration* and *recursion*. I discuss when each of these methods are appropriate, and how to convert back and forth between them.

3. *Algorithmic Generators* gives an intriguing look at a new way to think about algorithm design: that of turning algorithms into objects and classes of their own right. The generator concept that's developed here is used throughout the book.

4. *Primes, Factors, and Permutations* takes old algorithms—some of them dating back to the third century B.C.—and gives them new clothes, in the form of C++, and oriented around the generator concept of Chapter 3.

5. *Linear Congruential Generators* is a continuation of Chapter 4, putting primes, factors, and permutations into the use of building generators that produce *linear congruential number sequences*, the most well-known▲ of these being random number generators.

6. *Random Combinations and Permutations* is a continuation of Chapter 5, putting random number generators to use in sampling applications and image dissolve algorithms, among others.

7. *Hashing Algorithms* also builds on the techniques of the previous chapters, discussing techniques that use prime numbers and randomization to build highly efficient search tables known as *hash tables*.

8. *Heaps* discusses how to build contiguous priority queues using data structures known as *heaps*. Also shown is how to build double-ended priority queues with data structures known as *deaps*. Both of these data structures come into use in later chapters.

9. *Basic Sorting Algorithms* covers the most popular sorting algorithms, such as *selection sort, insertion sort, shell sort, heap sort,* and *quick sort.* Comparisons of these sorting routines is also given.

10. *More Sorting Algorithms* is a continuation of Chapter 9, where sorting algorithms such as *merge sort, external quick sort,* and *external bubble sort* are discussed.

11. *Text Searching* shows the most popular algorithms used in searching for patterns in strings of text. The main focus of the chapter is the *Boyer-Moore* algorithm, an exceedingly fast way to search text. Numerous variations of the algorithm are discussed with the focus on peak performance and optimization.

12. *Shift-OR Text Searching* discusses a relatively new algorithm for searching for more complex patterns in text. These patterns can contain "don't care" characters and "wild cards," among others.

13. *Graph Data Structures* lays the groundwork for the rest of the book, discussing the basics of *graphs,* the most general of all data structures, and shows how to implement graphs in C++.

14. *Basic Graph Traversals* is a continuation of the discussion of graphs, and talks about the various methods used to process graphs. Covered are depth-first and breadth-first traversals, along with algorithms than can be spawned directly from these traversals.

15. *Graph Minimization Algorithms* takes the traversals of Chapter 14 one step further, and discusses an even more general method of traversing graphs, *priority-first traversal.* Many interesting algorithms can be developed directly from priority-first traversal, such as finding shortest-paths through networks. The culmination of the chapter is the A* algorithm, a prime example of an artificial intelligence technique.

16. *Finite State Machines* is the final chapter of the book. It gives yet another application of graphs, in the form of *finite state machines,* which are devices that can be used in such diverse applications as parsing, pattern matching, controlling hardware devices, and so forth. We show an example of a finite state machine known as the *Aho-Corasick* machine, a highly practical way of searching for multiple patterns in parallel.

WHAT YOU NEED

The code in this book was written to work on any platform supported by C++, with little that relies on any particular machine or compiler vendor. The C++ 3.0 standard was used as the language base, and you'll need a compiler up to this specification. The code was developed using Borland C++ 3.1.

Since the time the companion volume was released (1993), I had hoped that the implementation of templates in the current crop of compilers would have matured. Unfortunately, that is happening ever so slowly. Templates are still supported only by a relatively few number of vendors, and no compilers that I know of can handle them flawlessly. Since templates are an excellent way of presenting code for a book, I decided to keep using them anyway. If you do have trouble compiling the template code, I provide a text file on disk that shows you how to strip the template syntax out of the code.

The disk that's included with this book is formatted for MS-DOS, and Borland C++ 3.1 make files are supplied. You can take these make files and convert them for your own compiler's use. Unfortunately, there are no standards as far as make files go, so there is no way I could create the make files in a generic way.

ACKNOWLEDGMENTS

I would like to thank Holly Mathews for her support and understanding in seeing me through yet another long book project. Thanks go also to my editor Paul Farrell at Wiley for the seemingly infinite amount of patience my books seem to require. Special thanks go to technical reviewer Michael Mohle for his insights and encouragements in seeing the subject of algorithms presented in a more practical slant, and for providing a "sanity check" of the code and techniques used. Finally, my apologies to Dog Zilla, for once again making him miss many walks while I was writing this book.

CONTACTING THE AUTHOR

If you have questions concerning the contents of this book, you may contact me in writing by either postal mail (Azarona Software, *Practical Algorithms in C++*, P.O. Box 768, Conifer, CO 80433), or via the Internet (73057.3172@compuserve.com).

Bryan Flamig
Conifer, Colorado
October, 1994

O N E

Basic Concepts

An *algorithm* is a procedure or method that describes how to accomplish some task in a finite number of well-defined steps. The word *algorithm* is derived from the name of a ninth-century Persian mathematician, Mohammed al-Khowarizmi, who wrote down procedures for performing basic arithmetic on decimal numbers. But the concept of an algorithm goes back farther, to the time of the Greek mathematician Euclid, between 300 and 400 B.C. At that time, Euclid described what many consider to be the first algorithm, one that finds the greatest common divisor between two numbers. You'll see this algorithm, modernized into the C++ language, in Chapter 5.

ALGORITHMS: RECIPES FOR ACTION

In this modern age, algorithms are usually associated with methods for solving tasks on the computer. However, any well-defined, step-by-step task can be considered an algorithm, whether it involves a computer or not. For instance, the procedure for making chocolate chip cookies is an algorithm in its more general sense, although we are more likely to use the term *recipe*. Taking some ingredients, such as eggs, sugar, flour, and, of course, chocolate chips, the recipe describes, step by step, the procedure used to turn these ingredients into (it is hoped) delicious cookies. Figure 1.1 shows a sample recipe, or algorithm if you will, for making chocolate chip cookies.

There is a difference between the chocolate chip cookie recipe in the abstract and the recipe as written down on a piece of paper. Our cookie recipe might be translated and written down in French, Spanish, or some other language. Still, each one of these written procedures describes the same

1

Figure 1.1 A sample recipe.

Dena's Famous Chocolate Chip Cookies

1 *Mix 2 eggs, 1 cup sugar, 1/2 cup brown sugar, 1 cup Crisco, 1/6 cup flour, and 1 teaspoon of each of the following: baking soda, salt, baking powder, and vanilla.*

2 *Add 4 and 1/2 cups of quick oats. Mix lightly. You want the mixture clumpy.*

3 *Add 12 oz. of chocolate chips. Again, mix lightly.*

4 *Drop into blobs onto a cookie sheet. There's no need to form the blobs.*

5 *Bake at 325 degrees for 12 minutes, or until the cookies are slightly brown around the edges. They'll look raw but they are really done. Yum!*

---

method for baking chocolate chip cookies. We say that the method, in the abstract sense, is the *algorithm* for making cookies. The written down part *implements* or gives an *expression* of the algorithm.

A similar relationship exists in computer programs. An algorithm can be described in an abstract sense, and then a particular implementation of the algorithm can be given, written as a procedure in some computer language, such as C++.

FROM ALGORITHMS TO C++: SELECTION SORTING

Suppose you are dealt five cards of a poker hand, and you wish to determine what kind of hand you have. Most of us would proceed by sorting the hand, ordering the cards first by suit and then within each suit, by the face value of each card. To keep things simple, suppose that all of your cards are in the same suit. Ordering the hand becomes a simple task of sorting the cards by face value. Some card players would use a procedure similar to the one given in Figure 1.2.

The method given in Figure 1.2 is called the *selection sort algorithm* because we are selecting, from the remaining cards to be processed, the next card to be placed in sorted order. Let's see a concrete example. Suppose the cards you are dealt are all in the same suit and have the face values of Ten,

Figure 1.2 A card sorting algorithm.

Selection Sort Algorithm for Cards

1 *Scan the cards, looking for the highest. Make it the leftmost card.*

2 *With the cards that remain, repeat Step 1, placing the second highest card immediately to the right of the highest card. Continue this process until only one card hasn't been processed. Go to Step 2.*

3 *The hand is now sorted.*

Queen, Jack, King, and Ace. We'll abbreviate these as T, Q, J, K, and A, respectively, and represent the hand as

```
[T Q J K A]
```

The first step in the sorting process is to scan all five cards, selecting the highest, and placing that card on the left. We do this for the Ace card, and obtain the new hand

```
[A  T Q J K]
```

Now, considering the remaining cards, we find the King high, and move it just to the right of the Ace card, ending up with

```
[A K  T Q J]
```

This process is repeated, selecting and moving the Queen, and then the Jack, whereupon we end up with a sorted hand:

```
[A K Q  T J]
```

```
[A K Q J  T]
```

```
[A K Q J T]  <-- Hand sorted
```

An Alternate Selection Sort Algorithm

The selection sort algorithm, as given in Figure 1.2, is stated rather loosely. We don't say exactly how the highest card is placed on the left. You probably assumed that the selected card was to be removed from its original location and inserted in front of all the remaining ones, shifting them down, as we did in the example. However, if we were to represent the cards in a computer program as an array of numbers, this process of removal, insertion, and shifting would involve a lot of data movement.

There is another way to do the selection sorting—one that is more suited to efficient computer implementation. We could swap the highest card found with the leftmost of the remaining cards. In that case, our sequence of hands using the swapping technique becomes the following:

```
[T Q J K A] — swap A with T —> [A   Q J K T]
[A   Q J K T] — swap K with Q —> [A K   J Q T]
[A K   J Q T] — swap Q with J —> [A K Q   J T]
[A K Q   J T] — swap J with J —> [A K Q J   T]
[A K Q J   T] — swap T with T —> [A K Q J T]
```

This eliminates a lot of the card movement. Admittedly, the sequence is different from what was obtained with the original selection sorting technique, but the final result still is the same sorted hand. When implemented using arrays in C++, the swapping method for selection sort is more efficient. So, while it's nice to discuss selection sort in the abstract, we can't ignore the consequences of how we decide to implement the algorithm.

From English to Pseudo Code

Stating an algorithm in English is often too vague to allow the algorithm to be implemented. This is due in large part to the fact that English can be very ambiguous. In order to remove some of the ambiguity for programmers, many books present algorithms in what is called *pseudo code*, which is halfway between English and real computer code. Often this pseudo code has ALGOL- or Pascal-like syntax, with a little natural language sprinkled in. For example, Figure 1.3 gives pseudo code for the "swapping version" of the selection sort

Figure 1.3 An example of pseudo code.

```
Selection sort:

let i, j, and m be array indices into an
array A of length N

for each value of i from 1 to N-1
    m <-- i
    for each value of j from i+1 to N
       if A[j] > A[m] then m <-- j
    if m <> i then
       tmp <-- A[i], A[i] <-- A[m], A[m] <-- tmp
```

algorithm, written with arrays in mind and written to perform the sort in descending order.

Your author invented the pseudo code in this figure on the fly, and on purpose, to bring up an important point. The main problem with pseudo code is that because there aren't any standards for it, numerous versions abound. As a result, someone might take the pseudo code and interpret it all wrong. For instance, in our pseudo code, we assumed that arrays are indexed using 1 as the starting index, but C++ uses 0 as the first index. Also, the pseudo code assumes that the reader is familiar with the convention of using indentation to determine where the extents of the *for-loop* and *if-then-else* constructs are. It's easy for the reader to get it wrong, especially in more complicated *if-then-else* expressions.

Perhaps the worst assumption being made is in the last line, where the three assignments are meant to be done in order sequentially, from left to right. A reader might misinterpret that and assume that the assignments are to be done in parallel. Or the converse may happen. Where statements are meant to be accomplished in parallel, a reader might assume that they happen sequentially. Your author encountered such problems when researching algorithms for this book, finding vague and misleading pseudo code in otherwise respectable computer journals.

Since you can't actually run pseudo code, you can never know for sure what the author of that code intended. For this reason, you won't see much pseudo code in this book. Instead, we'll first give the algorithms in high-level terms using the manuscript, and then we'll code the algorithms directly in C++. Sometimes, however, both the English descriptions and the C++ implementations can be so complex that the underlying algorithms are obscured. In such cases, we'll use pseudo code too. That is, we'll choose whatever approach leads to the greatest clarity.

From Pseudo Code to C++

Let's now implement the selection sort algorithm in C++, assuming that we are sorting arrays of integers. For completeness, we've wrapped a complete program around the algorithm.

```cpp
#include <iostream.h>

void SelnSort(int a[], int n)
// Using selection sort, this function sorts the integer
// array 'a', of length n, in descending order.
{
  int i, j, m, tmp;

  for (i = 0; i<n-1; i++) { // For each array position, except last
      m = i; // m will hold index of highest remaining element.
```

```
      for (j = i+1; j<n; j++) { // Select highest of remaining elements.
          if (a[j] > a[m]) m = j;
      }
      if (m != i) { // Swap highest with leftmost of remaining elements.
          tmp = a[i]; a[i] = a[m]; a[m] = tmp;
      }
   }
}

//               'T' 'Q' 'J' 'K' 'A'
int hand[5] = {1,  3,  2,  4,  5};

main()
{
   SelnSort(hand, 5);
   for(int i = 0; i<5; i++) cout << hand[i] << ' ';
   cout << '\n';
   return 0;
}
```

The code for this sorting example can be found on disk in the file *ch1_1.cpp*.

Can We Do Better?

When choosing an algorithm, we usually want to know how efficient it is. The selection sort algorithm just given takes about 0.32 seconds to sort 1,000 integers on your author's machine. But what does this really tell us? The time of 0.32 seconds is valid only for one machine, for one type of array element, for one size of array, and just as important, for only one implementation of the algorithm.

With all of these factors to consider, how can we go about analyzing the effectiveness of the selection sort algorithm? What can we say about the algorithm that is independent of implementation, machine, and input data? What can we say about selection sort in comparison to other sorting algorithms? Are there better sorting algorithms?

In the next section, we'll discuss the standard techniques that are used to analyze the effectiveness of algorithm, and we'll see just how good selection sort is (or isn't).

COMPLEXITY FUNCTIONS

The effectiveness of an algorithm often is expressed using what are known as *complexity functions*. These functions indicate how the time and space requirements of the algorithm vary given different amounts of input. The idea is to express the complexity functions so that as much detail as possible about the implementation and running environment is thrown away. This allows us to compare the essence of the algorithm without confusing the issue as to what machine environment or computer language we're using.

There are two types of complexity functions. The *time complexity* gives a measure of how long the algorithm takes. The *space complexity* gives a measure of how much memory overhead is needed for the algorithm to operate. Both time and space complexities are based on the "size" of the input, n, which reflects in some way how much data is to be processed.

Consider the following array printing algorithm. (It's almost too trivial to be called an algorithm, but let's use it as an example anyway.)

```
int arr[i] = {4, 7, 3, 6, 2, 5, 1};
int n = 7;

for(int i = 0; i<n; i++) cout << arr[i] << ' ';
cout << '\n';
```

Here the size of the input is the number of elements to be printed. Given n elements to process, the time complexity of the algorithm can be written as

$$t(n) = an + b$$

In this equation, a is a constant multiplier that represents the time required to execute each pass through the loop. The constant b represents the setup time required by the algorithm. In this case, the time complexity is linear, but in general it may be some arbitrary function.

You might wonder what we mean by "time" in this example. Is it to be the number of seconds needed by the algorithm? It could be, but remember that the idea is to make the complexity function independent of the actual hardware being used. We want the "time" to be a little more abstract. Often the time is expressed as the number of *steps* taken by the algorithm. What constitutes a step varies depending on what we wish to measure. In our array printing algorithm, it makes sense to define the process of printing out an array element as one step, since that's the main function of the algorithm.

Using a printing step as the unit of time, we can write our time complexity function as

$$t(n) = n$$

since for n array elements, n steps are needed to print the array.

The space complexity of an algorithm involves the amount of extra memory required by the algorithm. In the array printing case, only a constant amount is needed, regardless of the number of elements to be printed. The amount is made up of the space required by the code and a few extra memory locations for loop indices, stack locations, and so on. We could write the space complexity for array printing as

$$s(n) = c$$

where c denotes the constant amount of space needed. Note that the space complexity represents only the overhead needed to make the algorithm work, not the space for the data being processed. We don't include the size of the array being printed, for instance.

Rather than using machine-dependent units such as bytes, the space complexity is given in units of "memory cells." What constitutes a memory cell depends on what it is we are trying to measure. Often it is in units of whatever data was being stored in the first place. For instance, if we're storing integers, a memory cell might be the amount of memory needed to store one integer. Thus, we talk of the space complexity in relation to how many extra integer-size chunks of memory are needed. This allows us easily to calculate the percentage overhead required by the algorithm, since the same units are being used for both the storage of the data being processed and the overhead of the algorithm.

In the case of printing an array, our space complexity is independent of the size of the input, but like time complexity, space complexity can be some arbitrary function. For example, the quick sort algorithm in Chapter 9 has space overhead proportional to $\log n$.

The space complexity of an algorithm is often easy to determine, and usually it turns out to either be constant or linear. In contrast determining the time complexity is often difficult, and the result can be highly non-linear. Thus, the time complexity usually receives the most attention. Because of this, the term *time complexity* often is shortened to just *complexity*. That is, when we speak of the complexity of an algorithm, we're usually talking about the time complexity, unless otherwise stated.

Defining the Steps in an Algorithm

You want to define the steps of an algorithm in terms of the overall goal of the algorithm or those portions that are known to dominate the running time. In any case, the definition of a step should depend in some way on the size of the input if possible; otherwise the complexity function won't mean much. In our array printing algorithm, we defined a step as the act of printing one of the array elements. Certainly, this definition of a step depends on the size of the input—the number of elements to be printed.

For a more complicated example, consider a sorting algorithm. Usually two basic processes are carried out at each sorting step: a comparison, followed by an exchange of items. Often these are lumped together for the purpose of analysis, and the time complexity is couched in terms of how many comparisons are needed to sort n items. If the data exchanges dominate the runtime—perhaps a lot of exchanges take place for each corresponding comparison, or the items are large in size—then you might want to consider a data exchange as the algorithmic step. In some cases, two time complexities are stated for sorting algorithms, one for comparisons and one for data exchanges. You'll see an example of this later.

An algorithmic step can range from a single machine instruction to hundreds of instructions. For example, in comparing two strings during a sort, many instructions may be executed as the strings are compared character by character. This level of detail may impact how you define a step. Consider an algorithm for multiplying the numbers 42 and 17 together. On a machine capable of doing multiplication as a single instruction, the algorithm involves just one step. A less sophisticated machine may not have a multiply instruction and instead just has instructions for adding. For this machine, the multiplication algorithm might be "take the number 42 and add it to the result 17 times." Here, one step becomes an addition, and the number of steps required is a function of the magnitude of the numbers being multiplied.

Finding Time Complexities

In our array printing algorithm, we could determine the complexity by inspection, since the algorithm was so simple. For more complicated cases, a rule of thumb sometimes helps: If the algorithm involves a single loop, chances are it's linear. If it involves two nested loops, chances are it's quadratic. An algorithm with a triply nested loop is probably cubic, and so forth. If the algorithm consists of several (possibly nested) loops that execute in sequence, then the complexity is determined by the nature of the most dominant loop.

This rule of thumb works best when each loop iterates fully n times. If that's not the case, then you must do a more careful analysis. Next, we'll do such an analysis for the selection sort algorithm. As a reminder, here is a sketch of the selection sort algorithm.

```
// Sketch of selection sort.
for (i = 0; i<n-1; i++) {
    for (j = i+1; j<n; j++) {
        // Do comparison here.
    }
    // Do item exchange here.
}
```

Since this algorithm consists of a nested loop, you might guess that selection sort is quadratic in nature. You are half right. It depends on what you define as the algorithmic step. If you choose an item exchange as one step (basically implying that you think it will dominate the runtime), the complexity is actually linear, since there are $n-1$ exchanges needed to sort n items. That is,

$$t(n) = n - 1$$

Suppose you choose a comparison as one step. Note that all the comparisons occur in the inner loop. By working through the loop indices, you can see that the inner loop iterates $n-1$ times during the first iteration of the outer loop, and then $n-2$ times, and so on, down to the last pass, where it executes once. Thus, the time complexity, using a comparison as the step, becomes

$$t(n) = (n-1) + (n-2) + \ldots 1 = \frac{n^2}{2} - \frac{n}{2}$$

In other words, the complexity becomes quadratic. In this case our rule of thumb worked. We have a doubly nested loop and expected a quadratic result. But that won't always be the case. In Chapter 9, you'll see another sorting algorithm, known as *shell sort*, that also involves two nested loops, but its complexity for comparisons is somewhere between being linear and being quadratic. In fact, the shell sort algorithm is interesting because no one has been able to find a precise, analytical formulation for its complexity. Instead, the complexity has been determined empirically, by timing the results of many runs of the algorithm and taking averages.

The complexity of an algorithm also can depend on the type of data you give it. For example, later we'll show you yet another sorting algorithm, *insertion sort*, which involves a doubly nested loop and has a quadratic complexity on average. If, however, the data is already or nearly sorted, the complexity is linear.

Let's take a closer look at a case where you have multiple loops in an algorithm, but they aren't nested, as in

```
for (i=0; i<n1; i++) { // Do something. }
for (j=0; j<n2; j++) { // Do something else. }
```

What is the time complexity for this algorithm? It could be written as

$$t(n) = a_1 n + a_2 n + c$$

where a_1 represents the time factor of the first loop, a_2 the time factor of the second loop, and c the setup time for both loops. This function could be rewritten as

$$t(n) = (a_1 + a_2)n + c = a_3 n + c$$

and you can see we have a linear function, rather than a quadratic, as might have been the case for two nested loops.

Suppose the algorithm uses recursion rather than iteration. How do we determine the complexity then? Basically, you use the same technique by treating simple recursion as a single loop. If the algorithm involves a single, straightforward recursion, it will probably have a linear time complexity. If the algorithm involves one recursion nested inside another, then it will most likely have a quadratic complexity, similar to a nested loop, and so forth. Again, careful analysis is needed to find the precise complexity.

COMPARING ALGORITHMS: THE BIG-OH NOTATION

The purpose of finding the time and space complexities of an algorithm is to be able to compare the algorithm to others. When doing these comparisons, you must be careful to ensure that you use the same definition for *step* and *memory cell* in each complexity function. Even given identical step and memory cell definitions, comparing algorithms can be a tricky process. The complexities are often non-linear, so the size of the input greatly affects the results. Consider, for example, the following time complexity functions from two competing algorithms.

$$t_1(n) = 17n^2 + 42n$$
$$t_2(n) = 200n + 25$$

Take the case when n is 5. The $t_1()$ function yields 635 steps, whereas $t_2()$ yields 1,025 steps. In this case, we would choose the first algorithm since it yields fewer steps. However, suppose n is 15. Here $t_1()$ yields 4,455 steps, whereas $t_2()$ yields 3,025 steps. Now the second algorithm is faster. The break-even point is at approximately 9.45 steps, as illustrated in Figure 1.4. Thus, the fastest algorithm depends on which n we pick. What's a programmer to do?

One method that's used involves comparing the algorithms as n goes to infinity. This is known as *asymptotic analysis*. In this type of analysis, the complexity functions are given as the upper bound of the asymptotic behavior (as n approaches infinity) of the algorithm. To represent these upper bounds, a notation called the *big-Oh* notation is used. Here are the big-Oh notations for our two respective complexity functions:

$$t_1(n) = O(n^2)$$
$$t_2(n) = O(n)$$

Figure 1.4 Comparing two complexity functions.

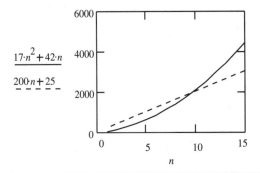

$17 \cdot n^2 + 42 \cdot n$

$200 \cdot n + 25$
- - - -

The first equation is read "t_1 is order n^2", or "t_1 is quadratic." The second equation is read "t_2 is order n", or "t_2 is linear."

Note how our complexities are now couched in terms of whether they are linear or quadratic, and so forth. That's exactly the behavior we wish to know about the functions, because it's what determines how the functions respond as n gets large.

Formally, the big-Oh notation is defined as follows. (Get ready to put your math hat on.)

Definition 1.1. A function $f(n)$ is $O(g(n))$ if and only if there exists two positive constants c and n_0 such that for all $n \geq n_0$, $f(n) \leq cg(n)$.

What does all this mean? Using a little more English, $f(n) = O(g(n))$ means the upper bound of $f(n)$ is $g(n)$. Earlier we stated the order of our two sample complexity functions as

$$t_1(n) = O(n^2)$$
$$t_2(n) = O(n)$$

How do we know this? For the case of $t_1()$, we need to prove that $t(n) \leq cn^2$, for some c and for all $n \geq n_0$. For example, if $c = 31$ and $n_0 = 3$, then

$$17 \times 3^2 + 42 \times 3 \leq 31 \times 3^2$$
$$279 \leq 279$$

The values that we chose for n_0 and c aren't the only ones that could be used. For instance, $c = 20$ and $n_0 = 17$ would work as well. What values we pick isn't really important. What is important is that such values exist. For instance, is $t_1(n)$ order n? In other words, can we find some constant c and some value of n_0 such that, for all values of n greater than or equal to n_0, $t_1(n)$ is less than or equal to cn? The answer is no, if we assume the c and n are positive. In contrast, it's easy to prove that our second complexity function, $t_2(n) = 200n + 25$, is linear. Here we could pick $c = 300$ and $n = 1$.

There is more than one upper bound for any given function. For instance, not only is $t_1(n)$ an $O(n^2)$ function, but it is also $O(n^3)$, $O(n^4)$, and even $O(2^n)$. To see the latter, try the constants $c = 3$ and $n_0 = 10$. Even so, we wouldn't normally treat $t_1(n)$ as being of these higher orders. It's better to pick the smallest upper bound that will work, since this will make comparisons fairer and more meaningful. Hence we would pick $O(n^2)$ as the representative upper bound for $t_1(n) = 17n^2 + 42n$.

The following rule will help you find the smallest upper bound.

Rule 1.1: If $f(n) = a_m n^m + \ldots + a_1 n + a_0$, and $a_m > 0$, then $f(n) = O(n^m)$.

Rule 1.1 means that, to find the smallest upper bound of a polynomial function $f(n)$, you should keep only the term with the highest exponent, throwing away its coefficient as well as all other lower-order terms in the polynomial. Here are some examples.

$$3n^5 + 2n^2 + 1 = O(n^5)$$
$$17n^2 + 42n = O(n^2)$$
$$200n + 25 = O(n)$$
$$55 = O(1)$$

Why, in the last example, do we come up with $O(1)$? The term 55 can be rewritten as $55n^0$, so by Rule 1.1, we get $O(n^0)$, and since $n^0 = 1$, we obtain $O(1)$.

One consequence of Rule 1.1 is that it doesn't makes sense to say, for example, $O(n^2+n)$, or $O(17n^2)$. Both of these really mean $O(n^2)$. By throwing away all constants and lower-order terms, we are simplifying the complexity functions a great deal. This lets us hide the supposedly irrelevant details when making comparisons. We are losing precision in the process, though, which may be misleading. We'll have more to say about this later.

Combining Upper Bounds

In doing algorithm analysis, sometimes it's easier to consider parts of the algorithm separately and then to combine the results. Also, you may want to merge two algorithms together and treat them as one algorithm. The following rules will help you find upper bounds in cases like these.

> **Rule 1.2:** If $f_1(n) = O(g(n))$, and $f_2(n) = O(h(n))$, then
> (a) $f_1(n) + f_2(n) = \max(O(g(n)), O(h(n)))$,
> (b) $f_1(n) * f_2(n) = O(g(n) * h(n))$.

Rule 1.2a covers the case where one algorithm follows sequentially after another. That is, first we do algorithm 1 and then we do algorithm 2. In this case, the algorithm having the highest order (that is, the highest exponent in its complexity function) will dominate the runtime (or space requirements) as n gets larger. Thus, it makes sense to pick the highest order as the overall order for the algorithm. For example, suppose we sort an array using selection sort and then print the results, using code like the following.

```
SelectionSort(arr, n);
for(i = 0; i<n; i++) cout << arr[i] << '\n';
cout << '\n';
```

We know that the selection sort has a complexity of $f_1(n) = O(n^2)$ and the array printing a complexity of $f_2(n) = O(n)$. Using Rule 1.2a, we obtain:

$$f_1(n) + f_2(n) = \max(O(n^2), O(n))$$
$$= O(n^2)$$

This specific application of Rule 1.2a is easy to prove by applying Rule 1.1.

$$f_1(n) + f_2(n) = (a_1 n^2 + b_1 n + c_1) + (b_2 n + c_2)$$
$$= a_1 n^2 + (b_1 + b_2)n + c_1 + c_2$$
$$= O(n^2)$$

In contrast to Rule 1.2a, Rule 1.2b covers the case where one algorithm is nested inside another. Suppose, for example, that we wish to test how well a sorting algorithm works on a nearly sorted array. We might devise a test that takes an already sorted array, compares two elements, and, if they are in order, swaps them around and then calls the sort routine. Suppose we did this at every position in the array. The code might look like:

```
// Assume array arr, of size n, is sorted coming in.
for(i=0; i<n-1; i++) {
   if (arr[i] > arr[i+1]) {    // In descending order, so swap
      Swap(arr[i], arr[i+1]); // to make out of order.
   }
   SelnSort(arr, n); // Assume sort is in descending order.
}
```

This *for* loop has two algorithms combined. One of the "algorithms" compares and swaps adjacent elements in an array. This is like one sorting step, so this algorithm has a complexity of $f_1(n) = c_1 n$, if we're using sorting steps as our units. The second algorithm does a selection sort, which we know to have a complexity function of the form $f_2(n) = c_2 n^2$. Since the selection sort is nested inside the adjacent element swapping algorithm, we combine the complexity functions by multiplying them together and then find the upper bound.

$$f_1(n) * f_2(n) = c_1 n c_2 n^2$$
$$= c_1 c_2 n n^2 = c_3 n^3$$
$$= O(n^3)$$

HITTING A BRICK WALL

The most common complexity functions are given in Table 1.1, which lists results of these functions for a few selected values of n. Figure 1.5 plots these results. From the table and figure, you can see that functions like 2^n, n^3, and even n^2 grow rapidly. For large n, algorithms with these complexities can quickly become impractical. For example, even with a computer that could operate 1 billion algorithmic "steps" per second, it would take over 36 years to complete an algorithm that had a time complexity of 2^n, for the modest value of $n = 60$! It's like we hit a brick wall with such complexities, as illustrated in Figure 1.6.

Table 1.1 Complexity function comparisons

$f(n)$	Description	$n = 1$	$n = 2$	$n = 4$	$n = 8$	$n = 16$	$n = 32$
$O(1)$	Constant	c	c	c	c	c	c
$O(\log n)$	Logarithmic	0	1	2	3	4	5
$O(n)$	Linear	1	2	4	8	16	32
$O(n\log n)$	Logarithmic	0	2	8	24	64	160
$O(n^2)$	Quadratic	1	4	16	64	256	1024
$O(n^3)$	Cubic	1	8	64	512	4096	32768
$O(2^n)$	Exponential	2	4	16	256	65536	4294967296

Figure 1.5 Graph of typical complexity functions.

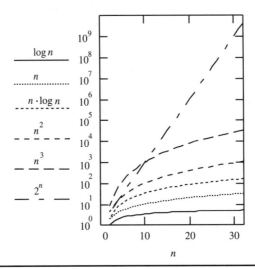

As you can see, when choosing algorithms, it behooves you to find those with complexities such as O(log n), O(n), or O($n$log$n$), as these complexities will work reasonably well even as n gets large. And by the way, not many non-trivial algorithms have a complexity of O(1), although it's too bad they don't! Often you won't even have the luxury of using an O(n) or O($n$log$n$) algorithm; you might find yourself stuck with a problem that seems to have an algorithmic solution that's O(n^2), or O(n^3), or even worse, O(2^n).

Note By *algorithmic solution*, we mean a solution that yields an exact answer using a well-defined step-by-step procedure. With some problems that can't be solved exactly, you might have to back off from hoping for exact solutions and settle for more approximate answers instead. You'll see an example of this in Chapter 15, with the A^* algorithm.

Figure 1.6 Hitting the proverbial brick wall.

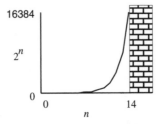

For some problems, no known algorithmic solutions have been found that have complexities less than some exponential form, such as 2^n. Such problems are called *intractable problems*; as you might guess, researchers in computer science have been trying for years to come up with better solutions for these problems. Unfortunately, it seems that no better solutions exist.

As a matter of fact, many hard problems fall into a category known as *NP-complete*. The *NP* stands for *non-deterministic polynomial time*. Such problems have benign polynomial-time algorithms that can solve them, but these algorithms are *non-deterministic*, meaning that they require a magical omnipotent being watching over the proceedings and making exactly the right choices at the right times to yield the correct answer. Of course, building such non-determinism in a machine would be difficult, and we're stuck using *deterministic algorithms*, where no magic is required. Unfortunately, for NP-complete problems, such algorithms cannot run in polynomial-time, and instead have exponential complexities, such as 2^n. Thus, for practical purposes, NP-complete problems are intractable.

The reason for the term *complete* is that, if someone should find a deterministic, polynomial-time algorithm for *any* NP-complete program, then there is a deterministic, polynomial-time algorithm for *all* problems. Of course, this would be highly desirable, for many now intractable problems would become practicable. Unfortunately, it is conjectured that no such solutions exist—but no one has been able to prove that! Proving or disproving this conjecture has turned out to be one of the hardest theoretical problems of computer science, and has yet to be resolved.

Note See [Harel 87] for an excellent discussion about NP-complete problems and algorithmic analysis in general.

In the discussion about complexities, you may have been a little bit confused when we used terms such as $n \log n$. What logarithm base are we talking about anyway? For many algorithms, the logarithms turn out to be base 2. That's due to the binary nature of the algorithms, since they are essentially making decisions involving two choices, or recursively splitting a problem in half. So, unless we say otherwise, assume the logarithm base is 2.

It turns out, though, that if you are using logarithms as part of a big-Oh complexity, then it doesn't matter at all what base you use. If you can recall from your high school algebra (your author couldn't, he had to look it up), a logarithm of some base n can be converted to a logarithm of another base m using the following equation.

$$\log_n x + \frac{1}{\log_m n} \log_m x$$

The term $\log_m n$ is a constant for a given m and n. Thus, the values of $\log_n x$ and $\log_m x$ differ only by a constant. But remember that constant factors are ignored in big-Oh complexities. Thus,

$$O(\log_n x) = O(\log_m x)$$

So you can use any logarithm base you wish when using big-Oh notation.

PRACTICAL CONSIDERATIONS IN ALGORITHM ANALYSIS

The big-Oh notation is useful because it allows us to compare algorithms without getting lost in the details. The notation gives us information on how the algorithms compare for *sufficiently large* n. The reason for considering large n is that we want algorithms that are robust, working reasonably well for both small and large n. That way, we can just plug the algorithm into our application and not worry about it afterward. We'll never have to say "Darn, my customer wants to sort 500,000 records, but my $O(n^2)$ algorithm wouldn't complete that sort for days! I sure wish I would have put in that $O(n\log n)$ algorithm I keep hearing about." (At this point, you may want to stop and compute the complexities for $n = 500,000$, assuming each sorting step takes one microsecond.)

Considering the Lost Terms

While the big-Oh notation is useful, the very reason it's useful also can make it misleading. The big-Oh notation works by ignoring all constants and lower-order terms. Yet these constants and lower-order terms can make a big difference in the actual running time of an algorithm. For example, consider two algorithms with the following time complexities.

$$t_1(n) = 17n$$
$$t_2(n) = 420n$$

Although these are both $O(n)$ algorithms, from an efficiency standpoint, the first algorithm is clearly better, being over *24 times faster* than the second. You'll see this same issue come up in the sorting algorithms in Chapters 9 and 10. Several sorting algorithms have $O(n\log n)$ complexities. However, the *quick sort algorithm* has the tightest inner loop of any, which means it has the smallest constant multiplier. In practice, quick sort is often twice as fast as its nearest competitor.

What can we conclude from all this? The big-Oh notation gives us a valid way to compare algorithms, but only when all other aspects are equal. The constants and lower-order terms must be the same. All things being equal, an O(n) algorithm will outperform an O(n^2) algorithm every time. However, if the step time of the O(n^2) algorithm is much smaller than the step time of the O(n) algorithm, you might actually be better off with the O(n^2) algorithm, at least for small n.

Obtaining the "Mileage" Estimate of an Algorithm

We've spent a lot of time considering how the size of the input affects algorithm performance, but other factors are involved as well. Not only does the size of the input matter, but many algorithms are also sensitive to the *values* of the data being processed. We can see this trait by examining another sorting algorithm, known as *insertion sort*.

```
void InsertionSort(int a[], int n)
// Insertion-sorts the array items in descending order.
{
  int i, j;
  int item_to_sort;

  // For each element of array, (except first):
  for(i = 1; i < n; i++) {
    item_to_sort = a[i];
    j = i;
    // Given item_to_sort, scan to the left until a
    // larger item is found. While scanning, shift
    // elements to the right to make room for the
    // resting place of item_to_sort.
    while(j > 0 && item_to_sort > a[j-1]) {
        a[j] = a[j-1];
        j -= 1;
    }
    a[j] = item_to_sort;
  }
}
```

 The code for this sorting algorithm is given on disk in the file *ch1_2.cpp*.

Let's walk through this algorithm, using a hand of poker cards as an example. Insertion sort works by taking each card, starting with the second, and scanning to the left, inserting the card so that it is ordered with respect to all of the other cards on the left. This is done for each card in the deck. Suppose you're dealt five cards of the same suit in this order:

[Q K T J A]

To sort these cards, you can take the second card, K, and insert it in front of Q, since K is higher than Q.

[K Q T J A]

The first two cards are now in order with respect to each other. Now you can take the third card, T, and scanning the cards to the left, find the spot where T belongs. In this case, with respect to the cards K and Q, the T card is already in the right spot. Thus, you can look next at the fourth card, J. Scanning to the left, you can see that it should come before the T, so you insert the J in front of the T, arriving at

[K Q J T A]

The last card to be inserted into its proper place is A. Scanning to the left, you see that it should come before all the other cards.

[A K Q J T]

At this point, you've arranged all the cards in sorted order.

If you inspect the code for insertion sort, you'll see that it involves a doubly nested loop. Thus, you might suspect that it has quadratic behavior. It does, in general. However, even though the outer loop is guaranteed to iterate n-1 times, the number of iterations in the inner loop will vary significantly depending on the data, and thus the complexity of the algorithm is affected.

Note that the inner loop of insertion sort terminates when the item to be inserted is less than the one immediately preceding it. In the worst case, when the item i is the largest, the inner loop will iterate i-1 times. In the best case, when the item is smaller than all those to the left, the loop won't execute at all. On average, assuming randomly ordered data, the inner loop will execute $i/2$ times. Thus, the average time complexity can be represented as.

$$t(n) = \tfrac{1}{2}(1+2+...n-1) = \tfrac{1}{4}n^2 = O(n^2) \text{ average case}$$

In the worst case, the time complexity also turns out to be $O(n^2)$. In the best cast, however, the inner loop never executes, so the time complexity depends solely on the outer loop, which executes n-1 times. We end up with

$$t(n) = O(n) \text{ best case}$$

Thus, insertion sort, in the best case, is linear. This happens when the array comes in already or nearly sorted.

The nature of this type of analysis is similar to the EPA (Environmental Protection Agency) estimates used to determine the gas consumption of automobiles. The EPA performs standardized mileage tests so that each vehicle is tested under the same kind of conditions. Given identical conditions, a car rated at 25 miles per gallon will use less gas than one that's rated at 20 miles per gallon. The EPA has chosen the categories of "city" and "highway" driving and gives two ratings. The city driving is the worst-case rating, and the highway driving is the average-case rating.

Algorithm analysts use a similar technique by categorizing an algorithm's behavior in terms of *worst, average,* and *best-case* behavior. You just saw where insertion sort is $O(n^2)$ in the worst and average cases, but $O(n)$ in the best case. Selection sort, on the other hand, is $O(n^2)$ for all cases.

The complexities we just gave for insertion sort and selection sort hold when we're using comparisons as the algorithmic step. If, however, a data exchange is the algorithmic step, then insertion sort is $O(n^2)$ for the average and worst cases and $O(1)$ in the best cast. (When the array is already sorted coming in, the inner loop of insertion sort does not execute at all; thus, no data exchanges take place.) For selection sort, the complexity for the average and worst cases is $O(n)$, and its best case complexity is $O(1)$. Thus, which algorithm is better depends both on what your needs are for the worst, average, and best cases and what you've chosen as the algorithmic step.

Generally it's easier to find the worst-case behavior analytically than it is the average case. The set of data that leads to worst-case behavior is often quite evident, and the analytical equations for it may be easy to solve. In contrast, analytically solving for the average-case behavior often is mathematically quite difficult, and it can be hard to pin down what an "average" set of data is.

For insertion and selection sort, we were lucky in that we could easily find the average complexities analytically. Often, though, an algorithm analyst will have to resort to empirical means to find the average complexity. Usually, this is done by running the algorithm many times for random inputs and then averaging the results. An example is the popular shell sort algorithm, covered in Chapter 9. No one really knows precisely what its complexity function is in the average case. Instead, shell sort's complexity has been determined empirically by running numerous tests. (By the way, the complexity is roughly $O(n^{1.5})$).

PRACTICAL DESIGN ISSUES

Choosing an efficient algorithm or data structure is just one part of the design process. Actually, there are three things you should consider in your algorithm design: the need to save

1. Time.
2. Space.
3. Face.

In almost all situations, an algorithm that runs faster is a better algorithm, so saving time is an obvious goal. Likewise, an algorithm that saves space over a competing algorithm is considered desirable. However, even the fastest, most memory-efficient algorithm is no good if it doesn't work. You want to "save face" and not be embarrassed by having your algorithm lock up, or generate reams of garbled data.

These three design goals are often mutually exclusive. The classic programming trade-off involves time versus space. Implementing a faster algorithm generally involves using more memory. There is also the trade-off in speed, memory, and comprehensibility. Generally, the faster or more memory efficient an algorithm is, the "smarter" is has to be. These smarts show up in more complicated code. And the more complicated the code, the harder it is to maintain. Sometimes, however, you can find an algorithm that is fast, memory efficient, and easy to understand. When you find such an algorithm, you know you've hit pay dirt.

Balancing the three design goals makes programming more of an engineering art than an exact science, for a large number of factors—many hard to quantify—come into play. In the companion to this book, *Practical Data Structures in C++* [Flamig 93], we outlined numerous design principles that can help when crafting data structures. These same principles apply for algorithm design. In particular, the following four principles are especially important.

Design Principle 1. *The 80-20 percent rule*. This rule comes in many variations. Here is one variation: Few designs work well in *all* situations, so make your designs work well at least 80 percent of the time. For the remaining 20 percent, the performance may suffer, but as long as your design doesn't totally stop working or produce the wrong results, don't worry about it too much. You've done the best you know how. Of course, if you know a design that works well 100 percent of the time, and that design is cost-effective, by all means, use it!

Design Principle 2. *Get the program working, then optimize it*. As programmers, probably we have all made the mistake of trying to optimize our code even before we know that it works correctly. Also, before optimizing the actual code, ask yourself if there is perhaps a better algorithm. Choosing a good algorithm often can make much more difference than merely optimizing the algorithm you already have. For instance, no matter how clever you are in optimizing a selection sort algorithm, a shell sort or quick sort algorithm will perform much better, without any optimization at all.

Design Principle 3. *Don't overoptimize*. Before putting the "fastest code known to man" into your design, ask yourself what impact that code will have. Suppose a sort routine you are using takes 100 milliseconds to complete, but the overall application takes several minutes. Would replacing the sort routine with one that runs twice as fast make a difference? Of course, if you have already coded the faster sort routine, perhaps for a previous application, by all means, go ahead and reuse it!

Design Principle 4. *The "time-doubling" rule*. For an algorithm to be noticeably faster than a competitor, you want it to be twice as fast. At the very least, you want it to be 50 percent faster. Smaller differences than that are often not noticeable to the user (unless the algorithm is run for days, then even a 10 percent difference is appreciated). Thus, working long and hard to optimize an algorithm to be 10 or 20 percent faster may not be worth it, but, getting it to work two or four times faster certainly is.

It's interesting that this time-doubling effect also appears in other disciplines. For example, in photography, the lens aperture settings (called *stops*) are often spaced so that each setting lets twice as much light through than its predecessor. That's because it takes twice as much exposure for the resulting photograph to be significantly lighter. The audiophonics field has another example. A sound level has to be twice as large to be significantly noticeable to the ear. Because of this, sound levels are often given in units (called *decibels*) that are based on powers of 2.

We've outlined just a few of the many principles in good algorithm design. For an excellent discussion of good programming techniques as they apply to writing "smarter" and "practical" algorithms, see [Bentley 86] and [Bentley 88].

ALGORITHM ANALYSIS IN THIS BOOK

This chapter may have scared you into thinking that we are going to be using a lot of mathematical analysis. You can rest easy, however. The purpose of this chapter is to expose you to how the analysis of algorithms is done and to give you enough information to know what the analysis means. By now, you should know what is meant when we say an algorithm is $O(n^2)$ or $O(n\log n)$.

Most of the algorithms in this book have been around for years and have been analyzed every which way by computer scientists. We'll merely be stating the results of these analyses, so that you can make the proper decisions when choosing one algorithm over another. The focus of this book is in working with the algorithms—in particular, how to apply C++ to its best advantage.

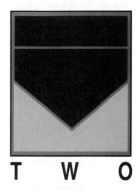

T W O

Algorithm Construction in C++

In this chapter, we take a look at how to construct algorithms using the features of C++. Here we'll focus more on the conventional, procedural style of programming. However, the fact that C++ supports object-oriented programming will not be ignored. In Chapter 3, we'll take a look at how you can use object-oriented programming to design algorithms in new intriguing ways.

All algorithms are made up of a finite series of steps that are supposedly progressing toward some goal. Implied in this process is some kind of looping or sequencing. There are two basic ways to achieve this looping: iteration and recursion. The major portion of this chapter will cover the use of these methods. Then, at the end of the chapter, we'll switch gears and take a look at how we can use *templates*, a relatively new feature of C++, to define algorithms generically, so that they can work for different types of data.

ITERATIVE STYLES

Iteration involves sequencing through a series of statements and then jumping back to the start of the sequence and repeating. This continues until some test indicates that it's time to terminate the process. C++ provides three high-level constructs to implement iteration, as shown in Figure 2.1.

These three looping constructs follow structured programming principles in controlling the ways that loops are exited. The test for exiting is specified either at the beginning of the loop, as in the *for* and *while* loops, or at the end, as in the *do-while* loop.

Figure 2.1 High-level iterative constructs of C++.

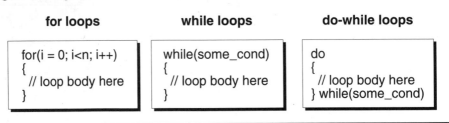

for loops	while loops	do-while loops

```
for(i = 0; i<n; i++)
{
  // loop body here
}
```

```
while(some_cond)
{
  // loop body here
}
```

```
do
{
  // loop body here
} while(some_cond)
```

The rationale behind controlling where the test takes place is that, presumably, it leads to greater clarity. But that's true only when it makes sense for the exit test to be at the beginning or the end. In many algorithms, the exit test belongs more naturally somewhere in the middle. In some instances, you need multiple tests in multiple places. Forcing all the test conditions to one place can lead to awkward code and often involves using auxiliary flags.

Exits from the Middle

As an example of a loop where exiting from the middle is natural, take the *binary search algorithm*. This algorithm is used for searching an array of sorted items in an efficient manner. Suppose the array is sorted in ascending order. The idea is to test the middle item. If it's greater than the target, then repeat, using the half of the array with smaller keys. If the middle item is smaller than the target, try the half of the array with larger keys. As Figure 2.2 illustrates, the array is repeatedly divided into halves until the middle item of the appropriate half matches the target or there are no more items to try.

Figure 2.2 Binary searching.

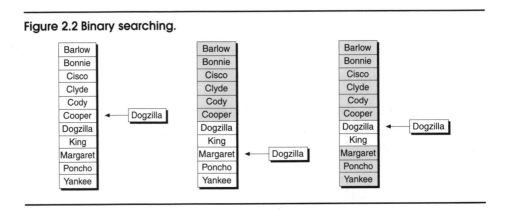

Using the convention of putting all of the exit conditions in one place (at the beginning of the loop, for instance), the following *BinarySearch()* function shows how the binary search algorithm might be implemented:

```
// bs1.cpp: "Structured" binary search algorithm

int BinarySearch(float *arr, int n, float x)
// Searches the sorted array arr of size n for the value x.
// Returns the index of the matching element, or -1 if not found.
// NOTE: The array MUST be sorted in ascending order.
{
  int btm, top, mid, found, match_indx;
  btm = 0; top = n-1; found = 0;

  while(top >= btm && !found) { // Perform the binary search.
    mid = (top + btm) / 2;
    if (arr[mid] == x) { // We have a match.
      match_indx = mid; found = 1;
    }
    else {
      if (arr[mid] < x)
          btm = mid+1;        // Narrow search to upper half
          else top = mid-1; // Narrow search to lower half
    }
  }

  if (found) return match_indx; else return -1;
}
```

 Running programs involving the binary search algorithm can be found on disk in the files *bs1.cpp, bs2.cpp, bs3.cpp, bs4.cpp,* and *bs5.cpp*. These files correspond to the versions of the algorithm given in this chapter, using the names given in the comments at the top of each version.

The *BinarySearch()* function was adapted from an algorithms book that espouses the strict use of structured programming principles. Take note of the use of the auxiliary flag variable *found*, which must be checked along with the test *top >= bottom* to determine when the loop is to terminate. The reason for the *found* flag is that the test for a matching item can only done conveniently only in the middle of the loop. Since the test really occurs there, why not exit the loop right at that point? A C++ programmer could use a *break* statement for that purpose and recode the *BinarySearch()* function as follows:

```
// bs2.cpp: Modified binary search algorithm using a break

int BinarySearch(float *arr, int n, float x)
{
  int btm, top, mid, match_indx;
  btm = 0; top = n-1;
  match_indx = -1; // Default the return value to "not found".
```

```
while(top >= btm) { // Perform the binary search.
  mid = (top + btm) / 2;
  if (arr[mid] == x) { // We have a match.
     match_indx = mid; // Record matching index.
     break;            // Exit loop.
  }
  // Otherwise, narrow the search.
  if (arr[mid] < x) btm = mid+1; else top = mid-1;
}

return match_indx;
}
```

Here we've defaulted *match_indx* to –1, which represents the "not found" case, and then set the variable to the real matching index if a match is found. When a match is found, the loop is exited immediately. Structured programming purists would say that this is bad, because someone reading the code must look in two places to see how the loop might be exited. But is the use of the *found* flag any better? Surely not, for the reader must then check for all the places where the flag is set and modified, and also note the values used. And it's easy for the programmer to forget to initialize the flag, not to mention the fact that the flag is adding extra overhead for no good reason.

It's usually better to use *break* statements (and their companion *continue* statements) than to use a bunch of auxiliary flags. That's the style we'll be using in this book. Sometimes, as in the case of the *BinarySearch()* function, we can even get rid of the *break* statements and simply do a function return instead. For example, here's yet another way to code *BinarySearch()*, which allows us to get rid of both the *found* flag and the *match_indx* variable as well.

```
// bs3.cpp: Another modified binary search algorithm using returns

int BinarySearch(float *arr, int n, float x)
{
  int btm, top, mid;
  btm = 0; top = n-1;

  while(top >= btm) {
    mid = (top + btm) / 2;
    if (arr[mid] == x) return mid; // Match found.
    if (arr[mid] < x) btm = mid+1; else top = mid-1;
  }

  return -1; // No match found
}
```

Again, structured programming purists would say that it is bad to return from a function in more the one place. In some cases that may be true, but in the simple case here, it's no worse than maintaining a potentially buggy auxiliary flag. It's better to choose clarity instead of dogmatically following a style simply for that style's sake. And if possible, it's better to use efficient code as well, as long as clarity doesn't suffer too much in the process.

When initially developing an algorithm, it's best not to worry about placing exits in one common place. Instead, place the exits where they fall naturally. Only after getting the algorithm to work should you see if the exits can be rearranged easily to fit a structured style.

Forcing loops to rigorous structured programming standards at the initial design stage can actually induce errors. For example, in one study involving beginning programming students, a majority failed to code the binary search algorithm correctly when forced to place the loop exit condition at either the beginning or the end of the loop. Had the programmers been allowed to place the exits where they fell naturally, it's likely there would have been fewer mistakes.

For the aforementioned reasons, in this book we'll often write loops with explicit break statements. This author likes to use the following style:

```
while(1) { // Don't even try to test conditions here.
  ...
  if (c) break; // Test them in the middle instead.
  ...
}
```

Here we use an indefinite *while* loop whose formal test condition always evaluates to "true." Then, inside the *while* loop, we perform tests and use *break* statements as needed. If after coding the algorithm, one test condition is found, and that condition falls either at the end or the beginning, we use the corresponding *while* loop or *do-while* loop as appropriate. Also, if it makes sense to use a *for* loop, we will.

With some compilers, using *while(1)* can actually lead to inefficiencies. Some compilers generate code that test the *1* for being true each time through the loop, even though that's entirely unnecessary. (That's mostly true with older compilers. Newer compilers usually optimize this condition.) Thus, another common technique is to use the following *for* loop style.

```
for(;;) { // Most compilers will compile this efficiently.
  ...
  if (c) break;
  ...
}
```

This author likes the *while* loop style better—the syntax is a little less ugly—but feel free to use whatever style you like. Another idea is to use a macro, such as in the following code.

```
#define loop for(;;)

loop {
  ...
```

```
    if (c) break;
    ...
}
```

Gotos Not Always Harmful

Sometimes it's not convenient to use a *break* statement. For example, if you have a nested loop, there's no way to exit the outer loop from within the inner loop using a *break*, since the break can take effect only for one level. Another problem encountered is the need to exit a loop while in a *switch* statement residing in the loop. For example:

```
while(1) {
  ...
  switch(x) {
    ...
    case xxx:
      if (c) goto end_of_loop; // Need to exit loop.
    break;
    ...
  }
  ...
}

end_of_loop:
```

Here we can't use a *break* statement to exit the loop, since inside the *switch* statement, it would exit the switch statement, not the loop. You have three basic ways to solve this problem: One is to use a *return* statement, if you can arrange it. Another is to maintain a flag variable, as mentioned before. That's not really very satisfactory, but sometimes it's the best way. A third way is to use—you guessed it—a *goto*, jumping to a label at the end of the loop, as we show in our example.

Many programmers are taught to think of *gotos* as unforgivably bad, to be avoided no matter what. But is using them really any worse than using flags? The answer is no, as long as you use them within reason. Like almost any programming construct, *gotos* can be abused. If you stylize their use and use *gotos* only when warranted, they are fine. For example, later in the chapter you'll see a technique for turning recursion into iteration. This technique involves using *gotos* that don't always lend themselves to being transformed into a structured programming style.

Use *Gotos* with Caution

Uncontrolled use of *gotos* can wreak havoc to code that uses objects needing constructors and destructors. For example, what happens if you jump past a constructor call for an object and then try to use that object? The following code illustrates the problem.

```
    goto MyLabel;
    Stack stk(42); // Construct a stack object.
MyLabel:
    stk.Push(17);   // Is this safe to use?
```

Obviously, it doesn't make sense to use an object that hasn't been constructed. It's similar to using a integer variable that hasn't been initialized. However, with uninitialized objects, it can be much worse than simply accessing a random value, for the object's constructor might be allocating additional memory, opening files, and so forth. For that reason, it's an error in C++ to jump past the initialization of an object that might be used after the jump.

It's possible for an initialization skip to occur even when doing a backward jump. Consider the following.

```
void Func()
{
    ...
    {
        Stack<int> stk(42); // Stack constructor called.
        ...
BadBoy:
        stk.Push(17);

        ...
        // Stack destructor implicitly called.
    }
    ...
    goto BadBoy;
    ...
}
```

Here we have a nested block with a local stack object. This stack is constructed on entry to the block and destructed on exit. At the point the backward jump to *BadBoy* occurs, the stack is not in a constructed state, yet we try to use it after the jump. Fortunately, the compiler will catch jumps like this and give you a compiler error.

Note, however, that if you don't have local variables with constructors in a block, the compiler *will* allow jumps into the middle of the block. Of course, any sane programmer would avoid doing this, but later in the chapter, you'll see an algorithm transformation process that may lead to such jumps.

A more subtle case happens when you jump backward past an initialization of an object, as in the following code.

```
BackLabel:
    if (some_cond) goto ExitLoop;
    Stack stk(42); // Construct a stack object.
    stk.Push(17);
```

```
    ...
    goto BackLabel; // Causes destructor to be called.
ExitLoop:
```

It turns out that this is legal, but the compiler must ensure that a destructor call takes place before the jump, so that the constructor and destructor calls can balance. Allowing jumps like this does make sense, as you can see if the code is transformed into a *while* loop.

```
while(some_cond) {
    Stack stk(42); // Constructor call
    stk.Push(17);
    ...
    // Implicit destructor call
}
```

Each time through the loop the stack is constructed, and it is destructed before the loop cycles again.

Another scenario to consider is jumping out of a block, as with the following *break* statement inside a *while* loop.

```
while(some_cond) {
    Stack stk(42); // Constructor call
    ...
    if (some_other_cond) goto ExitLoop; // Implicit destructor call
    ...
    // Implicit destructor call
}
ExitLoop:
```

Will the destructor get called properly for the stack? The answer is yes, fortunately. The compiler will see to it that the necessary destructors are called should you jump out of the middle of a block, with a *goto, break,* or *return* statement. Thus, while jumps into a block may cause compiler errors, jumps out of a block won't (unless the jump is into the middle of another block with local variables).

In summary, we are saying that it's okay to use *goto*s in your C++ code if warranted, but you must be very careful when objects are involved, particularly when you are jumping into the middle of a block. Note that, if you use highly structured code, it's difficult, if not impossible, to come up with a scenario where you can jump past the initialization of an object improperly, since you'll never be jumping into the middle of a block. (I'm sure some of you are going to try doing it, just to see if you can.)

RECURSION

Another way to loop through the sequence of steps in an algorithm is to use *recursion.* In C++, recursion can be accomplished by using recursive function calls—that is, by defining functions that call themselves. Recursion is a natural way of implementing algorithms known as *divide-and-conquer* algorithms. The idea behind divide and conquer is to take a task and recursively split it into smaller steps, until you get tasks that are simple enough to be handled directly.

An example of divide and conquer is the binary search algorithm given earlier. We are splitting an array to be searched in half and then searching the appropriate half, recursively splitting that half into two smaller halves, and so on. Using this idea, we can rewrite the binary search algorithm using a recursive function.

```
// bs4.cpp: Recursive binary search algorithm

int BinarySearch(float *arr, int n, float x, int btm, int top)
// Assumes arr is sorted in ascending order.
{
  if(top >= btm) {
    int mid = (top + btm) / 2;
    if (arr[mid] == x) return mid; // Return matching index.
    if (arr[mid] < x) btm = mid+1; else top = mid-1;
    return BinarySearch(arr, n, x, btm, top); // Recurse.
  }
  else return -1; // Signal no match found.
}
```

This algorithm is a direct adaption of the algorithm given earlier in *bs3.cpp*. In essence, the iterative *while* loop is replaced by a recursive function call. To start the recursion, we can supply an additional *driver function*, which takes care of the details of setting up the recursion. In particular, we need to initialize *btm* to 0 and *top* to *n*–1. Because C++ allows function names to be overloaded, we can use the same name, *BinarySearch*, for our driver function. And for utmost speed, we can make this function inline. For example:

```
inline int BinarySearch(float *arr, int n, float x)
// Driver function for recursive binary searching.
{
  return BinarySearch(arr, n, x, 0, n-1); // Start the recursion.
}

float arr[7] = {0, 1, 2, 3, 4, 5, 6};
float x = 3;
int i = BinarySearch(arr, 7, x);
```

Well-defined Recursion

Recursion can be quite unnerving if you are not used to the technique, for it seems you are defining something in terms of itself, in a hopelessly circular fashion. However, a properly designed recursive function isn't circular at all.

The trick is to define the recursion to have the following three important properties.

1. *Base cases.* Somewhere in the recursive function, there must be one or more tests that check for cases of the problem that can be handled without recursion. These cases are known as *base cases.* In our recursive binary search algorithm, we have two base cases: *arr[mid]* == *x* and *top* < *bottom.* With these cases we know that either we have a match, or we don't, and we can return the result immediately.

2. *Making progess.* The other cases of the recursion must be solved by arranging things so that the recursion always makes progress toward one of the base cases. In our example, we manipulate *top* and *btm* to progressively search a smaller portion of the array, until eventually one of the base cases is reached.

3. *Induction.* Many recursive algorithms can be defined using *induction.* This means to assume that the algorithm works for any arbitrary case *n* (with the base cases being axioms and thus proven by definition), and then reason about how to make the algorithm work for the next case *n*+1.

Another Divide-and-Conquer Example

As another example of recursion and the divide-and-conquer technique, consider the task of raising a number to an integral power. That is, given a floating pointer number *x* and an integer *n*, we wish to compute x^n. The obvious method is to multiply *x* to itself *n*–1 times. Thus, it would seem that integer exponentiation is an $O(n)$ algorithm (with multiplication by *x* considered one step). However, using a divide-and-conquer strategy, we can actually create a $O(\log n)$ algorithm, by arranging the calculations so that the number of multiplies is proportional to $\log n$. The strategy is based on the following properties of integer exponentiation.

$$x^0 = 1$$
$$x^1 = x$$
$$x^n = \left(x^2\right)^{\frac{n}{2}}, \text{ for even } n$$
$$x^n = x^{n-1}x, \text{ for odd } n$$

The cases where *n*=0 and *n*=1 represent the base of our recursion. Otherwise, when *n* is even, we can multiply *x* to itself (that is, compute x^2) and then recursively raise the result to the power *n*/2. When *n* is odd, we can first compute x^{n-1} and then multiply the result by *x*. Notice how we progress toward smaller values of n, until eventually one of the base cases is reached. This analysis leads us directly to the following recursive C++ function for computing integral powers.

```cpp
// pow1.cpp: Recursive power function

float Power(float x, int n)
// Raises x to the nth power, returning the result.
{
  if (n == 0) return 1;
  if (n == 1) return x;
  if (n & 1)
     return Power(x, n-1) * x;    // For odd n
     else return Power(x*x, n/2); // For even n
}
```

Running programs involving the algorithm for computing integral powers can be found on disk in the files *pow1.cpp, pow2.cpp,* and *pow3.cpp.* These files correspond to the versions of the algorithm given in this chapter, using the names given in the comments at the top of each version.

To compute x^{42}, the following series of calculations are made by the *Power()* function.

$$x^{42} = \left(x^2\right)^{21}, \quad x^{21} = x^{20}x, \quad x^{20} = \left(x^2\right)^{10}, \quad x^{10} = \left(x^2\right)^5,$$

$$x^5 = x^4x, \quad x^4 = \left(x^2\right)^2, \quad x^2 = \left(x^2\right)^1, \quad x^1 = x$$

From this it can be readily seen that 7 multiplies are needed to compute x^{42} (x^2 can be computed as $x*x$), instead of 41 multiplies as would be required using a naive algorithm. The divide-and-conquer technique can be quite useful indeed.

There is actually a simplification we can make to *Power()* that will get rid of one of the recursive calls. The following observations are used.

$$x^n = x^{n-1}x = x^m x, \text{ for odd } n, \text{ with } m = n - 1$$

$$x^m = \left(x^2\right)^{\frac{m}{2}}$$

$$\frac{n-1}{2} = \frac{n}{2}, \text{ if integer division is used with odd } n$$

From these observations, we can deduce that for odd n, we can use the same squaring of x and halving of n that we use for even n. The only difference is the need to multiply by an extra x. We also don't need to trap for the case *n* = 1, for when it is divided by 2, a 0 will result, which can be handled by the *if* statement. This leads to the following simplified *Power()* function.

```
// pow2.cpp: Simplified recursive power function

float Power(float x, int n)
// Raises x to the nth power, returning the result.
{
   if (n == 0) return 1;
   float v = (n & 1) ? x : 1.0;
   return Power(x*x, n/2) * v;
}
```

Implicit Recursion Stacks

In order for recursive functions to work, a stack must be employed. This stack is used to preserve copies of the local variables of the functions (including parameters), so that when a recursive call returns, the old values of the variables are still intact. The return address for the function is saved on the stack as well.

Thus, while recursion is a natural way to describe some algorithms, it does come with a cost. That cost is storage space in the form of a stack. One problem is that the amount of stack space varies and depends on the depth of the recursion. Often this is not a known quantity, and usually it depends on the input in some way. In the case of our *Power()* function, the stack space required is roughly $O(\log n)$.

When designing recursive functions, you must be aware of how many local variables are being used, and their sizes, since this affects the amount of stack space needed. When the local variables are just a few simple numbers, as with the *Power()* function, the stack space is not usually of concern. However, in C++, we may have local variables that are actually complicated objects. These objects can be fairly large, and may have constructors and destructors as well. An even worse scenario is when the objects come from some deeply nested class hierarchy. Then, when the constructors and destructors are invoked, a whole chain of base class functions are invoked.

For example, we might have a function similiar to the following:

```
void F(int n)
{
   ...
   SomeLargeDerivedObject x(a, b, c, d); // Constructor call
   ...
   F(n-5); // Make a recursive call to F().
   ...
   // Destructor for x called implicitly here.
}
```

At each call to *F()*, including the calls made during recursion, a new copy of *x* is created. During the creation of this copy, the constructor for *x* is called, which in turn might set off a chain reaction of base class constructor calls. Then, upon return from the function *F()*, the destructor for *x* is called, again possibly setting off a chain reaction of base class destructor calls. Recursive functions like this can become very expensive both in terms of space and speed.

REDUCING PARAMETER LISTS

One aspect of recursion that can chew up stack space is the parameter list. At each recursive call, these parameters are pushed onto the stack. If these parameters are large objects, or if there are many of them, stack space can be consumed in a hurry.

The standard way of reducing this overhead is to pass the parameters by reference, either by using pointers or by using reference arguments. We did this in the recursive *BinarySearch()* algorithm (*bs4.cpp*) when passing in the array elements. However, we got away with this only because we didn't need to save the values of the array elements across the function call, because the array was never modified. Pass-by-reference won't work if we need to pre-serve the arguments on the stack during the recursion.

Another technique that's used instead of pass-by-reference involves turn-ing arguments that don't need to be on the stack into global variables. How-ever, you will be "polluting" the global name space with these variables. Another way is to take advantage of C++'s object-oriented features. The secret lies in recognizing that member functions can be recursive just as ordinary functions can. Instead of using global variables or passing parameters that don't really need to be passed, store them in an object and then make the recursive function a member function. For example, if we have the following recursive function

```
void f(int a, int b, int c, int d, int e)
{
  ...
  f(a, b, c, d+1, e-2);
  ..
}
```

where *a*, *b*, and *c* don't really need to be on the stack, we can create a class with a modified form of the function as a member.

```
class MyClass {
private:
  int a, int b, int c;
  ...
public:
  ...
```

```
   void f(int d, int e);
};

void MyClass::f(int d, int e)
{
   ...
   f(d+1, e-2);
   ...
}
```

Now we are explicitly passing only two arguments, not five. Actually, there is a third argument, the *this* pointer, that's hidden. In effect, we are trading three arguments (*a*, *b*, and *c*), for one argument, *this*. We access *a*, *b*, and *c* through the *this* pointer, since they are now data members of the object. Again, remember that the only variables you can remove from the argument list and store in the object are those that don't need to be on the stack. In our example, we still had to pass *d* and *e*.

RECURSION REMOVAL

Besides playing games with the parameter lists, there are other ways to control the amount of stack space used. One way is to use an explicit, user-defined stack, where you control what gets pushed on the stack and turn the function into an iterative form rather than a recursive one. This is in lieu of using the techniques given in the last section. For some recursive functions, it's possible to eliminate the stack altogether and transform the recursion into simple iteration. We'll now take a look at these techniques.

Tail Recursion Optimization

Let's review the simplified recursive *Power()* function.

```
// pow2.cpp: Recursive power function

float Power(float x, int n)
{
   if (n == 0) return 1;
   float v = (n & 1) ? x : 0.0;
   return Power(x*x, n/2) * v;
}
```

Notice the last statement. When recursion occurs at the end of a function, it's known as *tail recursion* or *end recursion*. When a tail-recursive function call returns, the values popped off the stack can be discarded immediately, since there are no further statements in the function that require them. For example, in *Power()*, we no longer need *x* or *n* after the call. This suggests that these variables should not have been placed on the stack in the first place. In fact, you can perform an optimization that eliminates this stack pushing. The optimization involves wrapping a *while* loop around the code and, instead of doing the tail-recursive call, merely making assignments to the local variables, giving them the values they would have had during the recursive call. For example, we can remove the tail recursion from *Power()* as follows.

```
// pow3.cpp: Tail-recursion optimized Power function

float Power(float x, int n)
{
  float v = 1;
  while(1) {
    if (n == 0) return v;
    if (n & 1) v *= x;
    x *= x; n /= 2;
  }
}
```

We had to make some modifications to how *v* was used in order to make this function work. Specifically, *v* has to accumulate multiplications by *x* for each odd value of *n*, and it's *v* that we return from the function. The resulting code will produce integral powers quite efficiently, with no recursive stack space needed.

You can have recursive functions that, at first glance, don't look tail recursive, even though they are. You may have to experiment by rearranging the code. For example, the recursive form of the binary search algorithm that we gave earlier in the chapter in *bs4.cpp* is actually tail recursive. Of course, if we apply tail recursion optimization to *bs4.cpp*, we'll end up with iterative code similar to that given earlier in *bs3.cpp*.

Stack-based Iteration

Even when you can't apply tail recursion optimization, you can still convert recursion into iteration, using an explicit stack. You can do this conversion in an ad hoc fashion, but we don't recommend it. You're likely to end up with an algorithm that doesn't always work, and you may not know why, and worse, you may not discover the problem for a long time.

Next we'll describe a mechanical conversion process that's much less error prone than the ad hoc approach. This process involves using labels and *gotos*, and explicitly pushing parameters onto a user-defined stack. In effect, you'll be playing the role of a compiler. The resulting code will be quite ugly and low level, and will probably result in jumps into the middle of blocks, but it is hoped that you can clean up the code afterward. There are nine steps, as follows.

Step 1. Figure out what data needs to be saved between function calls. This will include some, but possibly not all, of the parameters and local variables. You also need an integer code that simulates the "return address" of a recursive call. We'll call this code the *place number*. Each recursive call gets a unique place number that represents the place where the call resides in the function. Once you've determined what variables need saved, collect them up into a structure that will serve as an element in a user-defined stack.

For example, suppose you have the following recursive function.

```
int MyFunc(int a, int b, int c)
{
  int d, e;
  ...
  MyFunc(...); // Place 1
  ...
  MyFunc(...); // Place 2
  ...
}
```

Let's assume that variables *a*, *b*, and *d* need to be saved across recursive calls, but *c* and *e* don't. We can define a structure that we'll call *Parms* to be used for the critical variables, remembering also to include the place number.

```
struct Parms {
  int a, b, d; // Variables that need to be preserved
  int place;   // The recursive call place number
  Parms() { }  // To make stack construction easier
  Parms(int a_, int b_) { a = a_; b = b_; } // For initial setup
};
```

We provide two constructors. The default constructor allows the stack class to allocate an array of *Parms*. The other constructor is used when making a call to the transformed function. We initialize only those variables that were parameters to the original function. Thus, *a* and *b* are initialized, but not *d* or *place*. These variables will be initialized during the simulated recursion.

Step 2. Modify the parameter list of the function to pass in a *Parms* structure and remove from the parameter list and function body all parameters and local variables that will be stored in the structure instead. For example:

```
int MyFunc(Parms &parms, int c)
{
  int e;
  ...
}
```

By making this change, we'll have to alter the way *MyFunc()* is called. For example:

```
MyFunc(Parms(a, b), c); // Instead of MyFunc(a, b, c);
```

Of course, a driver function could be provided to hide the change. Another possibility is to leave the parameter list in its original form and instead construct the *Parms* structure as a local variable. Here's an example.

```
int MyFunc(int a, int b, int c)
{
  int e;
  Parms parms(a, b); // Construct parms locally.
  ...
  MyFunc(...);
  ...
  MyFunc(...);
  ...
}
```

Now a driver function isn't needed. However, note how we have two versions of *a* and *b* in use: those from the parameter list and those in the *Parms* structure. In the transformed code, it will be *parms.a* and *parms.b* that we want to use, for they will be manipulated on our user-defined stack. The variables *a* and *b* won't. Accidentally using *a* and *b* will cause subtle, hard-to-find errors. For that reason, we modify the parameter list and pass in the *Parms* structure instead.

Step 3. In case the function has a return type other than *void*, add a local variable that will keep track of the current status of the return value. For example:

```
int MyFunc(Parms &parms, int c)
{
  int e;
  int r; // To keep track of return value
  ..
}
```

Step 4. After the variable declarations at the beginning of the function, add a declaration for a stack that can hold elements of type *Parms*. The best scenario is to use a template-based stack class, such as the one shown in Chapter 9 of the companion to this book, *Practical Data Structures in C++*. For example:

```
int MyFunc(Parms &parms, int c)
{
  int e, r;
  Stack<Parms> stk(20);
  ...
}
```

Here it's assumed that *Stack* is a template-based class that implements a fixed-size stack. Or you might want to use a stack that has no fixed limits, such as a list-based stack. For a fixed-size stack, the size determines the maximum level of recursion you can support. In our example, we fixed the maximum recursion depth to 20 levels.

Step 5. Put a label after the stack and local variable declarations. This will provide a jumping point when you want to simulate a recursive call. For example, you might use a label called *RecurStart*.

```
int MyFunc(Parms &parms, int c)
{
  int e, r;
  Stack<Parms> stk(20);

  RecurStart: // Jump here to start next pass.
  ...
}
```

Step 6. At each point where a simulated recursive call is to be made, modify the place number in the *Parms* structure, so that the recursive call has a unique code. Then push the *Parms* structure onto the stack. After the stack push, change those variables that represent function parameters to the values they would have if you were making a recursive call. Then use a *goto* to jump to the *RecurStart* label. The stack push, modification of parameter variables, and the *goto* together make up a simulated recursive call. For example, if the recursive call is to be *MyFunc(a+17, b–42, c+1)*, you might use code similiar to the following.

```
// Simulate recursive call MyFunc(a+17, b-42, c+1)

parms.place = 3; // Record place. (Use a unique number.)
stk.Push(parms);
parms.a += 17; parms.b -= 42; c++; // Modify paramenters.
goto RecurStart;                   // Do next pass.
```

Since *c* presumably does not need to be saved across recursive calls, it's not in the *parms* structure, so we can set it directly using an expression such as *c++*.

Step 7. Immediately following each simulated recursive call, add a label that serves as the place to jump to when we've "returned" from the recursive call. Each label should be unique. One suggestion is to use a label name of *Placep*, where *p* represents the corresponding place number for the recursive call. Following this label should be the code that would have normally come after the recursive call. For example:

```
// Simulate recursive call MyFunc(a+17, b-42, c+1)

parms.place = 3; // Record place. (Use a unique number.)
stk.Push(parms);
parms.a += 17; parms.b -= 42; c++; // Modify paramenters.
goto RecurStart;                    // Do next pass.

Place3: // After return from simulated recursive call
```

Step 8. Provide another label, partitioned into a separate part of the function, that serves as the place to jump to when a function return is to be simulated. You might use a label called *RecurReturn*. Add code to pop the *Parms* structure off the stack. If the stack is empty, it means the recursion has completely unwound, and you should do a "real" return from the function. If the stack isn't empty, use the place number to determine where to jump to. A switch statement is convenient here. There will be a case for each unique place number. For example:

```
RecurReturn:

  if (stk.IsEmpty()) return r; // Really do a return.
  stk.Pop(parms);              // Otherwise, restore parameters.
  switch(parms.place) { // Determine where to jump to.
   case 1: goto Place1;
   case 2: goto Place2;
   ...
  }
```

Step 9. Everywhere a function return is used in the original recursive function, replace the *return* statement with a jump to *RecurReturn*. If a return value is being used, update the special local variable set aside for that purpose (we've used *r*) before doing the jump. For example:

```
// Simulate "return 42;"

r = 42; // r holds return value.
goto RecurReturn;
```

Conversion Example: The Exact-Sum KnapSack Problem

Next we'll do complete example of converting recursion into iteration. We'll use an algorithm that solves the exact-sum version of the classic *knapsack problem*. In the knapsack problem, a burglar has a knapsack that can hold up to a certain weight. The burglar has stolen a collection of items he'd like to put in the knapsack, each item having a value associated with it. The burglar would like to put as many items possible into the knapsack without overburdening it, such that the total value taken is maximized. The rest of the items will have to be left behind.

In the exact-sum variation of this problem, an item's value is proportional to its weight, and items are added until we reach the weight capacity of the knapsack exactly. This implies that there may not be a solution. The burglar will go home empty if he can't find the right combination of items to match the weight exactly.

The collection of items the burglar has to choose from can be represented by an array of integers, each integer representing a weight. One way to solve the knapsack problem is to add weights to a running total, keeping track of the candidates. If the exact target weight is reached, we have a complete solution. If we try all of the candidates and their sum doesn't reach the target weight, we can have no solution. If, while we're trying candidates, we happen to go over the target weight, we subtract the last weight added and then skip past it, trying the next candidate. We may have to subtract all of the candidates and go back to the beginning. Then we'll skip past the first candidate and start a new attempt using the second candidate.

Our solution to the exact-sum knapsack problem is an example of a *backtracking algorithm*. In such an algorithm, we charge ahead trying a solution and then, when we see that the current search path won't lead to a solution, we backtrack to an appropriate place, and then try in a new direction. Because this backtracking can be accomplished easily using a stack, this implies that a recursive function might solve the problem. Such a function is given as follows.

```cpp
// knap1.cpp: Recursive solution to the exact sum knapsack problem

int KnapSack(int wts[], int n, int tw, int ci)
// Given a set of n weights stored in wts, and a target weight
// tw, pick from the set wts, a subset that will total to the
// target weight, starting with the candidate weight indexed
// by ci.
// Returns 1 if there is a solution, 0 if not.
{
    // We might have the exact weight total, or a total that's
    // too big, or we may have run out of candidate weights.

    if (tw == 0) return 1;
    if (tw < 0 || ci == n) return 0;

    // Try a solution that includes the current candidate weight.

    if (KnapSack(wts, n, tw-wts[ci], ci+1) == 1) {
        // We have a good candidate, so report it.
        cout << wts[ci] << ' ';
        return 1; // Signal partial solution.
    }
    else {
        // Candidate won't work, so try solving w/o candidate.
        return KnapSack(wts, n, tw, ci+1);
    }
}
```

Code for the exact-sum knapsack problem can be found on disk in the files *knap1.cpp, knap2.cpp, knap3.cpp*, and *knap4.cpp*. These files correspond to the versions of the algorithm given in this chapter, using the names given in the comments at the top of each version.

In the recursive *KnapSack()* function, we solve the problem by successively reducing it to a series of smaller problems, starting with the desired target weight and then subtracting candidate weights from the target and recursing again. We start out with the first candidate weight, ($ci=0$) and then walk through the candidates, backtracking if needed. When we know we have a good candidate weight, the value is printed out. Here's a sample call to *KnapSack()*.

```
int wts[7] = {2, 2, 2, 5, 5, 9, 11}; // Candidate weights

if (KnapSack(wts, 7, 18, 0)) // Try target weight of 18.
   cout << "Solution found\n";
   else << "No solution found\n";
```

The output from this example would be

```
9 5 2 2
Solution found
```

In general, there are many possible solutions to any given instance of the exact-sum knapsack problem. The *KnapSack()* function merely finds the first one it can. Algorithms with this type of behavior are known as *greedy algorithms*, and you'll see them again in Chapters 14 and 15.

Let's see if we can transform the recursive *KnapSack()* function into an iterative one. The first step in the transformation process is to define the *Parms* structure. What values do we need to save on the stack? We need a place number and the parameters *tw* and *ci*. If you study the function closely, you'll notice that we never need to save the parameters *wts* and *n*, since their values never change. Thus, our *Parms* structure should be defined as

```
struct Parms {
   int tw, ci; // Saved values of the target wt and candidate index
   int place;  // Place number
   Parms() { } // to make stack creation easy
   Parms(int t, int c) { tw = t; ci = c; }
};
```

Given this *Parms* structure, here is the *KnapSack()* function transformed to use iteration, using the mechanical conversion process discussed earlier.

```
// knap2.cpp: First iterative form
```

```
int KnapSack(int wts[], int n, Parms &parms)
{
  Stack<Parms> stk(n); // Max recursion depth depends on n.
  int rv;

  RecurStart:

  if (parms.tw == 0) {
    rv = 1; goto RecurReturn; // Simulate "return 1;"
  }
  if (parms.tw < 0 || parms.ci == n) {
    rv = 0; goto RecurReturn; // Simulate "return 0;"
  }

  // Simulate call "KnapSack(wts, n, tw-wts[ci], ci+1)"

  parms.place = 1;
  stk.Push(parms);
  parms.tw -= wts[parms.ci]; parms.ci++;
  goto RecurStart;

  Place1: // After return from first recursive call
  if (rv == 1) {
    cout << wts[parms.ci] << ' ';
    rv = 1; goto RecurReturn; // Simulate "return 1;"
  }
  else {
    // Simulate call "KnapSack(wts, n, tw, ci+1)"
    parms.place = 2;
    stk.Push(parms);
    parms.ci++;
    goto RecurStart;

    Place2: // After return from second recursive call
    goto RecurReturn; // Simulate "return rv;"
  }

  RecurReturn: // Simulate return from recursive call.

  if (stk.IsEmpty()) return rv; // Really return from function.
  stk.Pop(parms); // Restore parameters, jump to appropriate section.
  if (parms.place == 1) goto Place1;
    goto Place2;
}
```

Here's how we would call this function, using the example given earlier.

```
KnapSack(wts, 7, Parms(18, 0));
```

Going Toward More Structure

The iterative form of *KnapSack()*, while identical in function to the recursive form, is quite ugly, involving lots of labels and *goto*s. Even worse, we have jumps into and out of the insides of *if* statements that would make any sensible programmer shudder. Our next task, then, is to take this mechanically derived code and try to eliminate the low-level *goto*s.

First, notice how the last statement, *goto Place2;*, can be simplified to just *goto RecurReturn;* (and we can remove the two statements at label *Place2*). The last part would then look like

```
RecurReturn:
if (stk.IsEmpty()) return rv;
stk.Pop(parms);
if (parms.place == 1) goto Place1;
goto RecurReturn;
```

When we pop *parms* off the stack, which includes members *tw*, *ci*, and *place*, notice that for *place == 2* (returning from the second recursive call), we never use the parameters *tw* and *ci*. That suggests that we should have never put them on the stack in the first place when making the second recursive call. If this reminds you of tail recursion, it should, for that's exactly what it is. (We should have noticed it before starting the general conversion process.) Thus, we can optimize the second simulated recursive call by not doing a stack push, as follows.

```
// Simulate call "KnapSack(wts, n, tw, ci+1)"
parms.ci++;
goto RecurStart;
```

Since we don't push on the stack at this point, that means when popping from the stack, *place* will always equal 1. In other words, we don't need to store the place number on the stack. Also, the *RecurReturn* code simplifies to

```
RecurReturn:
if (stk.IsEmpty()) return rv;
stk.Pop(parms);
goto Place1;
```

Here is a new version of *KnapSack()*, where we've optimized for the tail recursion we discovered and moved the *RecurReturn* code to just after the first simulated recursive call, for reasons to become apparent.

```
// knap3.cpp: Tail recursion optimized iterative form

struct Parms {
  int tw, ci; // Place number removed from here.
  Parms() { }
  Parms(int t, int c) { tw = t; ci = c; }
};

int KnapSack(int wts[], int n, Parms &parms)
{
```

```
    Stack<Parms> stk(n);
    int rv;

RecurStart:

    if (parms.tw == 0) {
        rv = 1; goto RecurReturn; // Simulate "return 1;"
    }
    if (parms.tw < 0 || parms.ci == n) {
        rv = 0; goto RecurReturn; // Simulate "return 2;"
    }

    // Simulate call "KnapSack(wts, n, tw-wts[ci], ci+1)"

    stk.Push(parms);
    parms.tw -= wts[parms.ci]; parms.ci++;
    goto RecurStart;

RecurReturn: // Simulate return from recursive call.

    if (stk.IsEmpty()) return rv; // Really return from function.
    stk.Pop(parms); // Restore parameters, jump to appropriate section.
    goto Place1;

Place1: // After return from first recursive call
    if (rv == 1) {
        cout << wts[parms.ci] << ' ';
        rv = 1; goto RecurReturn; // Simulate "return 1;"
    }
    else {
        // Simulate call "KnapSack(wts, n, tw, ci+1)"
        parms.ci++;
        goto RecurStart;
    }

}
```

At this stage, we can try converting jumps into *while* loops. For example, the first part of the code can be transformed into

```
RecurStart:

while(1) {
    if (parms.tw == 0)               { rv = 1; break; }
    if (parms.tw < 0 || parms.ci == n) { rv = 0; break; }
    stk.Push(parms);
    parms.tw -= wts[parms.ci]; parms.ci++;
}
```

Next, the second half can be transformed into a *while* loop. We can take out the jump to *Place1* and the label as well.

```
while(1) {
  if (stk.IsEmpty()) return rv;
  stk.Pop(parms);
  if (rv == 1) {
    cout << wts[parms.ci] << ' ';
    rv = 1;
  }
  else {
    parms.ci++; break;
  }
}

goto RecurStart;
```

Finally, we can remove the *RecurStart* label and associated *goto*, and replace it with a while loop. Our final, simplified function is as follows.

```
// knap4.cpp: Final iterative form

int KnapSack(int wts[], int n, Parms &parms)
{
  Stack<Parms> stk(n);
  int rv;

  while(1) {
    while(1) {
      if (parms.tw == 0)               { rv = 1; break; }
      if (parms.tw < 0 || parms.ci == n) { rv = 0; break; }
      stk.Push(parms);
      parms.tw -= wts[parms.ci++];
    }
    while(1) {
      if (stk.IsEmpty()) return rv;
      stk.Pop(parms);
      if (rv == 1) {
        cout << wts[parms.ci] << ' ';
      }
      else {
        parms.ci++;
        break;
      }
    }
  }
}
```

Is Iteration Worth It?

The simplified iterative *KnapSack()* function given in *knap4.cpp* is certainly better than the original iterative version given in *knap2.cpp*. However, it's not nearly as clear as the original recursive function in *knap1.cpp*. Someone reading the iterative code who wasn't aware of how the code was derived might go nuts trying to figure out what it does.

So what did we gain by converting the recursion to iteration? In this example, not much. We're using slightly less stack space, because all we push on the stack are intermediate values of *tw* and *ci*, and not the values for *wts*, *n*, the return value, and the return address.

Keep in mind, though, that we're using a user-defined stack, as opposed to the built-in processor stack. Instead of doing simple machine instructions such as *push* and *pop* for stack manipulation, we have to make function calls to *stk.Push()* and *stk.Pop()*. Because of this, the iterative version may run slower than the recursive version.

However, in cases where the recursion involves large local variables, perhaps with sophisticated derived objects having constructors and destructors and so forth, the iterative version may be more efficient, if we can trim down the amount of data stored on the recursion stack. That's the main advantage of going to an iterative approach.

There is one other advantage, but to appreciate it, you'll have to study Chapter 3. In that chapter we discuss a way to turn algorithms inside out, so that we can use them in more flexible ways. Part of this process involves starting with iterative code, so if our algorithms are recursive, we must first make them iterative to take advantage of this new approach.

When to Leave Things Recursive

Some algorithms are inherently recursive, such as the *KnapSack()* function, and are best implemented that way, since the code often is much clearer than an iterative form. Other examples of inherently recursive algorithms are those that process recursive data structures. For example, in the companion book, we describe algorithms for walking binary trees. Since binary trees are recursive in nature, the algorithms for walking them can be described concisely in a recursive form. In later chapters you'll see other algorithms that lend themselves well to recursion, such as algorithms to traverse the nodes of a graph, which are generalizations of the tree-walking algorithms.

In the *KnapSack()* conversion example, we were lucky in that we were able to clean up the iterative code and remove all the *goto*s. In general, though, you won't be so lucky. Many times the code produced by the mechanical translation process will be in a form that makes it nearly impossible to get rid of all the *goto*s, unless you use auxiliary flags or duplicate some of the code in several places. And you may have jumps into the middle of blocks that you can't get rid of, something that would make anyone queasy. In cases like this, it's best to leave the algorithm in recursive form, unless you have some strong overriding reason to make it iterative.

When Not to Leave Things Recursive

Leaving recursive algorithms recursive is a good rule to follow, but it's not infallible. You saw where tail-recursive functions such as *Power()* and *BinarySearch()* are best put into an iterative form. Also, some non-tail-recursive algorithms have horrible recursive implementations but quite efficient iterative implementations. A classic example stems from the *Fibonacci* numbers. The sequence of Fibonacci numbers can be produced by the following definition, which happens to be a special kind of definition known as a *recurrence relation.*

```
Fib(0) = 0; // Base case
Fib(1) = 1; // Base case
Fib(n) = Fib(n-1) + Fib(n-2); // Recursive, or inductive case
```

Using this recurrence relation, we can produce the sequence of Fibonnaci numbers, which is 0, 1, 1, 2, 3, 5, 8, 13, 21, and so on.

You might be tempted to code this recurrence relation using a recursive function. For example:

```
// fib1.cpp: Recursive implementation for Fibonnaci numbers

long Fib(long n)
// Return the nth Fibonacci number, where n >= 0;
{
  if (n <= 1) return n;
  else return Fib(n-1) + Fib(n-2);
}
```

One the surface, this seems quite elegant. However, note that in calculating *Fib(n–1)* we calculate *Fib(n–2)*. Unfortunately, we don't store the result of *Fib(n–2)*, and as a consequence we have to turn right around and calculate it again. This happens recursively. It turns out the time complexity of this function is $O(O^n)$, where $O = 1.61803 \ldots$ is the "golden ratio," one of the magical constants of mathematics. The value of O isn't important here. What's important is that we have an exponential algorithm, something to be avoided if at all possible.

Fortunately, we can avoid this exponential algorithm, for there is an $O(n)$ algorithm that happens to be iterative.

```
// fib2.cpp: Iterative implementation for Fibonnaci numbers

long Fib(long n)
{
  if (n <= 1) return n;
  long curr, old = 1, older = 0;
```

```
while(n-- > 1) {
  curr = old + older;
  older = old;
  old = curr;
}
return curr;
}
```

 Sample programs for the Fibonacci numbers can be found on disk in the files *fib1.cpp* and *fib2.cpp*.

QUEUE-BASED ITERATION

The iterative version of *Fib()*, as given in *fib2.cpp*, explicitly keeps track of previous values to avoid lengthy recomputations. Note that this iterative form was not derived from doing tail recursion optimization or from the general transformation process given earlier. There is a reason for that. Take note of the variables *old* and *older*. Together, these in effect make up a queue of length 2. That is, our iterative form is using a queue, not a stack!

This suggests that some types of recursions are better handled with a queue. Recurrence relations are an example. While a stack *can* be used for recurrence relations, as we did in *fib1.cpp*, it's likely to be highly inefficient. It's simply the wrong data structure for the job.

INDIRECT RECURSION

The recursion examples you've seen up to this point involve a function directly calling itself. This is known as *direct recursion*. It's possible to have *indirect recursion*, where a function might call another function that, in turn, calls the first function. The following code illustrates this scenario.

```
void f()
{
  ...
  f(); // Direct recursion
  ...
  g(); // Possible indirect recursion
  ...
}

void g()
{
  ...
  g(); // Direct recursion
  ...
  f(); // Possible indirect recursion
  ...
}
```

In the case of *f()* and *g()*, it's possible that indirect recursion can take place for both functions. Functions like this are said to be *co-recursive*, with *co-recursion* taking place.

Most people have a hard enough time with simple direct recursion but uses of indirect recursion can be truly mind boggling. Fortunately, many of the recursive algorithms in everyday use involve only direct recursion. Examples of algorithms that use indirect recursion can be found in recursive descent parsers, such as those discussed in [Aho 88].

It's beyond the scope of this book to show how to turn indirect recursion into iteration. Suffice it to say that it can be done (but you may not want to—it's messy), and the technique is similar to that given in this chapter for direct recursion.

TEMPLATE-BASED ALGORITHMS

Before we close this chapter, one feature of C++ that lends itself well to algorithm design needs to be discussed. *Templates* allow you to parameterize both classes and functions so that they can work with different types of data. In the companion book, we use templates extensively when designing classes for data structures. For example, we have template-based trees, stacks, queues, and so forth. This allows us to use the same code for implementing, say, a stack of integers and a stack of recursive function parameters. We merely have to plug in the appropriate type when declaring a stack object.

Templates are also a natural for implementing algorithms that process homogenous data. For example, take the selection sort algorithm given in Chapter 1. We wrote the algorithm to work specifically for arrays of integers. By using templates, we can write the selection sort algorithm so that it can work for other types of array elements. Here is an example.

```
template<class TYPE>
void SelnSort(TYPE a[], int n)
// Descending selection sort
// ASSUMES TYPE has '>' and '=' properly defined.
{
  TYPE tmp;
  int i, j, m;

  for (i = 0; i<n-1; i++) {
      m = i;
      for (j = i+1; j<n; j++) {
          if (a[j] > a[m]) m = j;
      }
      if (m != i) {
          tmp = a[i]; a[i] = a[m]; a[m] = tmp;
      }
  }
}
```

The body of this function is almost identical to that given in Chapter 1. All we've done is parameterized the array element type using the template parameter *TYPE*. Any type can be substituted for *TYPE*, as long as it has the operations for '>' and '=' properly defined. Here is an example of using the template-based selection sort algorithm for two different types of data.

```
char charr[5] = {'J', 'K', 'A', 'Q', 'T' };
float farr[5] = {42.0, 17.0, 55.0, 21.0, 7.0};

SelnSort(charr, 5); // Calls character version
SelnSort(farr, 5);  // Calls floating point version
```

When the compiler sees a call to a template-based function, it determines what version to use by inspecting the parameter types. In the first call of our example, since an array of characters is used, a character array version of *SelnSort* is called. In the second case, a floating point version of *SelnSort()* is called. If the compiler hasn't already done so, it will see to it that specific versions of *SelnSort()* are generated. That is, there will be code for two *SelnSort()* functions. These automatically generated versions will be just as efficient as if we coded them by hand. And therein lies the main advantage to using templates. The disadvantage, of course, is that we are generating more source code. This is a classic example of the speed versus space trade-off.

Whenever it's appropriate, we'll be showing the algorithms in this book using templates. This gives us a perfect way of clearly showing an algorithm yet, at the same time, allowing you to see the algorithm adapted for different types of data, staying efficient in the process.

It is not our intent to discuss the details of templates here. If you are not familiar with using templates, an introduction to them can be found in Chapter 2 of the companion book. You may also want to consult a good C++ language book, such as [Stroustrup 91].

The text file *template.txt* is provided on disk, which contains a discussion of templates excerpted from Chapter 2 of the companion book. This file describes how to declare and use templates and discusses how to cope with the many different implementations of templates as found in various C++ compilers.

T H R E E

Algorithmic Generators

n Chapter 2 we discussed how to construct algorithms from the normal procedural viewpoint of iterative loops and function recursion. In this chapter we'll take a fresh, object-oriented perspective, and see how we can reorganize the code making up an algorithm so that it becomes, in effect, an "algorithmic object." We'll call such objects *algorithmic generators* (or *generators* for short). These objects are used to *generate* the steps of an algorithm.

Note The idea of generator objects was inspired by similar techniques used in the languages Snobol, Icon (see [Griswold 83]), and Lucid (see [AshWa 85]).

TURNING ALGORITHMS INSIDE OUT

When algorithms are constructed in the traditional procedural fashion, it's often the case that the code to output the results of an algorithm often becomes awkwardly embedded in the middle of a loop. For example, let's review the *KnapSack()* function, as given in the *knap4.cpp* program of Chapter 2.

```
// knap4.cpp: Final iterative form

int KnapSack(int wts[], int n, Parms &parms)
{
  Stack<Parms> stk(n);
  int rv;

  while(1) {
    while(1) {
      if (parms.tw == 0)                    { rv = 1; break; }
```

```
         if (parms.tw < 0 || parms.ci == n) { rv = 0; break; }
         stk.Push(parms);
         parms.tw -= wts[parms.ci++];
      }
    while(1) {
        if (stk.IsEmpty()) return rv;
        stk.Pop(parms);
        if (rv == 1) {
           cout << wts[parms.ci] << ' '; // Output partial result
        }
        else {
          parms.ci++;
          break;
        }
      }
    }
  }
}
```

The *KnapSack()* function produces a sequence of numbers that make up the solution to a given instance of the exact-sum knapsack problem. Unfortunately, we're restricted on how that sequence is output. In this case, we sent the numbers to the *cout* stream. But what if we wanted to add the numbers to a linked list or an array, or perhaps send them to other algorithms? We'd have to write additional versions of the algorithm, mostly duplicating the code.

But there's a way out of our dilemma. We can turn the algorithm "inside out," bringing the results of the algorithm to the "outside." To do this, we'll invent a new type of class, which incorporates as a member a function that, when repeatedly called, will generate the sequence of numbers that make up the solution to the knapsack problem. Then it's the user, not the algorithm itself, that determines what to do with those numbers. We'll call such classes *algorithmic generators*. Here's a sketch of the technique.

```
class KnapSacker { // A knapsack generator.
    ...
    KnapSacker(int wts[], int n, int tw); // Initialize the algorithm.
    int Step(int &good_wt); // Perform one step of the algorithm.
    ...
};
...
int wts[7] = { 2, 2, 2, 5, 5, 9, 11 };
KnapSacker kns(wts, 7, 18); // Create a generator.

int good_wt;
while(kns.Step(good_wt)) cout << good_wt << ' ';
```

Instead of having the algorithm control when sequencing takes place, the user controls it. The generator has enough smarts to perform one step of the algorithm and to return the result of the step. Figure 3.1 illustrates the concept of a knapsack generator.

Figure 3.1 A knapsack generator.

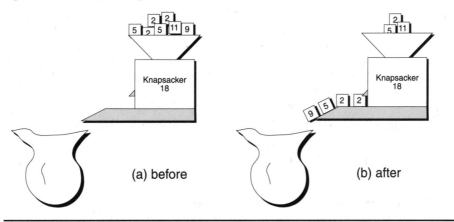

(a) before   (b) after

Creating Generators

Next we'll describe, step by step, how you can take an algorithm and make a generator from it. We'll use the knapsack algorithm as an example. In the following steps, we assume that the algorithm can be implemented in a single function. Later on we'll talk about how to remove this restriction.

Step 1. If the algorithm is recursive, you must turn the recursion into iteration, perhaps using the techniques given in Chapter 2.

Step 2. Create a new class for the generator. In our example, we'll call the class *KnapSacker*. Note that with generators, the objects aren't to be viewed as passive data structures, but instead, as more active agents. The name of a generator class should reflect the fact that the generator implements an algorithm.

Step 3. Remove the parameters and local variables from the function constituting the algorithm, and make them members of the generator class. Not all of the local variables need to be taken out of the function, but only those that must be preserved across calls to the *Step()* function given later. Next, add a variable, called *state*, which keeps track of the state of the generator (for example, to indicate when the algorithm is finished). Depending on the algorithm, you may not need this extra variable. The original set of variables may suffice. Here's a sketch of the idea for the *KnapSacker* generator.

```
class KnapSacker {
private:
  Stack<Parms> stk; // Was a local variable
```

```
    Parms parms;      // Was a parameter
    int *wts;         // Was a parameter
    int n;            // Was a parameter
    int rv;           // Was a local variable
    int state;        // Newly added variable
    ...
};
```

Step 4. Add a constructor that is passed enough information to describe the input to the algorithm. The constructor's task is to initialize the generator so that it is ready to run the algorithm. Often the constructor parameters are borrowed from the original function. For example:

```
KnapSacker(int wts[], int n, int tw);
```

Step 5. Add a member function that returns the result of the last step executed. We'll call that function *Curr()*, which stands for "current value." In general, *Curr()* returns a "snapshot" of the algorithm. You might need to add one or more member variables to record this snapshot. Here's the *Curr()* function for the KnapSacker class.

```
class KnapSacker {
  ...
  int curr_wt;
  ...
  int Curr();
  ...
};
...
int KnapSacker::Curr()
{
  return curr_wt;
}
```

After the constructor is called, you might want the generator to be in a state such that the result of the first step is available, through the *Curr()* function. That is, you may want the constructor to cause the first step of the algorithm to be executed—to prime the pump, so to speak. In other cases, you may want the user to command the first step explicitly. As we've written *Curr()*, it doesn't check for the generator being in a valid state, so you must either ensure that the generator will be in a valid state any time *Curr()* is called, or alternatively, modify *Curr()* to add a check, as follows.

```
int Curr(int &cv)
// Returns 1 and loads cv if in valid state, otherwise
// returns 0 and leaves cv untouched.
{
  if (not_in_valid_state) return 0;
  cv = curr_wt;
```

```
    return 1;
}
```

Step 6. Remove all parameters and local variables from the original function, and rename the function with the name *Step()*. Then, at the place where you would normally return the result of a step, record the result in the designated member variables instead and return a 1. If past the end of the algorithm (with no result to report), then a zero should be returned.

In order to pick up where you left off the next time *Step()* is called, you'll have to record in the *state* variable an indication of where to return to. The state variable serves a similar purpose as the place number of Chapter 2. For example:

```
int KnapSacker::Step()
{
  if (state == 1) goto State1;
  State0: // Start of a step.
  ...
  // At the end of one step
  curr_wt = wts[ci]; // Record result.
  state =1; // Record where to come back to.
  return 1; // Indicate algorithm not finished yet.

  State1: // Come back here next time Step() is called.
  state = 0;
  ...
  goto State0; // Back to the top of a step.
  ...
  // Past end of algorithm.
  return 0;
  ...
}
```

Note that the *state* variable is in addition to any place numbers you might be using for a recursive algorithm turned iterative. Remember that the place numbers will most likely be stored in some *Parms* structure and maintained on a recursion stack.

Step 7. You may want to add another *Step()* function that behaves much like a "post-incrementer." That is, it passes back the result of the last step and then advances the algorithm one step farther. The result can be passed as a parameter, and then the function can return a code indicating the status of the algorithm, as before. For example:

```
int Step(int &cv)
{
  if (past_end_of_algorithm) return 0;
  cv = curr_wt;
  Step();
```

```
    return 1;
}
```

This function is provided mainly for convenience. You may want to add other convenience functions, such as one to reset the generator to start the algorithm over and another to indicate whether the algorithm as ended. Also, you may want to add a destructor and any other functions needed to make the generator complete.

A KnapSack Generator

Here's an example of applying the techniques of the last section to create a generator that can step through the exact-sum knapsack algorithm:

```
// knap5.cpp: Exact sum knapsack algorithm generator

struct Parms { // Recursion stack parameters
  int tw, ci;
  Parms() { }
  Parms(int t, int c) { tw = t; ci = c; }
};

class KnapSacker {
private:
  Stack<Parms> stk; // Was a local variable to KnapSack() function
  Parms parms;      // Was a parameter to KnapSack() function
  int *wts;         // Was a parameter to KnapSack() function
  int n;            // Was a parameter to KnapSack() function
  int rv;           // Was a local variable to KnapSack() function
  int state;        // New variable to keep track of generator state
  int curr_wt;      // New variable that saves current output result
public:
  KnapSacker(int w[], int sz, int tw);
  void Reset(int tw=0);
  int Eos() const;
  int Status() const;
  int Curr() const;
  int Curr(int &cv) const;
  int Step();
  int Step(int &pv);
};

KnapSacker::KnapSacker(int w[], int sz, int tw)
// Constructor to set up a knapsack problem solver
: stk(sz) // Max recursion depth depends on sz.
{
  wts = w;  n = sz;
  Reset(tw);
}
```

```
void KnapSacker::Reset(int tw)
// Resets the generator and if tw != 0, uses a new target weight.
{
  if (tw) parms.tw = tw;
  parms.ci = 0;
  state = 0; curr_wt = 0;
  stk.Clear();
  Step(); // Prime the pump
}

int KnapSacker::Eos() const
// Returns 1 if past all steps, else 0.
{
  return state > 1; // Signals end of all of steps.
}

int KnapSacker::Status() const
// Returns state variable:
// 0 = still working
// 1 = still working, picking up where we left off
// 2 = done, no solution
// 3 = solution complete
{
  return state;
}

int KnapSacker::Curr() const
// Returns current output result. Does not test for being
// at the end of all steps.
{
  return curr_wt;
}

int KnapSacker::Curr(int &cv) const
// If at end of all steps, nothing is passed back, and a 0 is
// returned, else c is loaded with the current output value,
// and a 1 is returned.
{
  if (Eos()) return 0;
  cv = curr_wt;
  return 1;
}

int KnapSacker::Step()
// Advances the generator one step, returning a 1 if not at
// end of all steps, else 0.
{
  if (state >= 1) {
    if (state == 1) goto State1; // Pick up where we left off
    return 0;                    // Or, we're already done
  }
  while(1) {
    while(1) {
```

```
       if (parms.tw == 0)                 { rv = 1; break; }
       if (parms.tw < 0 || parms.ci == n) { rv = 0; break; }
       stk.Push(parms);
       parms.tw -= wts[parms.ci++];
     }
     while(1) {
       if (stk.IsEmpty()) {
         // We're done, for one reason or another.
         // state = 2: end of all steps, no solution
         // state = 3: end of all steps, solution complete
         state = 2 + rv;
         return 0; // Return end of steps indicator
       }
       stk.Pop(parms);
       if (rv == 1) {
         state = 1; // I.e.: We broke out at this point.
         curr_wt = wts[parms.ci];
         return 1;  // More steps to come, possibly.
         State1: // Jump back here upon next call to Step().
         state = 0;
       }
       else {
         parms.ci++;
         break;
       }
     }
   }
 }

int KnapSacker::Step(int &pv)
// Like first Step() function, except c is loaded with the
// current output result, before the advance (similar
// to post-increment). If already at end of all steps,
// c is not loaded, and a 0 is returned. Else a 1 is returned.
{
  if (Eos()) return 0;
  pv = curr_wt; // Grab current value before advancing.
  Step();
  return 1;
}
```

Complete code for the knapsack generator, along with a test program, can be found on disk in the file *knap5.cpp*.

Pay particular attention to the first *Step()* function. This function is similar to the *KnapSack()* function, except all the parameters and local variables are now members of the generator. Also note how we have code that forces jumps into the middle of blocks. That's one of the few drawbacks to the generator technique. Sometimes it's possible to get rid of potentially dangerous jumps by using flags, *switch* statements, rearranging code, and so on. For example, the *Fibber, SelnSorter,* and *TowerSolver* generators you'll see later in the chapter were derived using the techniques presented here, but the code was rearranged to get rid of explicit jumps. We didn't do that here, so that you could see directly how the code was derived.

Here's an example of using our newly defined knapsack generator.

```
const int SIZE = 7;
int wts[SIZE] = { 2, 2, 2, 5, 5, 9, 11 };

KnapSacker kns(wts, 7, 18);

int good_wt;
while(kns.Step(good_wt)) cout << good_wt << ' ';

if (kns.Status() == 2)
   cout << "No solution possible\n";
   else cout << '\n';
```

With our knapsack generator, we have a lot of flexibility in what we do with the results. For example, we can collect the results into an array *without changing the generator class at all*.

```
kns.Reset(); // Reset the generator to the start of the algorithm.

int rn = 0;
int results[SIZE];
while(kns.Step(good_wt)) results[rn++] = good_wt;
```

Generators are valuable because they can be used for many different situations. Using generators may cause you to shift your way of thinking about algorithm design, a topic we'll pick up again toward the end of the chapter.

A Canonical Form for Generators

Next we'll take a look at a suggested canonical form for generators. The purpose behind this form is to allow the generator to be as flexible and efficient as possible. We'll show the canonical form as a pseudo-class template, but please note we are not suggesting that you make your generators templates. You may want to do this, but in this book, we'll be showing the generators as independent classes. Note that the *KnapSacker* generator follows the canonical form, as do many of the generators in this book. However, we won't use *all* of the suggested member functions every time. For some generators, not all of the functions make sense.

```
// Canonical generator. This is a pseudo-class template and is not
// valid C++ as it stands.

template<class TYPE>
class Generator {
```

```
private:
  // All member data here, including a record of
  // the results of the last step, plus one or more
  // optional state variables
public:
  Generator(algorithm_parameters);
  ~Generator();
  void Reset(optional_new_algorithm_parameters);
  int Eos() const;
  int Status() const;
  TYPE Curr() const;
  int Curr(TYPE &curr_val);
  int Step();
  int Step(TYPE &prev_val);
  TYPE Run();
};
```

Here, *TYPE* is the type used to represent the results of a step. In general, the result type will be a structure containing whatever values are necessary to record one snapshot of the algorithm. You also may have more than one type of result that can be returned. Different member functions could be provided for this.

The meaning of most of the member functions can be found by examining the comments for the *KnapSacker* generator. We've added a destructor to point out that, in general, your generators may need destructors. In some cases, you may not want to see the intermediate steps but only the final result. The *Run()* function is for this purpose; it steps through the algorithm until completion and returns the final result. In most cases, *Run()* can easily be coded as:

```
template<class TYPE>
TYPE Generator::Run()
{
  while(Step());
  return Curr();
}
```

For those algorithms that may not have results if given the wrong kind of inputs, you may want *Run()* to return a status instead. For example:

```
template<class TYPE>
int Generator::Run(TYPE &fv)
// If algorithm completes in a valid state, the final result
// is passed back in fv, and a 1 is returned. Else, fv remains
// untouched and a 0 is returned.
{
  while(Step());
  if (in_good_state) { fv = Curr(); return 1; }
  return 0;
}
```

Multifunction Algorithms

Now we'll show you how to convert an algorithm that consists of a collection of data structures and functions to generator form. It's actually rather simple in theory, but messy in implementation. All of the data structures and functions should be members of the generator class. The *Step()* function becomes a driver of sorts. Through the use of state variables, *Step()* will determine which of the other functions to call, in order to pick up where the generator left off. For example:

```
class MyGenerator {
  ...
  int state_variable_1; // Which function were you in?
  int state_variable_2; // Which place in the function?
  ...
};


int MyGenerator::Step()
{
  switch(state_variable_1) {
    case 1: Func1();
    case 2: Func2();
    ...
}

int MyGenerator::Func1()
{
  switch(state_variable_2) {
    case 1: goto State1;
    case 2: goto State2;
    ...
}
...
```

If the original functions were recursive or, in general, co-recursive, you may need to keep track of multiple stacks and so forth. We won't be covering generators of this complexity in this book. Fortunately, many algorithms can be implemented with a single function (perhaps with some simple helper functions).

NUMBER GENERATORS

An important class of algorithms produce sequences of numbers. Examples are the prime number generator of Chapter 4 and the random number generators of Chapter 5. These sequences are often produced by recurrence relations. Next we'll show a simple number sequence generator that produces the Fibonacci sequence. The generator is based on the iterative function given in *fib2.cpp* in Chapter 2.

```
// fib3.cpp: Fibonnaci generator

class Fibber {
private:
   long m, n, curr, old, older;
public:
   Fibber(long m_);
   void Reset(long m_ = -1);
   int Eos();
   long Curr();
   int Step();
   int Step(long &pv);
   long Run();
};

Fibber::Fibber(long m_)
// Initialize a Fibonnaci generator to produce the
// first m Fibonnaci numbers, or if Run() is used,
// Fib(m).
{
   Reset(m_);
}

void Fibber::Reset(long m_)
// Reset the Fibonnaci generator, and optionally use a new
// sequence length if m_ >= 0.
{
   if (m_ >= 0) m = m_;
   n = 1; curr = 0; old = 0; older = 0; // Prime the pump
}

int Fibber::Eos()
// Returns 1 if after end of the sequence, else 0.
{
   return n > m;
}

long Fibber::Curr()
// Returns the current value of the sequence.
{
   return curr;
}

int Fibber::Step()
// Advance to the next number in the sequence. Returns 0 if
// past the end of the sequence, else 0.
{
   if (Eos()) return 0;
   older = old; old = curr;
   if (n++ == 1) curr = 1; else curr = old + older;
   return 1;
}
```

```
int Fibber::Step(long &pv)
{
  if (Eos()) return 0;
  pv = curr;
  Step();
  return 1;
}

long Fibber::Run()
// Runs the generator to the end of sequence, and returns the
// last value produced. Thus, it computes Fib(m), where
// Fib(0) = 0 is the first number in the sequence.
{
  while(Step());
  return curr;
}
```

The *Fibber* class can be used in two ways: Either it can produce the first *m* numbers of the sequence, using the *Step()* function, or it can produce the equivalent of *Fib(m)* using *Run()*. For example:

```
// Produce the first m numbers.

Fibber Fibber(m);
long x;
while(Fibber.Step(x)) cout << x << ' ';

// Compute Fib(m), where Fib(0) = 0 is the first number.

Fibber.Reset(m);
cout << Fibber.Run() << '\n';
```

The *Fibber* generator is fairly simple, but it's trickier than it looks. For instance, *Reset()* is responsible for computing the first number, and *Step()* is valid only for computing the second number onward. Also, *Step()* should return only a 0 when you are *past* the end of the algorithm, not when you've just reached the end during the step.

When creating generators, it's often tricky to get *Step(), Run(),* and *Curr()* to work uniformly, with no inconsistencies. You'll often encounter off-by-one errors, where one or more of these functions runs the algorithm too far, or not far enough, or gives a step result out of sync. It may take some time to get everything right.

GENERATORS AS DEBUGGING AIDS

In general, a generator returns a sequence of objects. Next we'll see an example of a selection sort generator that returns, at each step, the partially sorted array. We can use this generator as a demonstrative or debugging aid. Here is the generator, which is based on the selection sort algorithm of Chapter 1.

```
// selngtor.cpp: Selection sort generator

class SelnSorter {
private:
  int *a;     // Array we're sorting
  int i, n;   // Current step index i, and array size n
public:
  SelnSorter(int arr[], int size);
  void Reset();
  int Eos() const;
  const int *Curr() const;
  int Step();
  const int *Run();
};

SelnSorter::SelnSorter(int arr[], int size)
// Construct a selection sorter object.
{
  a = arr; n = size;
  Reset();
}

void SelnSorter::Reset()
// Reset the loop counter, so that we can sort the array
// again. (We don't put the array back to its original state
// or anything.)
{
  i = 0;
}

int SelnSorter::Eos() const
{
  return i >= n-1;
}

const int *SelnSorter::Curr() const
// Takes a snapshot of the sorting process by
// returning the current state of the array.
{
  return a;
}

int SelnSorter::Step()
// Advance one step in the sorting process.
{
  int tmp, j, m;

  if (Eos()) return 0;

  m = i;
  for (j = i+1; j<n; j++) {
      if (a[j] > a[m]) m = j;
  }
```

```
    if (m != i) {
       tmp = a[i]; a[i] = a[m]; a[m] = tmp;
    }

    ++i;

    return 1;
}

const int *SelnSorter::Run()
// Run the sorting process to completion.
{
   while(Step());
   return a;
}
```

This generator, like all generators, has a *Step()* function that computes one step of the algorithm, but in this case the result is the partially sorted array. In other words, we have a snapshot of how far the algorithm has progressed. The *Step()* function is the same function as the *SelnSort()* function of Chapter 1, except the local variables and parameters are now class members and, most important, the outer loop of the algorithm has been removed. To make the sorting take place, the user must provide an outer loop. The data member *i*, which was the outer loop index originally, serves as the state variable so the generator can pick up where it left off.

One advantage to having the user control the step execution is that the different stages of the sorting process can be printed out, making it easy to see how the sorting takes place. For example, the following code

```
int arr[7] = {5, 3, 6, 1, 4, 7, 2};

SelnSorter ss(arr, 7);

while(1) {
   const int *aip = ss.Curr(); // Grab sorted array in progress.
   for(int i = 0; i<7; i++) cout << aip[i] << ' ';
   cout << '\n';
   if (ss.Step() == 0) break; // Advance one step.
}
```

produces

```
5 3 6 1 4 7 2
7 3 6 1 4 5 2
7 6 3 1 4 5 2
7 6 5 1 4 3 2
7 6 5 4 1 3 2
7 6 5 4 3 1 2
7 6 5 4 3 2 1
```

Using the *Run()* function, it's easy to use the generator to sort an array without seeing the intermediate steps.

```
SelnSorter ss(arr, 7);
ss.Run(); // Run algorithm to completion.
```

INSTRUCTION GENERATORS

Besides returning a snapshot of the algorithm at each step, you also can create generators that generate *instructions*, either to the user or perhaps as input to another algorithm. To see an example of this, we'll take an algorithm to solve the classic Towers of Hanoi game and convert the algorithm into generator form. The generator will give us the sequence of moves we should make to solve an instance of the game.

The Towers of Hanoi game goes like this: Suppose we are given three "towers" (in the game, they are actually pegs), with n rings stacked on the first peg, such that the rings are sorted from smallest to largest, with the smallest at the top. The goal is to move all the rings to another peg, stacked in sorted order, with the restrictions that only one ring can be moved at a time and that a larger ring cannot be placed on top of a smaller ring. Figure 3.2 shows an example of the moves made for this game, for $n = 3$, where peg *a* is the starting peg and peg *b* is the ending peg, with peg *c* used as an intermediate storage area.

Figure 3.2 Moves for three-ring Towers of Hanoi game.

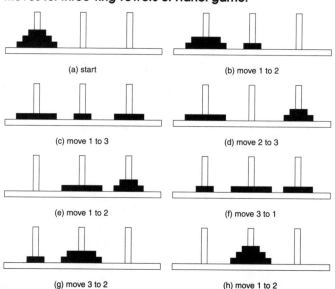

(a) start

(b) move 1 to 2

(c) move 1 to 3

(d) move 2 to 3

(e) move 1 to 2

(f) move 3 to 1

(g) move 3 to 2

(h) move 1 to 2

At first, it may seem impossible to construct an algorithm to solve this game for any *n*. However, if you think recursively, using induction, an algorithm can be derived easily. Let's represent the moves with the following function call

```
Move(n, a, b, c); // Move n rings from a to b using c.
```

which means "Move *n* rings from peg *a* to peg *b* using *c* as a temporary or working peg." Suppose we move all but one of the rings from peg *a* to peg *c* using the call

```
Move(n-1, a, c, b);
```

It's important at this stage to pretend that the *Move()* function works. Now, the last ring, the largest, will still reside on peg *a*. We can easily move the largest ring to *b*, without needing peg *c,* since peg *b* will be empty at this point. We could use the following call.

```
Move(1, a, b, b);
```

This is our base case, and we can write code inside the *Move()* function to handle this case.

```
if (n == 1) cout << "Move top ring from " << a << to << b << '\n';
```

With the largest ring where it belongs, we can then move the other rings from peg *c* to peg *b*, without disturbing the largest ring now on peg *b*, using the call

```
Move(n-1, c, b, a);
```

Believe it or not, these are all the ingredients we need to write the recursive *Move()* function, since we've handled both the base case and the inductive cases.

```
// tower1.cpp: Recursive Towers of Hanoi algorithm

void Move(int n, int a, int b, int c)
// Move n rings from peg a to peg b, using c as
// a working peg.
{
  if (n == 1) {
    cout << "Move top ring from " << a << " to " << b << '\n';
  }
  else {
    Move(n-1, a, c, b);
    Move(1, a, b, b);
    Move(n-1, c, b, a);
  }
}
```

Code for the Towers of Hanoi algorithm can be found on disk in the files *tower1.cpp, tower2.cpp, tower3.cpp, tower4.cpp, tower5.cpp,* and *tower6.cpp.* These files correspond to the versions of the algorithm given in this chapter, using the names given in the comments at the top of each version.

Here's how we would call the *Move()* function to solve the game for three rings.

```
// Move 3 rings from peg 1 to peg 2 using peg 3.
Move(3, 1, 2, 3);
```

The *Move()* function would give the following output.

```
Move top disk from 1 to 2
Move top disk from 1 to 3
Move top disk from 2 to 3
Move top disk from 1 to 2
Move top disk from 3 to 1
Move top disk from 3 to 2
Move top disk from 1 to 2
```

A Towers of Hanoi game with three rings needs seven moves to complete. This turns out to be optimal, although the sequence is not unique. The Towers of Hanoi algorithm in general takes 2^n-1 moves for a game of n rings. The algorithm is thus $O(2^n)$—in other words, an exponential algorithm. For a game involving 64 rings, it would take about 500 billion years to finish, even if a move could be made once a second!

We can simplify the original *Move()* function by replacing the second call to *Move()* with a single output statement, since all the call does is cause a move instruction to be printed. The resulting algorithm is

```
// tower2.cpp: Simplified recursive Towers of Hanoi algorithm

void Move(int n, int a, int b, int c)
// Move n rings from peg a to peg b, using c as
// a working peg.
{
  if (n <= 0) return; // Base of recursion
  Move(--n, a, c, b);
  cout << "Move top ring from " << a << " to " << b << '\n';
  Move(n, c, b, a);
}
```

We would like to get rid of the output statement in the middle of the function and pull it to the outside, so that we can sequence through the algorithm, collecting, storing, or reporting the moves any way we want. That is, we want to create a generator that generates move instructions. To do this, we first need to turn the recursion into iteration. Notice that the last recursive call is tail-recursive, so here is a tail-recursive optimized form:

```
// tower3.cpp: Towers of Hanoi with tail recursion optimization

void Move(int n, int a, int b, int c)
// Move n rings from peg a to peg b, using c as
// a working peg.
{
  while(1) {
    if (n <= 0) return;
    Move(--n, a, c, b);
    cout << "Move top ring from " << a << " to " << b << '\n';
    // Simulate: "Move(n, c, b, a);"
    int tmp = a; a = c; c = tmp;
  }
}
```

Next we need to turn the other recursive call into iteration. Here is a completely iterative form of the Towers of Hanoi algorithm, derived using the techniques of Chapter 2.

```
// tower4.cpp: Iterative Towers of Hanoi algorithm

struct Parms {
  int n, a, b, c;
  Parms() { }
  Parms(int n_, int a_, int b_, int c_);
};

Parms::Parms(int n_, int a_, int b_, int c_)
{
  n = n_; a = a_; b = b_; c = c_;
}

void Move(Parms parms)
// Move n rings from peg a to peg b, using c as
// a working peg. To start, use the call:
// Move(Parms(n, 1, 2, 3));
{
  Stack<Parms> stk(parms.n); // Need n stack elements
  int tmp;

  while(1) {
    while(1) {
      if (parms.n <= 0) break;
      // Simulate: "Move(--n, a, c, b);"
      parms.n--;
      stk.Push(parms);
      tmp = parms.b; parms.b = parms.c; parms.c = tmp;
    }
```

```
   // Returning from simulated recursive call
   if (stk.IsEmpty()) return;
   stk.Pop(parms);

   cout << "Move top ring from " << parms.a << " to " << parms.b << '\n';
   // Simulate: "Move(n, c, b, a);"
   tmp = parms.a; parms.a = parms.c; parms.c = tmp;
   }
}
```

The maximum depth of recursion at any given time is n; thus, we created a stack that can hold n sets of parameters. In other words, the Towers of Hanoi algorithm is $O(n)$ in space overhead.

The next task is to take the iterative form and turn it into a generator. Here is the result, using the techniques of this chapter and before simplifying.

```
// tower5.cpp: First Towers of Hanoi generator.
// Define Parms structure as before.

class TowerSolver {
private:
  Stack<Parms> stk;
  Parms parms; // Stores current move instructions
  int state; // 0 = ready for next step, 1 = done, 2 = continue step
public:
  TowerSolver(int n=3); // Default to three ring game.
  void Reset(int n = 0);
  int Eos() const;
  const Parms &Curr() const;
  int Step();
  int Step(Parms &pv);
};

TowerSolver::TowerSolver(int n)
// Set up a Towers of Hanoi generator, to solve the game
// for n rings.
: stk(n) // Stack size required = n
{
  Reset(n);
}

void TowerSolver::Reset(int n)
// Resets the Towers of Hanoi generator, with an optional
// new number of rings to solve (if n != 0).
{
  if (n) parms.n = n;
  state = 0;
  parms.a = 1; parms.b = 2; parms.c = 3;
  stk.Resize(n); // Might need new stack size
  Step(); // Prime the pump
}
```

```
int TowerSolver::Eos() const
// Returns 1 if all moves have been accomplished.
{
  return state == 1;
}

const Parms &TowerSolver::Curr() const
// Returns the current move instruction.
{
  return parms;
}

int TowerSolver::Step()
// Advances to the next move. Returns 1 if more moves
// to follow, else 0.
{
  int tmp;

  if (state == 1) return 0;    // No more moves
  if (state == 2) goto State2; // Pick up where you left off.

  while(1) {
    while(1) {
      if (parms.n <= 0) break;
      parms.n--;
      stk.Push(parms);
      tmp = parms.b; parms.b = parms.c; parms.c = tmp;
    }

    if (stk.IsEmpty()) { state = 1; return 0; } // No more moves
    stk.Pop(parms);
    state = 2;
    return 1; // At this point, the move is: "parms.a to parms.b"

    State2: // Picking up where we left off
    state = 0;
    tmp = parms.a; parms.a = parms.c; parms.c = tmp;
  }
}


int TowerSolver::Step(Parms &pv)
// Assigns pv with the current move, and then advances
// to the next move and returns a 1. If at end of moves,
// pv is not assigned and a 0 is returned.
{
  if (Eos()) return 0;
  pv = parms;
  Step();
  return 1;
}
```

Here's an example of using the *TowerSolver* generator.

```
TowerSolver ts(3); // Solve game for 3 rings.

Parms pm;
while(ts.Step(pm)) {
  cout << "Move top ring from " << pm.a << " to " << pm.b << '\n';
}
cout << "Finished.\n";
```

The generator in *tower5.cpp* was derived directly using the techniques given earlier in the chapter. If we rearrange the code in the *Step()* function by moving the statements following *State2* to the top of the function, we can simplify things by getting rid of the *goto*, the label *State2*, the *state = 2* value (so *state* has only the values 0 or 1), and by removing the outer *while* loop. Also, the step can be started by doing the swap of parameters that used to occur right before the second recursive call in the original *Move()* function. Because the swap takes place at the start of the step, instead of in the middle where it used to be, we need to modify the *Reset()* function to compensate. For example:

```
// tower6.cpp: Simplified Towers of Hanoi Generator
// All code as before, except the following functions.

int TowerSolver::Step()
{
  int tmp;

  if (state == 1) return 0; // No more moves
  tmp = parms.a; parms.a = parms.c; parms.c = tmp; // Swap

  while(1) {
    if (parms.n <= 0) break;
    parms.n--;
    stk.Push(parms);
    tmp = parms.b; parms.b = parms.c; parms.c = tmp;
  }

  if (stk.IsEmpty()) { state = 1; return 0; } // No more moves
  stk.Pop(parms);
  return 1; // The move is: "parms.a to parms.b"
}

void TowerSolver::Reset(int n)
{
  if (n) parms.n = n;
  state = 0;
  // Pegs a and b are swapped due to Step() starting "halfway"
  // through the original step of the algorithm.
  parms.a = 3; parms.b = 2; parms.c = 1;
  stk.Resize(n);
```

```
   Step();
}
```

GENERATOR-ITERATOR DUALITY

You may have noticed that generators are similar to another popular type of object, called *iterators*, which can be found abundantly in the C++ literature. Iterators are objects that iterate, or walk, through data structures, returning the elements found during the walk. For example, you might have a linked-list iterator that returns all the nodes in the list. Chapter 10 of the companion book contains examples of iterators that walk through binary trees in various orderings. You'll see similar iterators that traverse graphs in later chapters.

In a sense, iterators are the duals of generators. An iterator relies on stored data that's already built and ready to use, while a generator typically has to generate data. While an iterator may *use* algorithms to accomplish its goal, for all intents and purposes a generator *is* an algorithm.

Iterators and generators both can be used in the same program, and to the user (that is, to the outside), they look the same. For example, you could create a Fibonnaci iterator that has stored in an array, precomputed values of the Fibonnaci sequence, which are then returned by indexing through the array. Or you could create a Fibonnaci generator, like *Fibber*, and compute the values on the fly. To the outside, the functionality appears the same.

The main advantage to using a generator is that it allows *lazy evaluation,* when the values in the sequence aren't computed unless asked for. In contrast, with iterators, you must know and compute ahead of time all the values that might be needed. Of course, you could create hybrid objects that are a combination of iterator and generator. For instance, you might use a cache to store some of the values computed by a generator and then use an iterator to extract these. Also, having a cache in a generator would allow you to do something not otherwise possible: go backward through the sequence. The generators presented in this chapter can only go forward.

Using iterators doesn't seem to require the same shift in thinking that using generators do. Programs with iterators may use algorithms still couched in terms of the traditional, procedural point of view. To use a generator, however, you must shift your thinking deep into an object-oriented frame of mind, so much so that algorithms are thought of as objects themselves. At that point, your programs become collections of "algorithmic objects" communicating with each other using a network of data streams.

GENERATORS AND DATA FLOW

If you think of each generator as a node in a network, then you can visualize data flowing through the net, from one process to another. Figure 3.3 illustrates

Figure 3.3 A dataflow net.

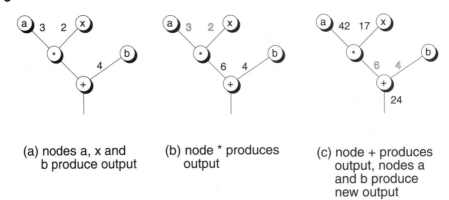

    (a) nodes a, x and   (b) node * produces   (c) node + produces
        b produce output   output   output, nodes a
                                                  and b produce
                                                  new output

the idea, using numerical data. Think of nodes *a, x,* and *b* as producers of data and nodes * and + as consumers.

In general, not only numbers can be flowing through the net, but the "instructions" of algorithms as well. For example, picture a *TowerSolver* object generating the moves to a Towers of Hanoi game and passing those instructions on to some other algorithm. (Your author does not know what use the Towers of Hanoi moves are to any other algorithm, but perhaps you can come up with some exotic purpose!)

This concept of data flowing through a network is nothing new. In fact, it is the programming paradigm call *dataflow programming,* often used to describe parallel programs. Attempts have been made to build dataflow languages (for example, the language Lucid, see [AshWa 85]), just as languages have been built around object-oriented paradigms. It's interesting that generators allow us to take the object-oriented paradigm and begin to implement the dataflow paradigm. Dataflow programming may be the next step beyond object-oriented programming. Generators seem to be an evolutionary bridge between the two styles of programming.

ALGORITHM INHERITANCE

Since generators are algorithms turned into objects, that leads to the idea that algorithms can have all the properties that other objects have. For example, we can create objects from a class that inherits from another class. Can we do the same with generators? That is, can one algorithm inherit from another? How do you inherit the "selection-sortedness" of the selection sort algorithm, for example?

Before attempting to answer this question, we'll review what we mean by inheritance. With inheritance, you start with an abstract or generalized class and then derive more specialized versions, going from general to specific. This directionality often is associated with the subtyping relationship called *IS-A*. For example, you might have a *Rectangle* class and derive from it a *Square* class. You are in effect saying that a square IS-A rectangle. The IS-A relationship is one-way, however. You wouldn't say that a rectangle IS-A square, for not all rectangles are squares. The question about algorithm inheritance is transformed, then, into the question of whether you can have two algorithms such that one algorithm IS-A subtype of the other.

There are counterpart relationships to IS-A. For example, the *HAS-A* relationship is used to indicate that one class incorporates objects from another. A stack may incorporate a linked list as part of its implementation, so a list-based stack HAS-A linked list. Closely related to HAS-A is *USES-A*. You also might say that a list-based stack USES-A linked list.

The USES-A relationship abounds in algorithm design. For example, you might have an algorithm that collects all the words in a document, throwing out duplicates. The words might be collected in a *hash table* (see Chapter 7), using a *hashing algorithm* to determine where the words go in the table. The hash table makes it convenient to check for duplicate words without having to sort them. In this case, the word-collecting algorithm USES-A hashing algorithm, but you probably wouldn't say that the word-collecting algorithm IS-A hashing algorithm.

The IS-A relationship fits naturally in designing data structures, as you probably realized if you have read the companion book. For example, in that book we show an abstract array class and, from that, derive arrays that allow static or dynamic memory allocation, variable-length arrays, resizable arrays, and so forth. The hierarchy is shown in Figure 3.4.

Figure 3.4 A data structure hierarchy.

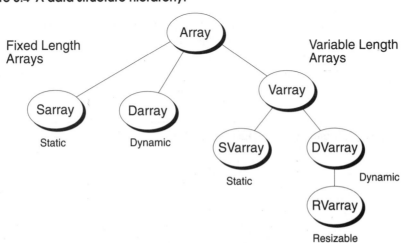

Since algorithms normally are written in a procedural style, it's hard to imagine how you go about "inheriting" from a procedure. It seems that the procedure must be broken up into pieces, to allow you to modify or replace some of those pieces. There are two possible ways a procedure or algorithm can be taken apart: laterally or serially.

Breaking Up Algorithms Laterally

Taking apart an algorithm laterally involves, in the extreme case, replacing every statement in the procedure implementing the algorithm with a function call. Then the procedure and all the tiny new functions can be made member functions of a class, and all the functions can be made virtual. At this point, you can then derive new classes, overriding the virtual functions as you see fit.

You probably wouldn't replace every statement in the original procedure by a function call. Most likely, you would use a coarser-grained approach. For example, you might break up the selection sort algorithm into two parts: the selection of a candidate element and the movement of the candidate element to the front. These can be driven by an overall driver function. You might create an abstract selection sorter class similiar to the following.

```
class Sorter {
  ...
  void Sort();              // Driver function
  virtual void Select() = 0; // Virtual selection function
  virtual void Move() = 0;   // Virtual movement of selected element
};
```

From this, you could derive a class that allows for sorting arrays, for instance, or one for sorting linked lists. A possible hierarchy, then, could be created, as shown in Figure 3.5.

Note that the *Sorter* class we present here is not a generator. You don't ask repeatedly for each step of the sort to occur. Instead, you simply call the *Sort()* function. This is in contrast to the *SelnSorter* generator we gave earlier. Of course, you could apply the same lateral breakup technique when creating a generator. For instance, you can replace the *Sort()* just given with a *Step()*

Figure 3.5 Selection sorting class hierarchy.

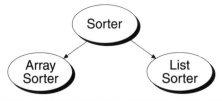

function, which presumably calls *Select()* and *Move()*. Now you can override the latter functions to provide different types of sorting generators.

Breaking Up Algorithms Serially

By laterally breaking up an algorithm, you are still, in many ways, adhering to the procedural point of view. You are controlling how each step is used internally. Another way to break up an algorithm is to manipulate the *sequence* of steps, in a serial fashion. You are then controlling how the steps are used, at a higher level. Of course, a generator allows you to do just that.

After breaking an algorithm down into serial steps, you can modify or replace these steps, coming up with a new, derived algorithm. In other words, you could take a generator class and derive from it a new generator class that overrides in some way how the steps are used.

Many such derived generators act like filters, filtering out the steps of the base class generator. Consider a generator, called *Factors*, that computes all the prime factors of a number. You might derive from this generator another generator, *UniqueFactors*, that generates all the *unique* prime factors. In this case, the overidden *Step()* function of *UniqueFactors* basically serves as a filter to remove all duplicate factors from the sequence that is produced by the *Factors* generator. In Chapter 4 you'll see this example in detail. There is real inheritance going on here, because you could certainly say that a unique factors sequence IS-A factors sequence. Thus, an algorithm inheritance hierarchy can be drawn for these algorithms, as shown in Figure 3.6.

Any time you have generators acting as filters, you have the potential for algorithm inheritance. That's because filters are a way to specialize sequences, and specialization is exactly the relationship represented by IS-A.

Filtering numbers is one thing, but what about generators that produce other types of objects, such as the partially sorted arrays of *SelnSorter* or the moves of *TowerSolver*? What algorithm could be derived from *SelnSorter*, for instance, or *TowerSolver*? The concept of algorithm inheritance using generators is

---

Figure 3.6 A factors algorithm hierarchy.

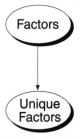

still too new to know what its ramifications are. Generators, and algorithm inheritance in general, do seem to be a fertile ground for research, though.

PERFORMANCE ISSUES

Generators allow you to use algorithms in more flexible ways. But any time you add generality or flexibility, performance is bound to suffer. So as you might expect, algorithms written in generator style probably won't run as fast as those written in procedural style.

The performance hit can be boiled down to the following: When designing a generator, basically you are removing the outer loop of an algorithm. It's up to the user then to provide the outer loop. Each pass through this loop becomes a call to *Step()*. In order to work properly, *Step()* must save the state of the algorithm between calls. Thus the overhead becomes one function call per outer loop, plus whatever is needed to maintain the state between calls. Whether this slowdown is noticeable is, of course, dependent on the application. Keep in mind that it is there, however.

The canonical form of generators we gave in this chapter was designed to minimize the overhead as much as possible, while still providing a lot of flexibility. For example, two *Step()* functions were given: one that takes doesn't return the step result and one that does. The parameterless *Step()* function should always be the one where the actual step code resides. That's so functions like *Run()* can work as fast as possible. Recall how *Run()* was defined:

```
TYPE Generator::Run()
{
  while(Step());
  return Curr();
}
```

It's hard to imagine a tighter loop for *Run()*. It's not as tight as actually including the step code directly, but then, if we did that we wouldn't be able to intercept the steps for their partial results, should they be desired. You can use *Run()* as a basis for comparing an algorithm written as a generator and a competing procedural version.

One primary disadvantage of generators is that they require you to code your algorithms in an iterative form. Some algorithms are more naturally expressed and implemented recursively. By making these algorithms iterative generators, you are gaining flexibility at the expense of clarity.

MORE EXAMPLES OF GENERATORS

You might get the impression that we are going to be writing all the algorithms in this book as generators. While Chapters 4, 5, and 6 will in fact cover more examples of generators, for the most part we'll be presenting algorithms in the conventional procedural style. That's because, admittedly, an algorithm written as a generator is harder to understand than one written procedurally. Written generator style, the workings of the algorithm are scattered across many function calls, such as in the constructor, the *Reset()* function, and the *Step()* function, whereas with a procedural algorithm, the code is mostly in one place.

There's an advantage to rearranging where the code of an algorithm resides, though. That reason lies right at the heart of the difference between procedural programming and the dataflow programming style suggested by generators. Both styles offer views into your program; they're just different views. Sometimes one view is better than the other. The procedural view lets you see the algorithm all at once, whereas the generator, dataflow view lets you think about the algorithm in more flexible ways.

As you learn about the algorithms presented in this book, we encourage you to think about how they might be implemented as generators and about how you might use the algorithms in new and exotic ways because of the technique. Who knows, maybe you'll invent some new algorithms in the process!

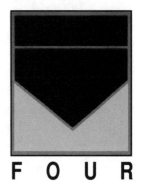

F O U R

Primes, Factors, and Permutations

In this chapter we'll take a look at several generators that produce primes, factors, permutations, and combinations. The sequences covered are at the very heart of number theory, and computer science, for that matter. However, they also have direct, practical applications, as you'll see in later chapters. Some of the algorithms presented here have been around since antiquity. It's interesting to see how we can modernize these classic algorithms using C++ and object-oriented techniques.

A PRIME NUMBER GENERATOR

Perhaps the most famous sequence of numbers is the prime number sequence. A *prime number* is an integer $p > 1$, such that no other integer $1 < x < p$ divides p evenly. Prime numbers are such an important foundation for mathematics that an algorithm to produce the primes, called the Sieve of Eratosthenes, was worked out as far back as the third century B.C. Next we take a reasonably efficient implementation of the algorithm, as given in Knuth [Knuth 69], and present it in generator form.

The Sieve of Eratosthenes works by "catching" those numbers that are not prime and letting the primes slip through. Consider the following sequence of integers.

```
2 3 4 5 6 7 8 9 10 11 12 13 14 15 16 17 18 19 20 21 22 23 24 ...
```

The number 2 is prime by inspection, and we also know that all multiples of 2 are *not* prime, by definition. Thus, we can output 2 as the first prime number and then "cross out" (remove from consideration) all other even numbers. We'll show the numbers output to the left of a vertical bar.

2 | 3 4 5 6 7 8 9 ~~10~~ 11 ~~12~~ 13 ~~14~~ 15 ~~16~~ 17 ~~18~~ 19 ~~20~~ 21 ~~22~~ 23 ~~24~~ ...

Next the number 3 is tried. Since it is not crossed out, we know it is prime, so it is output, and then all odd multiples of 3 are crossed out, starting with 3(3+0) = 9, and proceeding with 3(3 + 2) = 15, 3(3 + 4) = 21, 3(3 + 6)=27, and so on.

2 3 | 5 7 ~~9~~ 11 13 ~~15~~ 17 19 ~~21~~ 23 25 ~~27~~ 29 31 ~~33~~ 35 37 ~~39~~ 41 43 ~~45~~ ...

In general, given a prime p, we can cross out the multiples $p(p + 0)$, $p(p + 2)$, $p(p + 4)$, $p(p + 6)$, and so on. We start with the multiple p^2, because all multiples of p smaller than that would have already been crossed out during an earlier phase of the sieve process. For example, when $p = 3$, we don't need to cross out 3*2, since it was crossed out for $p = 2$. We still may end up crossing out numbers more than once, though. For instance, with the prime 5, we cross out 5(5 + 0), 5(5 + 2), 5(5 + 2), 5(5 + 4), 5(5 + 6), and so on. However, 5(5 + 4) = 45 was already crossed out as a multiple of 3. Here is the sequence for $p = 5$.

2 3 5 | 7 11 13 17 19 23 ~~25~~ 29 31 ~~35~~ 37 41 43 47 49 53 ~~55~~ ...

This process continues until we reach a prime p such that $p^2 > n$, where n is the largest number we wish to consider. At that point, all numbers $x \leq n$ have either been crossed out or are prime. Here is a list of the prime numbers less than 100 that was generated using the sieve algorithm.

 2 3 5 7 11 13 17 19 23 29 31 37 41
43 47 53 59 61 67 71 73 79 83 89 97

There are 25 primes less than 100, 168 less than 1,000, and 78,498 primes less than 1,000,000. In general, for any given x, there are roughly $x / \ln x$ primes less than x. The exact distribution is not known for all x, because no one has been able to discover any long-lasting patterns in the prime number sequence. Because of this, there are no known closed formulas for computing the primes. Most of the methods involve using some type of sieve algorithm similiar to the one shown.

How are numbers crossed out in the sieve algorithm? One way is to keep an array of flags, one for each number. We start out with all flags set to true. Multiples are crossed out by setting the corresponding flags to false. When we reach a prime $p^2 > n$, we know that all primes $1 < p$ n have been found, so the flags represent a table that tells us which numbers are prime.

Since there is one flag per number, the algorithm has a space complexity of $O(n)$. Thus, if each flag were stored in a byte, we would need a 1MB table to find the primes less than one million! We can reduce this overhead by a factor of 16 by doing the following. First, represent each flag as a bit, so eight flags per byte can be stored. Second, we need only store flags for the odd numbers, since we know that all even numbers besides 2 are not prime. We can use a bit vector (an array of bits) we'll call *bits*, of length $m = (n + 1)/2 - 1$, such that *bits*[j] is the flag for the number $2j + 3$, where $0 \leq j < m$. Note that a bit for the prime number 2 is not included in the table, only bits for the odd numbers > 1.

Here is a *Primes* generator that uses the Sieve of Eratosthenes in its *Step()* function. Shown are the most important functions *Reset()* and *Step()*. A *HugeBitVector* class, not shown, is used for storing the table of bits.

```
// The primes generator class

class Primes {
protected:
  HugeBitVector bits;
  unsigned long n, m, j, p, q, curr_prime;
  int state;
public:
  Primes(unsigned long max_n);
  void Reset(unsigned long max_n=0);
  int Eos() const            { return state == 3; }
  int Ok() const             { return bits.Ok();  }
  unsigned long Curr() const { return curr_prime; }
  int Step();
  int Step(unsigned long &p);
  const HugeBitVector *Run();
};

void Primes::Reset(unsigned long max_n)
// Resets this prime number generator to have the range
// 1 > p <= max_n, and sets the current prime to 2. If
// max_n <= 1 coming in (the default), n is left alone.
{
  if (max_n > 1) n = max_n;
  m = (n+1)/2-1;      // We have m odd numbers to work with.
  bits.SetTo(m, 1); // Set bit vector to length m, all 1's.
  state = 0;
  curr_prime = 2; j = 0; p = 3; q = 3; // Prime the pump
}

int Primes::Step()
// Advance to the next prime. Returns 1 if not at the end, else 0.
{
  if (state == 1) goto State1; // Still in sifting phase
  if (state == 2) goto State2; // In report-only phase
  if (state == 3) return 0;    // Finished.

  // Go from state 0 to state 1 to start the sifting phase.
  state = 1;

  while(1) {
    if (q >= m) break; // p^2 > n
    if (bits[j]) {
```

```
              // p = 2j+3 prime is prime, so record it and return
              curr_prime = p;
              return 1;
              State1: // Come back here at next call to Step()
              // Cross out all odd multiples of p, starting with p^2
              long k = q; // k = index of p^2
              while(k < m) { bits[k] = 0; k += p; }
          }
        j++;          // Get ready for next odd number.
        p += 2;       // p = 2j + 3, the next odd number
        q += 2*p-2; // q = 2j^2+6j+3, index corr. to p^2
    }

    // At this point, p^2 > n, so bits[] has all primes <= n recorded.
    // Now all we need to do is "report" the results.

    state = 2; // We are in the report-only phase.

    while(1) {
        if (j >= m) break; // No more bits to examine
        if (bits[j]) {       // p = 2j+3 is prime.
            curr_prime = p;
            return 1;
        }
        State2: // Come back here on next call to Step().
        j++; p += 2; // To next odd number
    }

    state = 3; // At the end of the sequence
    return 0;
}
```

 Complete code for the *Primes* generator is given on disk in the files *primes.h, primes.cpp, hbitvec.h,* and *hbitvec.cpp.* Two test programs are provided in *primtst1.cpp* and *primtst2.cpp.*

Here is some algebra to help you understand how the variables *j, p, q,* and *k* are being used. The variables are defined as follows.

j = index of bit for some odd number p, $0 \quad j < m$

$p_j = 2j + 3$

q_j = index of bit for $p_j^2 \Rightarrow 2q_j + 3 = p_j^2$

k_z = index of bit for $p_j(p_j + z)$, for $z = 0, 2, 4, 6, \ldots \Rightarrow 2k_z + 3 = p_j(p_j + z)$

A recurrence relation for p can be readily determined.

$p_0 = 3$

$p_j = p_{j-1} + 2$

We can solve for the bit index q and derive a recurrence relation as follows.

$$2q_j + 3 = p_j^2 = (2j+3)^2 = 4j^2 + 12j + 9$$
$$q_j = 2j^2 + 6j + 3$$

so,

$$q_0 = 3$$
$$q_j = q_{j-1} + d$$

where,

$$d = q_j - q_{j-1} = 2j^2 + 6j + 3 - 2(j-1)^2 - 6(j-1) - 3$$
$$d = 4j + 4 = 2(2j+3) - 2$$
$$d = 2p_j - 2$$

In a similar fashion, we can solve for the bit index k and derive a recurrence relation.

$$2k_z + 3 = p(p+z) = p^2 = pz$$
$$k_z = \frac{(2j+3)^2 + pz - 3}{2} = 2j^2 + 6j + 3 + \frac{pz}{2}$$
$$k_z = q + \frac{pz}{2}$$

so, remembering z equals 0, 2, 4, 6, ...

$$k_0 = q$$
$$k_z = k_{z-1} + p$$

The implementation given here carefully avoids multiplications and divisions. Only multiplications by 2 are needed, yielding a reasonably efficient algorithm. The time complexity is roughly $O(n)$, which was determined experimentally. (An analytical determination is difficult.) On a 33mhz 486 machine it takes a few seconds to find all primes less than $n = 1,000,000$. The space complexity, as mentioned before, is $O(n)$, needing 62,500 bytes for $n = 1,000,000$. The algorithm is not too bad for numbers up to several million, but for numbers larger than that, the algorithm becomes unwieldy, particularly in terms of storage.

Testing for Primes

Storing the bit table produced by the *Primes* generator can come in handy. Once the table is built, we can use it to test quickly whether an odd number 2 < x n is prime. For example:

```
unsigned long i = (x-3) / 2; // Compute bit index for odd x.
if (bits[i])
   cout << x << " is prime\n"; else cout << x << " is not prime\n";
```

If you are concerned only about producing the bit table, and not the sequence of primes, you can use the *Run()* function of the *Primes* generator, which builds the bit table silently.

```
const HugeBitVector *Primes::Run()
// Computes the table of bits indicating which numbers are
// prime and returns a pointer to the table. If there was a
// problem, then a 0 is returned.
{
  if (!Ok()) return 0;
  while(state <= 1) Step(); // Do only the sifting phase.
  return &bits;
}
```

Because *Primes* stores a bit table, this class is actually part generator, part iterator. During the sifting phase (*state*   1), *Primes* acts as a generator. During the report-only phase (*state* = 2), *Primes* acts more like an iterator.

Testing for prime numbers using *Primes* works as long as you don't mind paying the overhead for producing the bit table. For large numbers, such as a 20-digit number, it becomes highly impractical. Of course, on most machines today, integers are stored in 32 bits, so we couldn't represent a 20-digit number anyway without writing expensive multiprecision routines. Even if we could represent numbers greater than 2^{32}, there are no known ways of efficiently testing for the primality of large numbers. Instead, a number of ad hoc approaches can be used, none of which suffices on its own. If the numbers are less than 2^{32}, there are routines that will complete the testing in a reasonable amount of time. We'll take a look at one such technique in the next section.

FACTOR GENERATORS

One important use of prime numbers is to find the set of primes that make up the factors of a given number. Factoring numbers is for more than just theoretical interest. Many of today's cryptographic algorithms (methods for encrypting or scrambling messages and documents) have factoring as their basis. It turns out that no one knows how to factor large (such as 20-digit) numbers efficiently, and the encryption algorithms use this to their advantage. (Without the encryption key, large numbers must be factored to "crack" the encryption.)

Even though no efficient algorithms exist for factoring 20-digit numbers, there are algorithms that work reasonably well for numbers up to 10 digits (billions).We'll present the most simplistic and obvious algorithm next.

Suppose we have a number n that we wish to factor. One way to proceed is to try dividing n by each of the primes in succession, (using integer division). If we obtain a zero remainder when dividing by prime p_i, that prime is a factor. We can then use the quotient as the new n. If the remainder is not zero, and the quotient is smaller than p_i, then the current value of n is prime, and the sequence is finished. Otherwise, we repeat using prime p_{i+1} as the next divisor.

We can use this algorithm as the basis of a *Step()* function in a class that generates the finite sequence of factors for some integer n. We'll call this generator *Factors*. This generator could use the *Primes* generator to obtain a sequence of trial divisors. However, the *Primes* generator requires $O(n)$ storage for the bit table. It turns out that we can use another sequence for our trial divisors that requires only $O(1)$ storage, as suggested in Knuth [Knuth 69]. The trial divisor sequence is defined using the following recurrence relation.

$$d_0 = 2, \quad d_1 = 3, \quad d_2 = 5,$$
$$d_k = d_{k-1} + 2, \text{ for odd } k > 2$$
$$d_k = d_{k-1} + 4, \text{ for even } k > 2$$

The first few numbers of the trial divisor sequence are as follows:

```
2 3 5 7 11 13 17 19 23 25 29 31 35 37 41 43 47 49 53 55 ...
```

As you can see, the first nine numbers of this sequence are the first nine primes. After that, the numbers in the sequence include the primes, but they also include non-primes as well, such as 25, 35, 49, and so on. The factoring algorithm will still work even though some of these numbers are not prime. A non-prime divisor d_k will never be output as factor, because by the time d_k is tried, its own prime divisors will have already been factored out.

Here is the *Factors* generator that generates the sequence of factors for some number n. We show only the class definition for *Factors* and its *Reset()* and *Step()* functions. The *TrialDivisors* class is easy to implement from the corresponding recurrence relation.

```
// Prime factors generator

class Factors {
// Creates the sequence of prime factors (there may be repeats)
// for a given number n > 0, as specified in the constructor.
protected:
    TrialDivisors td;        // Sequence of trial divisors
    unsigned long num;       // The number we're trying to factor
    unsigned long curr_factor; // The current factor of the sequence
```

```
      unsigned long working_num; // Working number during factoring
public:
  Factors(unsigned long n);
  virtual void Reset(unsigned long n=0);
  int Eos()               const { return working_num <= 1; }
  unsigned long Curr() const { return curr_factor;        }
  virtual int Step();
  virtual int Step(unsigned long &pf);
};


void Factors::Reset(unsigned long n)
// Resets the generator so that it returns the prime factors of the
// number n. If n == 0, it means use the same number as before. The
// first factor will be > 1, unless n == 1, then 1 is returned.
{
  if (n) num = n;
  working_num = num; curr_factor = 1;
  td.Reset();
  Step(); // Prime the pump.
}

int Factors::Step()
// Advance to the next prime factor in the sequence.
// Returns a 0 if at eos, else 1.
{
  while(1) { // Repeatedly divide by the trial divisors
    if (working_num <= 1) return 0; // No more factors
    unsigned long quotient  = working_num / td.Curr();
    unsigned long remainder = working_num - td.Curr() * quotient;
    if (remainder == 0) {
       // The working number is a multiple of the trial divisor,
       // hence the trial divisor is a factor of num. And because of
       // the way this algorithm works, at this point, we know
       // the trial divisor is prime.
       curr_factor = td.Curr();
       working_num = quotient;
       return 1;
    }
    else if (quotient > td.Curr()) {
       // The trial divisor is not a factor of the working
       // number, and there are more candidate divisors worth
       // trying, so increment to next divisor and try again.
       td.Step();
    }
    else {
       // At this point, the working number is prime, so
       // it becomes the last factor in the sequence.
       curr_factor = working_num;
       working_num = 1;
       return 1;
    }
  }
}
```

Complete code for the *Factors* generator and *TrialDivisors* generator can be found on disk in the files *factors.h*, *factors.cpp*, *divisors.h*, and *divisors.cpp*. Also, the upcoming *UniqueFactors* class is given in the files *ufactors.h* and *ufactors.cpp*. Test programs are given in *divtest.cpp*, *factest.cpp*, and *ufactest.cpp*.

Here is an example of creating a *Factors* generator and producing the sequence of factors for the number 4,155,667,725.

```
Factors f(4155667725L);

while(1) {
  cout << f.Curr() << ' ';
  if (f.Step() == 0) break;
}
```

The sequence of factors produced is

```
3 5 5 11 53 101 941
```

The *Factors* generator may produce sequences that contain duplicate factors, such as the number 5 in this example. We can derive a new class, called *UniqueFactors*, from *Factors*, that filters out duplicates. Here is the class definition and its *Reset()* and *Step()* functions.

```
// Unique prime factors generator

class UniqueFactors : public Factors {
// Returns the sequence of unique prime factors (no repeats)
// for a given number n > 0, as specified in the constructor.
protected:
  unsigned long prev_factor; // Records last factor used
public:
  UniqueFactors(unsigned long n);
  virtual void Reset(unsigned long n=0);
  virtual int Step();
  virtual int Step(unsigned long &pf);
};


void UniqueFactors::Reset(unsigned long n)
// Resets the object so that it returns the prime factors
// of the number n. If n == 0, it means use the number
// that was used last.
{
  Factors::Reset(n);
  prev_factor = Curr();
}


int UniqueFactors::Step()
// Advance to the next unique factor. If no next unique
// factor, a 0 is returned, else a 1.
```

```
{
  while(1) {
    // Loop until we get a factor we didn't have before,
    // or we're at the end of the sequence.
    if (Factors::Step() == 0) return 0;
    if (Curr() != prev_factor) {
      prev_factor = Curr();
      return 1;
    }
  }
}
```

Now the sequence produced for 4,155,667,725 is

```
3 5 11 53 101 941
```

The *UniqueFactors* class shows an example of algorithm inheritance. There is a true IS-A relationship between *Factors* and *UniqueFactors*; a *UniqueFactors* sequence IS-A *Factors* sequence, but not all *Factors* sequences are *UniqueFactors* sequences.

Testing for Primes Revisited

We can use the *UniqueFactors* generator to test for prime numbers. If the only factor produced in the sequence for the number n is the number itself, we know that n is prime. Also, if you were to multiply the numbers in the unique factor sequence together and the result is n, then n is known as a *prime product*. Note that if there are duplicate factors, as produced by the *Factors* generator, then n won't be either a prime or a prime product. Here's an example of a program that tests for primes and prime products.

```
// ufactest.cpp: Test program for unique factors

#include <iostream.h>
#include "ufactors.h"

main()
{
  unsigned long n, fac, product = 1;
  int fac_cnt = 0;
  cout << "Enter number to factor: "; cin >> n;

  UniqueFactors uf(n);
  while(uf.Step(fac)) {
    fac_cnt++; product *= fac;
    cout << fac << ' ';
  }
  cout << '\n';
```

```
  if (product == n) {
     if (fac_cnt == 1)
        cout << "The number is prime\n";
        else cout << "The number is a prime product\n";
  }
  else cout << "The number isn't a prime or prime product\n";

  return 0;
}
```

The following text is a sample output from the program.

```
Enter number to factor: 47
47
The number is prime

Enter number to factor: 4239040639
65003 65213
The number is a prime product

Enter number to factor: 200
2 5
The number isn't a prime or prime product
```

When testing for primes, the main advantage of using the *UniqueFactors* generator over the *Primes* generator is that only $O(1)$ storage is needed, rather than $O(n)$.

Factoring Performance

The factoring algorithm given works fairly quickly on today's fast machines, although for numbers that are factors of large primes, such as 4,239,040,639, it takes bit longer. Keep in mind that the factoring algorithm works only for numbers up to 2^{32}. For numbers larger than this, the factoring algorithm slows down significantly, and the computation is burdened further by the need to use multiprecision arithmetic. There are better, albeit more complicated factoring algorithms, such as those given in Knuth [Knuth 69], but even these bog down for large numbers.

An exact analysis of the running time for factoring is hard to obtain, since the prime numbers themselves don't have any regular patterns that can be exploited. In the general case, the best overall time complexity you can obtain for factoring appears to be around $O(2^d)$, where d is the number of decimal digits in the number to factor. Thus factoring appears to be exponential in complexity. However, no one has been able to prove that there aren't more efficient factoring algorithms that have gone undiscovered.

PERMUTATION GENERATORS

Given an ordered set of n distinct objects, there are $n!$ *permutations*, or arrangements, of those objects. The expression $n!$ is computed using the equation $n! = n \times (n-1) \times (n-2) \times (n-3) \times ...1$. For example, if our set is the four numbers {1,2,3,4}, here are all $4! = 4 \times 3 \times 2 \times 1 = 24$ permutations possible:

```
(1) 1 2 3 4  ( 7) 2 1 3 4 (13) 3 1 2 4 (19) 4 1 2 3
(2) 1 2 4 3  ( 8) 2 1 4 3 (14) 3 1 4 2 (20) 4 1 3 2
(3) 1 3 2 4  ( 9) 2 3 1 4 (15) 3 2 1 4 (21) 4 2 1 3
(4) 1 3 4 2  (10) 2 3 4 1 (16) 3 2 4 1 (22) 4 2 3 1
(5) 1 4 2 3  (11) 2 4 1 3 (17) 3 4 1 2 (23) 4 3 1 2
(6) 1 4 3 2  (12) 2 4 3 1 (18) 3 4 2 1 (24) 4 3 2 1
```

Permutations are interesting to study, since they underlie many algorithms, such as those that perform sorting. For example, a sorting algorithm can be viewed as a process that generates permutations until one is found that keeps the objects in sorted order.

Next we'll construct a generator that generates, in sequence, all the possible permutations of n objects. We'll represent the objects to be permuted as the numbers from 1 to n. There is no loss in generality here, for the generator works even if the objects in the permutation array are not numbers.

There are many ways to go about generating permutations. In fact, since each sequence of permutations is in effect a permutation of permutations, this suggests that there are at least $(n!)!$ permutation algorithms possible for a given n, not counting those algorithms that generate the same sequence. The expression $(n!)!$ gets large quite rapidly, even for n as small as 4. In theory, there are at least $(4!)! = 24! = 6.205 \times 10^{23}$ algorithms possible for generating permutations of four numbers!

We'll use a special ordering for the sequences, called *lexicographic ordering*. This ordering can be understood by treating each permutation as an n-digit number, starting with $123...n$. Then the next permutation in the sequence is the one that yields the next highest n-digit number. The sequence we gave earlier for $n = 4$ is in lexicographic order.

An algorithm can be derived to generate permutations in lexicographic order by observing the sequence for $n = 4$. You'll notice that the sequence can be divided into four groups, with six permutations in a group. Figure 4.1 shows the permutations at the start of each group, and these permutations form an "outer sequence."

The outer sequence can be generated by first arranging the numbers in order, then doing a series of swaps as illustrated in the figure. The figure also shows the interesting fact that the first permutation can be restored by taking the last permutation and doing a left rotation, moving 4 to the end and 1 to the front, with the rest of the numbers shifting left as well.

Figure 4.1 Outer swaps and rotations for lexicographic permutations.

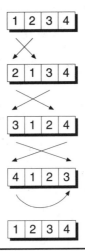

In general, for an array of *n* elements, you will have *n* groups of permutations in the outer sequence. The swapping and rotation can be applied recursively for each of the *n* groups. You permute the sets of *n* elements at the top level and then, while leaving the first number fixed, use recursion for subsets of $n-1$ elements, followed by recursion for subsets of $n-2$ elements, with two numbers fixed, and so on.

Even though it's easy to reason about this algorithm top-down, using a bottom-up implementation is more efficient, as shown in the following *Permute()* function. In this function, *s* is the offset to the start of the current subset, and *p* and *q* give the offsets to the two elements that are to be swapped. To reinforce the fact that this algorithm works for any type of object, we've written *Permute()* as a template function. The only restriction is that TYPE have assignment defined.

```
// Recursive, bottom-up lexicographic permutation algorithm

template<class TYPE>
void Permute(TYPE arr[], int n, int s=0)
// Generate all permutations, in lexicographic order, of the right-hand
// subset of the array starting at offset s. Using s = 0 generates
// permutations for the whole array of length n. When the function
// finishes, the array is restored to its original order.
{
  TYPE tmp;
  int t, p, q;

  Print(arr, n);        // Print out current permutation.
  if (s >= n-1) return; // Need at least two elements to permute.
```

```
    for (p = n-2; p>=s; p--) { // Work from right to left.
        for(q = p + 1; q<n; q++) { // Work from left to right.
          // Swap elements p and q.
          tmp = arr[p]; arr[p] = arr[q]; arr[q] = tmp;
          Permute(arr, n, p+1); // Permute subset starting at p+1.
        }
        // Use left rotation from p to n-1 to restore order.
        q--; t = p; tmp = arr[t];
        while(t != q) { arr[t] = arr[t+1]; t++; }
        arr[t] = tmp;
    }
}
```

In this algorithm, it's critical that at the end of the outer loop, a left rotation is done to restore the array to the same ordering it had before the inner loop. This restoration process is another example of *backtracking*. You saw backtracking being used in the knapsack algorithm in Chapter 2.

In the *Permute()* function, a lot of movement of the array elements takes place. This can become costly if the array elements are large. In such situations, you might want instead to permute an auxiliary array of indices, starting out with [0,1,2,...n–1]. Then you can use the index array to determine how to print out the actual array of data.

With this recursive algorithm in hand, a lexicographic permutation generator can be derived, as shown in the following *Permutator* template class. (Not all functions are shown.)

```
struct Parms {
   int p, q, s;
};

template<class TYPE>
class Permutator {
private:
   TYPE *arr;
   int n, eos;
   Stack<Parms> stk;
   Parms parms;
public:
   Permutator(TYPE arr_[], int n_);
   void Reset();
   int Eos() const;
   const TYPE *Curr() const;
   int Step();
};

template<class TYPE>
Permutator<TYPE>::Permutator(TYPE arr_[], int n_)
: stk(n_*n_) // More than enough stack space
{
   arr = arr_; n = n_; Reset();
}
```

```
template<class TYPE>
void Permutator<TYPE>::Reset()
{
  parms.s = 0; eos = 0; stk.Clear();
}

template<class TYPE>
int Permutator<TYPE>::Step()
// Generates the next permutation in lexicographic order.
{
  TYPE tmp;
  int t;

  if (eos == 1) return 0;

  if (parms.s < n-1) {
    for (parms.p = n-2; parms.p >= parms.s; parms.p--) {
        for (parms.q = parms.p + 1; parms.q < n; parms.q++) {
            // Swap elements p and q.
            tmp = arr[parms.p];
            arr[parms.p] = arr[parms.q];
            arr[parms.q] = tmp;
            // Simulate: Permute(arr, n, p+1);
            stk.Push(parms);
            parms.s = parms.p+1;
            return 1; // Have a new permutation to process

            Place1: // Back from simulated recursive call
        }
        // Use left rotation from p to n-1 to restore order.
        parms.q--; t = parms.p; tmp = arr[t];
        while(t != parms.q) { arr[t] = arr[t+1]; t++; }
        arr[t] = tmp;
    }
  }

  if (stk.Pop(parms)) goto Place1; // Return from simulated recursion.

  eos = 1;  // No more permutations, end of sequence
  return 0;
}
```

▼ **Note** Complete code for the *Permute()* function and *Permutator()* class is given on disk in the files *lexperm1.cpp, permtor.h, permtor.mth,* and *lexperm2.cpp.*

Here's an example of using the *Permutator* class:
```
const int size = 4;
int arr[size] = {1, 2, 3, 4};

Permutator<int> ptor(arr, size);

while(1) {
```

```
    Print(ptor.Curr(), size);    // Print the array.
    if (ptor.Step() == 0) break; // Next permutation
}
```

This generator is quite versatile, for you can easily generate the permutations in reverse lexicographic order (from the highest *n*-digit number to the lowest) by initializing the array in reverse order. For example, by starting the array as [4, 3, 2, 1], the following sequence is produced.

```
4 3 2 1 - 4 3 1 2 - 4 2 3 1 - 4 2 1 3 - 4 1 3 2 - 4 1 2 3 - 3 4 2 1 ...
```

It's possible to write permutation algorithms that don't use any backtracking or rotations, with just a simple swap each cycle. However, these algorithms don't generate the permutations in lexicographic order. In fact, it appears that backtracking is needed for lexicographic ordering, although your author doesn't know this for sure. (Perhaps *you* can discover a lexicographic permutation algorithm that needs no backtracking!) R. Sedgewick [Sedgewick 77] gives a survey of different permutation algorithms that you might be interested in reading.

Using Permutation Generators

Permutation generators are useful any time you need to have all the possible arrangements of a group of objects. Many algorithms can be viewed as generating permutations until some condition is reached. Unfortunately, since there are *n*! permutations for *n* objects, such exhaustive searching may lead to a worst-case complexity of $O(n!)$, which is practical only for very small *n*.

For example, one could concoct a sorting algorithm that uses a *Permutator* generator, and at each step, the current permutation can be checked to see if the objects are in sorted order. The code might look like the following.

```
template<class TYPE>
void PermutationSort(TYPE arr[], int n)
{
  Permutator<TYPE> ptor(arr, n);

  while(!InOrder(arr, n)) ptor.Step();
}
```

While this is nice conceptually, in practice it is not feasible. For example, in sorting an array of ten elements, in the worst case you'll need to generate 3,628,800 permutations! In contrast, even the inefficient $O(n^2)$ insertion sort and selection sort algorithms need generate only 100 permutations, and the most efficient $O(n \log n)$ sorting algorithms generate far less than that.

In essence, the design of certain algorithms can be thought of as the process of working out a way to avoid generating all permutations. You strive to cleverly generate only a small subset, and quickly converge on a solution. The sorting algorithms you've seen (and the ones you will see in Chapters 9 and 10) are nice examples of this.

COMBINATIONS

Related to permutations are combinations. A *combination* is formed by choosing k elements from a set of n elements. Unlike permutations, combinations are not concerned with orderings. For example, here are all the possible combinations of five elements choosing three:

```
1 2 3
1 2 4
1 2 5
1 3 4
1 3 5
1 4 5
2 3 4
2 3 5
2 4 5
3 4 5
```

Because combinations aren't ordered, the combination {3 2 1} is not considered different from {1 2 3}. We chose to list the combinations with a special ordering, which by now you should recognize as lexicographic ordering.

Let's compute how many combinations there are for n elements choosing k of them. There are n choices for the first element, $n - 1$ choices for the second, down to $n - k + 1$ choices for the kth element. This means that, if we're concerned about ordering, there are $n(n - 1)...(n - k + 1)$ choices. However, there are $k!$ ways of arranging k elements, so we must divide by $k!$ to reflect that we are in fact ignoring order. This leads to the following formula:

$$\binom{n}{k} = \frac{n(n-1)...(n-k+1)}{k(k-1)...(1)}$$

The symbol $\binom{n}{k}$ is called a *binomial coefficient* and is read "n choose k." It's called a binomial coefficient because it happens to appear in the expansion of $(x + y)^n$ as the coefficient for the term $x^k y^{n-k}$. The following function shows an easy way to compute binomial coefficients.

```
long Binomial(long n, long k)
// Compute the binomial coefficient for n choose k.
{
    if (k == 0) return 1;
    long np=n, kp=k;
```

```
    while(k > 1) { np *= --n; kp *= --k; }
    return np/kp;
}
```

By tabularizing the binomial coefficients, an amazing result emerges, in what is known as *Pascal's Triangle*, partially shown in Table 4.1. Note that the blank entries are actually zeros and that we've only shown the first portion of the triangle.

Pascal's Triangle is rife with patterns. For example, each row is symmetrical about its midpoint. Also, each number in the triangle can be obtained by summing the number above it and diagonally to the left. That's just two of the patterns. While studying Pascal's Triangle is beyond the scope of this book, it is fundamental to the study of number theory. For further information, you might start with Graham [Graham 92].

A Combination Generator

Now that you know to compute the number of combinations for n elements choosing k, how can you go about generating the combinations? Let's reexamine the lexicographic ordering for five elements choosing three:

```
1 2 3
1 2 4
1 2 5
1 3 4
1 3 5
1 4 5
2 3 4
2 3 5
2 4 5
3 4 5
```

Table 4.1 Pascal's Triangle

n	$\binom{n}{0}$	$\binom{n}{1}$	$\binom{n}{2}$	$\binom{n}{3}$	$\binom{n}{4}$	$\binom{n}{5}$	$\binom{n}{6}$	$\binom{n}{7}$
0	1							
1	1	1						
2	1	2	1					
3	1	3	3	1				
4	1	4	6	4	1			
5	1	5	10	10	5	1		
6	1	6	15	20	15	6	1	
7	1	7	21	35	35	21	7	1

You'll notice that we always increment the rightmost digit first. When it is as large as it can be, we back up one digit and increment it, and then set all digits to the right to form an increasing sequence. This process is repeated, working leftward, as we increment each digit to its maximum value.

But what is the maximum value of each digit? If we number the digits left to right starting with one, then the jth digit has a maximum value of $n - k + j$. For example, the first digit has a maximum value of $5 - 3 + 1 = 3$, the second a maximum of $5 - 3 + 2 = 4$, and the last digit $5 - 3 + 3 = 5$.

Using these facts leads us directly to an algorithm for generating combinations in lexicographic order. We present the algorithm in generator form, showing the *Reset()* and *Step()* functions.

```
class Combinator {
private:
  int *combo;
  int n, k, eos;
public:
  Combinator(int n_, int k_);
  ~Combinator();
  void Reset();
  int Eos() const;
  const int *Curr() const;
  int Step();
};


void Combinator::Reset()
// Resets the generator so that the first combination will
// be 0 1 ... k-1. If combo array wasn't allocated, then
// end-of-sequence will be indicated right from the start.
{
  if (combo) {
    eos = 0;
    for (int i = 0; i < k; i++) combo[i] = i;
  }
  else eos = 1;
}


int Combinator::Step()
// Given the current combination in combo, this advances to
// the next combination in lexicographic order. Returns 1 if
// not at end of combinations, else 0.
{
  if (eos) return 0;

  for(int j = k-1; j >= 0; j--) {
    if (combo[j] < n-k+j) {
      for (int v = combo[j]; j<k; j++) combo[j] = ++v;
      return 1;
    }
  }
}
```

```
   eos = 1;
   return 0;
}
```

Complete code for the *Combinator* class is given on disk in the files *combtor.h* and *combtor.cpp*. Test programs can be found in *combtst1.cpp* and *combtst2.cpp*.

The *Combinator* class stores the current combination in the array *combo*, which is initialized in the *Reset()* function to [0 1 2 ... k–1]. Note that we changed the notation and started from 0 rather than 1. This is to facilitate using *combo* as an index array. Unlike the *Permutator* class, the *Combinator* class works only with arrays of numbers. If you want to get combinations of an array of objects other than numbers, then you must use numbers in *combo* to serve as indices for the actual objects to select. The following program shows an example, using an array of characters.

```
// combtst2.cpp:
#include <iostream.h>
#include "combtor.h"

char charr[] = "abcde";
main()
{
  int i, k, n = 5;

  while(1) {
     cout << "Choose how many elements from 5? (0 to quit): "; cin >> k;
     if (k == 0) break;
     if (k > 5) { cout << "Sorry, maximum is 5\n";   k = 5; }
     Combinator cmbtor(n, k);
     do {
        for(i = 0; i<k; i++) cout << charr[cmbtor.Curr()[i]];
        cout << '\n';
     } while(cmbtor.Step());
  }

  return 0;
}
```

Here's sample output for five elements choosing four.

```
abcd
abce
abde
acde
bcde
```

F I V E

Linear Congruential Generators

In this chapter we'll take look at numerical sequence generators based on modulo arithmetic. These *linear congruential generators* (LCGs) can be used to produce permutations, and they form the basis of the most popular method for generating *random number sequences*, which we'll also discuss in this chapter.

LINEAR CONGRUENTIAL SEQUENCES

A *linear congruential sequence* is based on the following recurrence relation.

$$x_n = (ax_{n-1} + c) \bmod m, \text{ where } 0 \le x_0 < m$$

The variable a is known as the *multiplier*, c the *increment*, and m the *modulus*. Here is the sequence that is produced for $a = 3$, $c = 5$, $m = 11$, and $x_0 = 9$.

9 10 2 0 5 9 10 2 0 5 9 ...

A linear congruential generator can be easily implemented, as shown in the following *Lcg* class:

```
class Lcg {
protected:
  long x0, x, a, m, c;
public:
  Lcg(long x0_, long a_, long m_, long c_);
  Lcg() { }; // Default constructor does nothing
  void Configure(long x0_, long a_, long c_, long m_);
  void Reset(long x0_);
  void Reset()              { Reset(x0); }
```

```
    long InitialVal() const { return x0; }
    long Multiplier() const { return a;  }
    long Increment()  const { return c;  }
    long Modulus()    const { return m;  }
    long Curr()       const { return x;  }
    long Step();
};

Lcg::Lcg(long x0_, long a_, long c_, long m_)
// Constructs a linear congruential sequencer, using
// the configuration parameters given.
{
    Configure(x0_, a_, c_, m_);
}

void Lcg::Reset(long x0_)
// Resets the sequence using a new initial value x0.
{
    x0 = x0_ % m; // Keep x0 modulo m
    x = x0;
}

void Lcg::Configure(long x0_, long a_, long c_, long m_)
// Configures the LCG with a new set of parameters.
{
    a = a_; c = c_; m = m_;
    Reset(x0_);
}

long Lcg::Step()
// Advances to the next number in the sequence.
{
    return x = (a*x+c) % m;
}
```

Complete code for linear congruential generator class, and a test program, is given on disk in the files *lcg.h*, *lcg.cpp*, and *lcgtst.cpp*.

Full-Cycle Generators

Because of the modulo arithmetic used in an LCG, the sequence of numbers produced will cycle eventually. The length of the cycle is known as the *period*. For example, the sample sequence given earlier—9 10 2 0 5 9 10 2 0 5 9 . . .—has a period of 5.

The maximum period possible for a LCG with a modulus of m is m itself. An LCG with a maximum period is known as a *full-cycle LCG*. A full-cycle LCG produces, each cycle, a permutation (see Chapter 4) of the numbers between 0 and $m - 1$. Thus, full-cycle LCGs give us another way to generate permutations.

You can obtain a full cycle for any given m by choosing appropriate values for a, c, and x_0. The following rules, adapted from Knuth [Knuth 69], apply when c 0.

Rules to Guarantee a Full Cycle

- Let b_0 = the product of all the unique prime factors of m. If m is a multiple of 4, then multiply b_0 by 2, to guarantee that b_0 is a multiple of 4 as well. Then set $b = vb_0$, where v can be either 1 or any prime number, such that b m.
- Set the multiplier $a = b + 1$.
- The increment c can be any number such that $1 < c < m$, and with c *relatively prime* to m, or if no such relative prime exists, set $c = 1$. (Two numbers are relatively prime if they have no factors in common.)
- The starting value x_0 can be any number such that 0 $x_0 < m$.
- It's assumed that the calculations are arranged so as not to cause overflow. An example of this is given later in the chapter in the *Ran32* class.

Note   The full cycle rules apply only for c 0. LCGs with c 0 are sometimes called *mixed congruential generators*. In contrast, LCGs with $c = 0$ are sometimes called *pure multiplicative congruential generators*. We'll discuss the latter when we present random number generators later in the chapter.

Let's apply the rules for $m = 18$. Here are the possible values for b_0, b, a, c, and x_0.

b_0 = product of unique primes of $m = 2 \times 3 = 6$
$b = 6, 12, 18$, for $v = 1, 2, 3$
$a = 7, 13, 19$
$c = 5, 7, 11, 13, 17$
$x_0 = 0, 1, 2, 3, 4, 5, 6, 7, 8, 9, 10, 11, 12, 13, 14, 15, 16, 17$

Each combination of a, c, and x_0 will give us a unique permutation in one cycle. For $m = 18$, there are $3 \times 5 \times 18 = 270$ possible permutations we can achieve. However, this is a far cry from the total number of permutations of 18 objects, which is $18! = 6.4 \times 10^{15}$. In general, you'll only be able to produce only a small fraction of the $m!$ permutations using full-cycle LCGs of period m. Here is a list of 15 of the permutations for $m = 18$, generated by using a starting value of $x_0 = 1$.

```
m=18, x0=1, a= 7, c= 5: 1 12 17 16 9 14 13 6 11 10 3 8 7 0 5 4 15 2
m=18, x0=1, a= 7, c= 7: 1 14 15 4 17 0 7 2 3 10 5 6 13 8 9 16 11 12
m=18, x0=1, a= 7, c=11: 1 0 11 16 15 8 13 12 5 10 9 2 7 6 17 4 3 14
m=18, x0=1, a= 7, c=13: 1 2 9 4 5 12 7 8 15 10 11 0 13 14 3 16 17 6
```

```
m=18, x0=1, a= 7, c=17: 1 6 5 16 3 2 13 0 17 10 15 14 7 12 11 4 9 8
m=18, x0=1, a=13, c= 5: 1 0 5 16 15 2 13 12 17 10 9 14 7 6 11 4 3 8
m=18, x0=1, a=13, c= 7: 1 2 15 4 5 0 7 8 3 10 11 6 13 14 9 16 17 12
m=18, x0=1, a=13, c=11: 1 6 17 16 3 14 13 0 11 10 15 8 7 12 5 4 9 2
m=18, x0=1, a=13, c=13: 1 8 9 4 11 12 7 14 15 10 17 0 13 2 3 16 5 6
m=18, x0=1, a=13, c=17: 1 12 11 16 9 8 13 6 5 10 3 2 7 0 17 4 15 14
m=18, x0=1, a=19, c= 5: 1 6 11 16 3 8 13 0 5 10 15 2 7 12 17 4 9 14
m=18, x0=1, a=19, c= 7: 1 8 15 4 11 0 7 14 3 10 17 6 13 2 9 16 5 12
m=18, x0=1, a=19, c=11: 1 12 5 16 9 2 13 6 17 10 3 14 7 0 11 4 15 8
m=18, x0=1, a=19, c=13: 1 14 9 4 17 12 7 2 15 10 5 0 13 8 3 16 11 6
m=18, x0=1, a=19, c=17: 1 0 17 16 15 14 13 12 11 10 9 8 7 6 5 4 3 2
```

You should verify that each sequence indeed is a valid permutation, covering all numbers from 0 to 17, once and only once.

In the rules given earlier, we stated that b should not be greater than m. Actually, you can set $b > m$ and still get a full cycle, but the cycle obtained will be the same as if you took $b = b \bmod m$. For example, in the list of permutations just given, five of them used $a = 19$ (and thus $b = 18 = m$). For these cases, the same permutations would be generated had we used $a = 1$ (and thus $b = 0$). Also, you can use values of v other than prime numbers or the number 1, but again, you won't obtain unique sequences.

When m is a prime product, the number of unique permutations possible becomes limited. In effect, the only multiplier we can use is $a = 1$. That's because $b = m$ in this case, so that a will be $m + 1$, or 1 if b is taken modulo m. Here are seven of the possible 105 permutations for $m = 15$ (a prime product) and fixing $x_0 = 1$.

```
m=15, x0=1, a=1, c= 2: 1 3 5 7 9 11 13 0 2 4 6 8 10 12 14
m=15, x0=1, a=1, c= 4: 1 5 9 13 2 6 10 14 3 7 11 0 4 8 12
m=15, x0=1, a=1, c= 7: 1 8 0 7 14 6 13 5 12 4 11 3 10 2 9
m=15, x0=1, a=1, c= 8: 1 9 2 10 3 11 4 12 5 13 6 14 7 0 8
m=15, x0=1, a=1, c=11: 1 12 8 4 0 11 7 3 14 10 6 2 13 9 5
m=15, x0=1, a=1, c=13: 1 14 12 10 8 6 4 2 0 13 11 9 7 5 3
m=15, x0=1, a=1, c=14: 1 0 14 13 12 11 10 9 8 7 6 5 4 3 2
```

A Full-Cycle Generator Class

Using the techniques just outlined, we can define a full-cycle generator class. By inspecting the rules, you'll notice that we need to be able to do factoring, test for prime products, and determine when two numbers are relatively prime. Fortunately, you learned how to handle most of these tasks in Chapter 4. The following functions, which are built on the classes presented in that chapter, will help out further.

```
long ProductOfPrimes(long n)
// Finds all unique prime factors of n, computes the
// product of these primes, and returns the product.
```

```
{
  UniqueFactors uf(n);
  long product = 1;

  while(1) {
    product *= uf.Curr();
    if (uf.Step() == 0) break;
  }
  return product;
}

int IsPrimeProduct(long n)
// Returns 1 if n is a prime or prime product.
{
  return ProductOfPrimes(n) == n;
}
```

Along with these functions, we need to be able to test for relative primes (that is, when two numbers have no common factors). One way to do this is to find the greatest common divisor (GCD) of the two numbers. If the GCD is 1, then the numbers are relatively prime. Given a function *Gcd()* that returns the greatest common divisor, here is a function that tests for relative primeness.

```
int IsRelativePrime(long a, long b)
// Returns 1 if a and b have no common factors.
{
  return Gcd(a, b) == 1;
}
```

In turns out that the *Gcd()* function can be implemented by using what many consider to be the first algorithm ever invented: *Euclid's algorithm*, which came about in the third century B.C. That algorithm is reminiscent of the factoring algorithm given in Chapter 4, since we do repeated divisions until a zero remainder is obtained:

```
long Gcd(long n, long d)
// Returns the greatest common divisor of n and d, both
// ASSUMED to be positive and greater than 0.
{
  while(d > 0) { // Repeat until remainder of division is zero.
    long r = n % d;
    n = d;
    d = r;
  }
  return n;
}
```

An alternative way to implement the *Gcd()* function is to use subtraction rather than division. (Try working it out!) Since subtraction is cheaper than division, you may get a faster function, at least for some numbers. However, other times, using division will be faster. Knuth [Knuth 69] analyzes this in depth and gives other methods for computing the greatest common divisor.

Given these helper functions, here is a *FullCycler* class that can generate permutations using a full-cycle LCG.

```
class FullCycler : public Lcg {
protected:
  long first_b;
  int vi;
public:
  FullCycler(long p, long x0_);
  void NewPeriod(long p, long x0_);
  long Period() const { return m; }
  int NextSeq();
};
```

 Complete code for the *FullCycler* class is given on disk in the files *cycler.h*, *cycler.cpp*, and *cyctst.cpp*; the latter is a test program.

The *FullCycler* class is derived from *Lcg*, and the base LCG generator is configured to produce a full cycle, using the function *NewPeriod()*.

```
void FullCycler::NewPeriod(long p, long x0_)
// Sets up the linear congruential multiplier parms so
// that it will give you a modulo sequence that will be
// a complete permutation of the numbers 0..p-1, starting
// with the number x0_ and assuming 0 <= x0_ <= p-1.
{
  first_b = ProductOfPrimes(p);

  if (p - (p/4)*4 == 0) {
    // p is a multiple of 4, so guarantee first_b is a multiple of 4.
    first_b *= 2; // Note: first_b will already have 2 as a factor.
  }
  else {
    // If first_b == p, and p isn't a multiple of 4, then p is a prime
    // product. Because of this, we use first_b mod p = 0. All other
    // possible values for first_b won't produce unique sequences.
    if (first_b == p) first_b = 0;
  }

  vi = 0; // Initialize the index for the prime table.

  Configure(x0_, first_b + 1, FirstRelativePrime(p, 1), p);
}
```

The *NewPeriod()* function, called by the constructor, computes the first legal values of *a* and *c* that will produce a full cycle. The function *NextSeq()*, given later, will then compute the next legal values of these variables, if any. In order to produce the first and next values of *c*, the following helper function is used.

```
long FirstRelativePrime(long n, long low_bound)
// Finds the first number w such that low_bound < w < n,
// and such that w is relative prime to n.
// If no such number is found, a 1 is returned.
{
  for (long w = low_bound+1; w<n; w++) {
      if (Gcd(w, n) == 1) return w;
  }
  return 1;
}
```

The *FullCycler* generator is different from the generators you've seen in previous chapters, because it can generate sequences of sequences. The *NextSeq()* function is used to switch to a new permutation cycle having the same period as before.

```
int FullCycler::NextSeq()
// Use the next valid values of the multiplier and increment that will
// lead us to another linear congruential sequence having the same
// period as before. Returns 1 if there is a next sequence, else 0.
{
  // First, we'll look for another increment.

  long newc = FirstRelativePrime(m, c);
  if (newc != 1) { // We found one that will work.
     Configure(x0, a, newc, m);
     return 1;
  }

  // No more increments to try, so try a new multiplier. As long
  // as m isn't a prime product, we can take the original multiplier
  // and multiply it by a prime. We'll do this up to seven times.
  // (This is arbitrary; we could continue with as many primes as
  // we wished, until b > m.)

  if (IsPrimeProduct(m)) return 0; // It's no use.
  if (vi == 7) return 0; // Went through all seven choices.

  static int seven_primes[7] = {2, 3, 5, 7, 11, 13, 17};

  long b = first_b * seven_primes[vi++];
  if (b > m) return 0; // No point, we'll just repeat ourselves

  Configure(x0, b+1, FirstRelativePrime(m, 1), m);
  return 1;
}
```

The permutation lists given earlier were generated with the *FullCycler* class, using code similar to the following.

```
FullCycler cyc(p, x0); // FullCycler having period p, initial value x0
```

```
while(1) { // For each sequence
  cout << "m=" << p << ", x0=" << x0 << ", a=" << cyc.Multiplier();
  cout << ", c=" << cyc.Increment() << ": ";
  for(long i = 0; i<p; i++) { // For each number in sequence
    cout << cyc.curr() << ' ';
    cyc.Step();
  }
  cout << '\n';
  if (cyc.NextSeq() == 0) break;
}
```

RANDOM NUMBER SEQUENCES

Another type of sequence that is widely used in computer science is a *random number sequence*. By this we mean that each number in the sequence is, for all intents and purposes, independent of the numbers that came before it, in such a manner that we can't predict the next number that will turn up. Many techniques can be used for generating random sequences, but the most popular technique involves using LCGs. That's because LCGs are simple to implement, fast, and if set up properly, capable of producing quite random sequences.

Many people object philosophically to the idea of using LCGs—or any algorithm, for that matter—to produce random sequences. That's because, by definition, the numbers in the sequences are not supposed to depend in any way on the other numbers in the sequence. The only devices considered capable of producing such random events are physical ones, such as the rolling of a die, or the number of cosmic rays striking a surface during any given time interval.

If we use a device such as an LCG to generate the sequences, the numbers are far from being independent, since they follow the recurrence relation of the LCG. However, to someone who doesn't know the underlying relationship, the numbers do appear to be random. For this reason, the term *pseudo-random number sequence* is often used.

Randomness is truly in the eye of the beholder. For example, who is to say that the rolling of a die is random? It certainly follows the laws of physics, and is thus deterministic. It's just that we aren't capable of handling enough of the complexity to predict what number will turn up next. In a practical sense, all we care about is whether the numbers *appear* to be random and whether they are random enough for our application. For these reasons, we'll simply use the term *random number sequence*, dropping the *pseudo* qualifier.

Random number sequences can be generated using an LCG by carefully choosing values for m, a, and c. For example, if we choose $m = 18$, $a = 7$, $c = 5$, and start out with $x_0 = 12$, we get the fairly random sequence below:

```
12 17 16 9 14 13 6 11 10 3 8 7 0 5 4 15 2 1 12 17 16 ...
```

The only problem with this sequence is that it has a short period of only 18 numbers. Unless we are never going to use more than 18 numbers, a definite pattern emerges. Thus, one important criteria for using LCGs for random sequences is that they should have a long period. The way to achieve a long period is to use a large modulus, and follow the rules for generating a full cycle.

Using a large period does not guarantee a random sequence, however. Suppose we use a modulus of $m = 2^{31} - 1 = 2,147,483,647$, the largest number that can be obtained in a signed 32-bit representation. Following the rules for a full cycle, we can only use a multiplier of $a = 1$, since m is prime. Choosing an increment $c = 2$ and $x_0 = 1$, the following sequence emerges.

1 3 5 7 9 11 13 15 17 19 . . .

This is hardly random, since each number is just two added to the previous number. In fact, for any increment c that we use, the sequence produced will have c as the difference between two consecutive numbers. Here the rules for generating a full cycle get in the way. Any time we wish to get a full cycle with $c \neq 0$ and m as a prime or prime product, we'll run into this same problem.

We can work around this problem in three ways. One way is to give up trying to obtain a full cycle, and just use parameters that yield a random sequence, even if the period might be smaller. Another way is to choose the highest number possible for the modulus that isn't a prime or prime product. This may lead us to a sequence that's acceptable.

A third way is to abandon the use of a mixed generator altogether, and instead use a pure multiplicative generator, with $c = 0$. The recurrence relation then reduces to

$$x_n = ax_{n-1} \bmod m$$

Using a pure multiplicative generator involves one less computation (the addition) than with mixed generators. Because of this, pure multiplicative generators are often used in random number generators. But how do we insure the numbers are random and that the period is large?

No one knows how to guarantee randomness. The best you can do is try various values for a and m and test the resulting sequences. Ensuring a full cycle is easier. For pure multiplicative generators, a different set of rules is needed to ensure a full cycle from what is used for mixed generators. We won't go into these rules, since they require more sophisticated math than we wish to present here. You can read Knuth [Knuth 69] for more information.

Two important aspects of these rules need to be mentioned, however. One is that you *want* the modulus to be prime. That's opposite the case for a mixed generator, where using a prime or prime product modulus restricts your choices and can lead to non-random sequences. Another aspect of the rules is that it's not possible to get a full cycle with a pure multiplicative generator. However, we can get one less than a full cycle, which, practically speaking, is just as good. The only number that can't be used is the number 0. Once a 0 is output, all numbers to follow will be 0 as well.

THE MINIMAL STANDARD GENERATOR

Pure multiplicative generators have been studied extensively for use as random number generators. Picking the modulus m is fairly obvious. You want one that is as large as possible. And, unlike the case for mixed LCGs, you want a prime modulus. Picking the multiplier a is trickier. For the modulus 2,147,483,647, there are over 500 million values of a that will guarantee a full cycle. However, most of these will not generate good random numbers.

One set of parameters that has enjoyed wide use is m = 2,147,483,647 and a = 16,807. A generator with these particular set of parameters is called the *Minimal Standard Generator*. This term was coined by Park and Miller [ParkMill 88]. To paraphrase what they say in their paper, there might be better random number generators, but all generators ought to be at least as good as this one.

The Minimal Standard Generator has been tested extensively for randomness and is considered to produce acceptably random sequences. However, having that result in theory is not the same as having it in practice. One problem with implementing the Minimal Standard Generator is overflow in the calculations. For example, suppose the last number output is 424,217. To generate the next number, we first multiply by a = 16,807. The result is 7,129,815,119, which, unfortunately, requires more than 32 bits in its representation. The multiplication will overflow before we have a chance to take the result modulo $2^{31} - 1$. We end up with an erroneous sequence.

Note This problem of overflow applies to all LCGs, including the full-cycle generators given earlier in the chapter. Beware of the problem, especially as the modulus gets larger.

By rearranging the calculations to avoid overflow, Schrage [Bratley 87] developed a method that works using 32-bit arithmetic for the Minimal Standard Generator. We won't go into how the method was derived, but merely show it in the form of a function that produces the next random number.

```
long NextRandomNumber(long seed)
// Computes the next number in the Minimal Standard Random Number
// Sequence, given the previous number called the "seed."
{
  static const long a = 16807;
  static const long m = 2147483647L;
  static const long q = m / a;
  static const long r = m % a;
```

```
   seed = a * (seed - (seed/q) * q) - (seed/q) * r;
   if (seed < 0) seed += m;
   return seed;
}
```

Note Whenever we use the type *long* in this book, it's understood that a signed integer 32-bits wide is needed, and it's assumed that *long* is a signed 32-bit integer. This will be true with most C++ compilers in use today.

The method shown in *NextRandomNumber()* can be used to get around overflow problems any time $r < q$. This relationship can be obtained by making judicious choices for a and m, such as the ones used in the Minimal Standard Generator, with the result of $r = 2,836$ and $q = 127,773$.

Because you must keep the previous number (called the *seed*) around in order to compute the next number, it works best to encapsulate the random number generator into a class, such as the following *Ran32* class.

```
// A 32-bit Minimal Standard Random Number Generator

class Ran32 {
protected:
  static const long a, m, q, r;
  long seed;
public:
  Ran32(long s);            { Reset(s);        }
  void Reset(long s);
  long Curr()      const { return seed;    }
  long Step()              { return Next(); }
  long Next();
  long Curr(long n) const;
  long Next(long n);
  long Curr(long lo, long hi) const;
  long Next(long lo, long hi);
};

const long Ran32::m = 2147483647L; // "Minimal Standard" constants
const long Ran32::a = 16807L;
const long Ran32::q = m / a;
const long Ran32::r = m % a;
```

Complete code for the *Ran32* class is given on disk in the files *ran32.h* and *ran32.cpp*. Test programs reside in the files *ran32tst.cpp* and *ran32tst2.cpp*.

The *Reset()* and *Next()* functions are at the heart of the *Ran32* class:

```
void Ran32::Reset(long s)
// Resets the random number sequence by starting with a new seed,
// assumed to be between 1 and m-1.
{
```

```
  seed = s;
  if (seed < 1) seed = 1; // Keep seed in range.
  else if (seed >= m) seed = m-1;
}

long Ran32::Next()
// Advances to the next random number, carefully avoiding overflows.
{
  seed = a * (seed - (seed/q) * q) - (seed/q) * r;
  if (seed < 0) seed += m;
  return seed;
}
```

We've kept *Ran32* in canonical generator form by adding functions like *Curr()* and *Step()*, as shown inline in the class definition.

The numbers produced by *Ran32* range from 1 to 2,147,483,646 ($2^{31} - 2$). To start the sequence, you must supply a non-zero initial value, called the seed. Presumably, this seed should be chosen at random, using some other method, such as from the ticks on the system clock or by flipping coins to construct a random binary number. Any time you use the same seed, you'll get the same sequence. This can actually be an advantage, for you can rerun a program involving random numbers using the same sequence as before, perhaps to allow for debugging.

As long as you don't request more than 2,147,483,646 numbers from the *Ran32* sequence, the numbers won't repeat during any one run. In effect, using the *Ran32* generator is like drawing numbers from a bucket of 2,147,483,646 unique integers. Once you've drawn a number, it won't be replaced. This process is called *random draw without replacement*.

Scaled Random Numbers

More often than not, you don't want a sequence of numbers that range from 1 to $2^{31} - 2$, but rather, you want a much smaller range. For example, if you want to choose an element randomly from an array of ten elements, you'll want random numbers that range from 0 to 9. The following variation of the *Next()* function works by using modulo arithmetic to scale the next random number to be between 0 and $n - 1$.

```
long Ran32::Next(long n)
// Advances to the next random number nv, and
// scales its value such that 0 <= nv < n.
{
  return Next() % n;
}
```

Another, more general variation of *Next()* is the following function, which returns the next random number scaled and adjusted to be between *lo* and *hi*, inclusively.

```
long Ran32::Next(long lo, long hi)
// Returns the next value of the sequence, nv, and
// restricts its range to be lo <= nv <= hi.
{
  return lo + (Next() % (hi-lo+1));
}
```

When using the raw, unscaled output of *Ran32*, the numbers aren't re-peated until you've completed one cycle. That is, you are doing random draws without replacement. However, when scaling the output, you will get numbers showing up more than once, even before the next cycle. Of course, the smaller the scale, the more often numbers are repeated. For example, here's the sequence produced when using a seed of 7 and repeated calls of *Next(10)*.

7 9 3 6 5 9 7 8 8 3 1 6 4 9 0 1 3 6 4 7 ...

The numbers produced by the *Ran32* generator are *uniformly distributed*. This means that given a scale of n, each number from 0 to $n-1$ has, in theory, a $1/n$ chance of being the next number output, regardless of what the previous number was.

Local vs Global Randomness

When using scaled output from the *Ran32* class, you'll often see sequences that look anything but random. For example, here are the first few numbers produced by using a seed of 17 and calling *Next(10)*.

9 9 7 7 7 3 8 9 ...

The first few numbers of this sequence don't look very random, with both 9s and 7s appearing too often. Actually, the randomness depends on your point of view. When using random sequences, you must keep in mind the difference between *local randomness*, where only a short finite sequence is considered, and *global randomness*, where the theoretically infinite sequence is considered.

From a local point of view, the first five numbers of the sequence don't seem very random. The most bothersome aspect is the three 7s in a row. (This is known as a *run* of 7s). Intuitively, you would expect a sequence more like 5 7 1 3 9, where there are no runs. However, from a global point of view, runs can exist. For example, the probability that the first number is a 7 is 1/10, assuming a uniform distribution. The probability that the next number is also a 7 is 1/100. The probability of having three 7s in a row is 1/1000, and so on. Given a long enough sequence, there is a chance, however small—($10^{-1,000,000}$)—that you'll have a million 7s in a row.

In practical terms, local non-randomness can be a problem, especially when you are using a short sequences of a small range of numbers. For example, if you are using scaled output from *Ran32* to move a spaceship at random in a video game, you may be frustrated by the occasional lack of randomness in the spaceship's movements. The problem might very well be a case of local non-randomness. A better choice of seed, or a different method of using the random numbers to control the spaceship might alleviate the problem.

UNIFORM DEVIATES

In some cases, the desired output from a random number sequence isn't a sequence of integers, but rather a sequence of floating point numbers. This is particularly true in simulations, where, by convention, the random number generators usually output floating point numbers between 0 and 1. These types of random numbers are sometimes called *uniform deviates*, because a given number between 0 and 1 is just as likely to appear as any other.

Uniform deviates can be obtained easily from the sequence of integers produced by the *Ran32* class by first converting the integers to floating point and then dividing by the modulus *m*. The following *fRan32* class, derived from *Ran32*, does just that. We show the class definition and the *Next()* function.

```
class fRan32 : private Ran32 {
protected:
  static const float d; // d = 1/m
public:
  fRan32(long s);
  Ran32::Reset; // Make base function public.
  long Seed();  // To get direct access to seed
  float Curr() const;
  float Step();
  float Next();
  float Curr(float n) const;
  float Next(float n);
  float Curr(float lo, float hi) const;
  float Next(float lo, float hi);
};

const float fRan32::d = 1.0 / float(Ran32::m);

float fRan32::Next()
// Advances to the next random number in floating point
// form, and limits the range from 0 < nv < 1.
{
  return float(Ran32::Next()) * d;
}
```

 Complete code for the *fRan32* class is given on disk in the files *fran32.h* and *fran32.cpp*. A test program can be found in *fr32tst.cpp*.

Keep in mind that the numbers 0 and 1 cannot be output using *fRan32*, and there is a limit to the precision of the numbers produced. The smallest number possible is $1/m$, and the largest is $1 - 1/m$. The numbers from *fRan32* will be uniformly distributed, with each number having a probability of occurrence of approximately $1/m$.

Non-Uniform Deviates

When using random numbers for simulations, what's often desired is not uniform deviates, but rather random numbers following some other probability distribution, such as the familiar bell-shaped gaussian distribution. Numerous algorithms exist that can take uniform deviates and derive random variables with other forms of distribution. For further information, here are some references to get you started: [Bratley 87], [Knuth69], and [Press 88].

THE SPECTRAL TEST

How can you tell if a random sequence generator is truly producing random sequences? The answer is that it's not easy. However, many statistical tests have been developed. As usual, Knuth gives a good survey [Knuth 69], so that's a good place to start. These tests are beyond the scope of this book. Besides, the random number generators given here have already been tested extensively.

One criticism of the Minimal Standard Generator and actually of most random number generators based on LCGs, is that they fail what is known as the *spectral test*. In general, this test looks for correlations between the numbers produced, in *k*-space. Many random number generators will tend to cause the numbers to cluster in hyperplanes.

For a more down-to-earth example, take the case for $k = 2$. Here consecutive numbers are tested for correlations. A graphical way to perform this test is to take the random numbers from the sequence in pairs, the first being treated as an *x* coordinate and the second as the *y* coordinate. Dots are then plotted at these coordinates. This process is repeated many times until you get a picture full of dots.

If the numbers are truly random, there won't be any patterns noticeable in the dots. If there are correlations, you'll see streaks in the picture where the dots line up. For example, Figure 5.1 shows a spectral test picture produced for the Minimal Standard Generator. It certainly has noticeable streaks, meaning that there are correlations between consecutive numbers.

Code to produce the spectral test picture is given on disk in the file *spectral.cpp*. This code was written for BC++ 3.1 in a DOS environment but can be adapted easily for other environments.

Figure 5.1 Spectral plot for the Minimal Standard Generator.

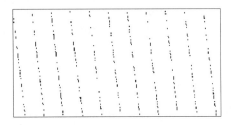

While it's interesting to perform the spectral test graphically, it's usually done with mathematics alone. In theory, you want to perform spectral tests on a random number generator for higher dimensions, not just $k = 2$. A perfect generator would not have any significant correlations in any dimension, but you would be hard pressed to find one.

COMBINED RANDOM NUMBER GENERATORS

If the two-dimensional correlations produced by the Minimal Standard Generator bother you, there is another type of generator, known as a *combined random number generator*, that doesn't have the correlations. This type of generator is based on the common technique of combining the outputs of two or more random number generators, in the hopes of producing a more random sequence.

One way of combining generators using two LCG generators is given in a paper by L'Ecuyer [L'Ecuyer 88]. We've incorporated his technique in the following combined generator class called *CRan32*.

```
class CRan32 {
protected:
  static const long a1, m1, q1, r1;
  static const long a2, m2, q2, r2;
  long seed1, seed2, curr;
public:
  CRan32(long s1, long s2);
  void Reset(long s1, long s2);
  long Curr() const;
  long Step();
  long Next();
  long Curr(long n) const;
  long Next(long n);
  long Curr(long lo, long hi) const;
  long Next(long lo, long hi);
};
```

Complete code for the *CRan32* class is given on disk in the files *cran32.h* and *cran32.cpp*. A test program can be found in *cr32tst.cpp*. Also, a companion class, *fCRan32*, which produces uniform deviates in floating point, is given in the files *fcran32.h* and *fcran32.cpp*, with a test program in *fcr32tst.cpp*.

L'Ecuyer tested many combinations of constants for the combined generator, and suggests that the following constants be used.

```
// Constants for first generator
const long CRan32::m1 = 2147483563L;
const long CRan32::a1 = 40014L;
const long CRan32::q1 = m1 / a1;
const long CRan32::r1 = m1 % a1;

// Constants for second generator
const long CRan32::m2 = 2147483399L;
const long CRan32::a2 = 40692L;
const long CRan32::q2 = m2 / a2;
const long CRan32::r2 = m2 % a2;
```

The *Next()* function shows how the two generators are combined. First, the two LCGs are computed separately, and then their results are merged together.

```
long CRan32::Next()
// Advances to the next random number with uniform distribution,
// with a range of [1..m1-1]. The period is (m1-1)*(m2-1)/2,
// or roughly 2.3x10^18 for the constants as given.
{
  long k;

  // Run the two generators independently.

  k = seed1 / q1;
  seed1 = a1 * (seed1 - k*q1) - k*r1;
  if (seed1 < 0) seed1 += m1;

  k = seed2 / q2;
  seed2 = a2 * (seed2 - k*q2) - k*r2;
  if (seed2 < 0) seed2 += m2;

  // Combine their results.

  curr = seed1 - seed2;
  if (curr < 1) curr += m1-1;

  return curr;
}
```

You'll notice that the calculations are arranged to alleviate overflow problems, using the same technique we gave for the Minimal Standard Generator.

As you can see, the *CRan32* generator takes two seeds rather than one. These seeds must be in the proper range, as shown in the following *Reset()* function.

```
void CRan32::Reset(long s1, long s2)
// Resets the random number generator by setting
// the two seeds. The seeds must be between
// [1 .. respective_modulo - 1].
{
  seed1 = s1;
  if (seed1 < 1) seed1 = 1;
  else if (seed1 >= m1) seed1 = m1-1;

  seed2 = s2;
  if (seed2 < 1) seed2 = 1;
  else if (seed2 >= m2) seed2 = m2-1;

  // Combine the seeds to get the current value.

  curr = seed1 - seed2;
  if (curr < 1) curr += m1-1;
}
```

The constants chosen by L'Ecuyer produce a combined generator that performs well on the spectral test, for dimensions as high as 6. (Only extreme purists and skeptics would look any higher than that.) Figure 5.2 plots the spectral map for $k = 2$, using the same program we gave earlier for *Ran32*.

In the literature, you'll find numerous schemes of combining generators, using other random number generators besides LCGs. Some of these are quite complex and ad hoc, involving, for example, the use of *shuffle arrays* (see Chapter 6) and other cumbersome devices. We advocate the use of two combined LCGs instead. As can be seen from the *Next()* function of *CRan32*, it's an extremely simple approach that takes very little memory. The *CRan32* generator also is quite fast; it's only twice as slow as the less random Minimal Standard Generator.

Figure 5.2 Spectral plot for the combined generator.

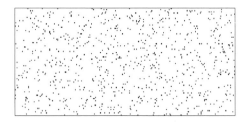

The *CRan32* class produces sequences with a long period of approximately 2.3×10^{18}. The sequences are random enough to satisfy all but the most hardened critics.

RANDOM BIT SEQUENCES

Sometimes all that you want are random numbers chosen from 0 and 1, which simulate a coin toss. You could generate these random bits using the *Ran32* class. Simply use only one of the bits of each the number generated. (This trick only works with good random number generators.) However, a faster way to generate random bit sequences, which involves only shifts and XORs. This technique utilizes what are known as *primitive polynomials modulo 2*. These are polynomials whose coefficients are either zero or one and that have certain properties beyond the scope of what we can discuss here. Here is a primitive polynomial modulo 2 with a degree of 32. (That is, its highest power is 32.)

$$x^{32} + x^7 + x^5 + x^2 + x + 1$$

You can represent this polynomial using a shorthand notation of (32, 7, 5, 2, 1, 0), or with a bit mask as well, of (1,00000000,00000000,00000000,10100111). It turns out that we can take this bit mask (ignoring the high bit at position 33), and use it in operations applied to a running, unsigned 32-bit number to produce random bits at the most significant position. The random sequence has a long period of $2^{32}-1$. The algorithm is quite simple to implement.

```
int RandomBit(unsigned long seed)
// Using the seed, to be passed repeatedly to this routine, a
// random bit sequence is produced, with a cycle length of 2^32-1.
// We assume that longs are 32 bits.
{
  // Note: Bit 33 is not actually used.
  const unsigned long mask = 0x000000AF; // Bits 7,5,3,2,1,0
  if (seed & 0x80000000L) {
    seed <<= 1;
    seed ^= mask;
    return 1;
  }
  else {
    seed <<= 1;
    return 0;
  }
}
```

Understanding how this works takes some sophisticated mathematics that we won't describe here. You can find the theory in Knuth [Knuth 69]. During operation, if the most significant bit is 1, then an XOR with *mask* is performed.

A left-shift follows in any case. The value of the most significant prior to the operations is returned. Here's a list of the first 10 values of *seed* produced when an initial value of 2,222,222,222 is used:

```
10000100011101000110101110001110 - High bit: 1
00001000111010001101011110110011 - High bit: 0
00010001110100011010111101100110 - High bit: 0
00100011101000110101111011001100 - High bit: 0
01000111010001101011110110011000 - High bit: 0
10001110100011010111101100110000 - High bit: 1
00011101000110101111011011001111 - High bit: 0
00111010001101011110110110011110 - High bit: 0
01110100011010111101101100111100 - High bit: 0
11101000110101111011011001111000 - High bit: 1
```

This results in a bit sequence of 1000010001. Note that *seed* must not be zero, or there will never be any 1-bits produced.

No XORing is done until the first time the most significant bit is a 1. Because of this, you may encounter some local non-randomness upon startup with seeds that don't have any 1s in the high positions. For example, here is the sequence using a seed of 42.

```
00000000000000000000000000101010 - High bit: 0
00000000000000000000000001010100 - High bit: 0
00000000000000000000000010101000 - High bit: 0
00000000000000000000000101010000 - High bit: 0
00000000000000000000001010100000 - High bit: 0
00000000000000000000010101000000 - High bit: 0
00000000000000000000101010000000 - High bit: 0
00000000000000000001010100000000 - High bit: 0
00000000000000000010101000000000 - High bit: 0
00000000000000000101010000000000 - High bit: 0
...
```

Here the first ten bits produced are all zeros. In fact, the first 1-bit doesn't appear until the 27th time. Any time you start out with a seed that has few or no 1-bits in the high positions, it takes a while for the generator to kick in gear, and start producing 1-bits. In most applications, this is not desirable, and is probably unexpected by anyone using the random bit generator. For this reason, you might prime the generator by running it awhile before actually using the resulting bits. You might discard the first 32 bits, for instance.

Keep in mind the difference between local and global randomness. Locally, a sequence may not appear very random, but globally, it might. You may find long runs of 1s or 0s, but that's to be expected. In theory, in a long enough sequence, there is even a small chance (very small) of having runs of a million 1s or 0s.

One way to test the randomness of bit sequences is to produce a fairly long sequence, and then keep track of how often runs of 1s and 0s appear.

Since the probability that any given number in the sequence is a 0 is 1/2, then the probability that the next number is a 0 is 1/4. A run of three 0s has a probability of 1/8. In general, a run of n 0s has a probability of $1/2^n$. The same is true for runs of 1s. Thus you could devise a program that computes and compares the actual frequency of runs of different lengths with the expected or theoretical probability.

Complete code for a random bit generator class, called *RanBit32*, which is based on the *RandomBit()* function, can be found on disk in the files *ranbit32.h* and *ranbit32.cpp*. Test programs can be found in *rb32tst.cpp*, *rb32tst2.cpp*, and *rb32tst3.cpp*.

CAVEAT EMPTOR

There are other techniques for generating random numbers, but few match the simplicity of LCG-based generators. The random number generators you've seen in this chapter are simple, fast, and capable of producing reasonably random sequences. However, the constants used in the underlying LCGs were chosen carefully, and have been tested extensively by various people. Unfortunately, that is not always true with other random number generators. Computer science lore is filled with horror stories related to the use of bad random number generators, many of which made it to standard libraries used by millions of people.

As Knuth cautions [Knuth 69], don't assume anything when it comes to random number generators. If you are going to use one, you should test it thoroughly, or use only those generators that *have* been tested thoroughly and are known to be good. It's best to treat any unknown random number generator you see with suspicion, particularly if you are going to use it in simulations that require statistical accuracy.

This rule applies to random number generators found in standard libraries. For example, C has the standard library function *rand()*. Different implementations of C use different generators for *rand()*, and many of them are no good. If you are going to do serious simulation work where randomness is crucial, you should check out your version of *rand()* thoroughly or, better yet, avoid using *rand()* altogether and use a generator that's known to be good.

Another common mistake people make is inventing their own random number generators. One word of advice: Don't! Unless you understand the underlying math and have a good suite of tests, it's doubtful you can come up with a good generator. More likely you'll produce a horribly bad one. You are far better off sticking to generators that are well known and thoroughly tested.

Random Combinations and Permutations

I n this chapter we'll put the linear congruential generators and random number generators of Chapter 5 to use in obtaining random combinations and permutations. Algorithms like the ones presented here are integral to many statistical applications, but they have other uses as well, as you'll see in the last part of the chapter.

RANDOM COMBINATIONS

In this section we'll take a look at some of the techniques for obtaining random samples, useful as input to statistical applications. Random sampling is based on the problem of choosing k of items at random given a population size of n items of data. If you've just finished studying Chapter 4, you may realize that what actually is being obtained is a *random combination*. The three techniques to be presented now are adaptations of those found in Knuth [Knuth 69] and Bratley [Bratley 87].

The type of algorithm used in sampling depends on how the data is organized. The first method we'll show is useful when you have random access to the data. The second technique is useful when you have only sequential access and where the input size is known. The third technique is used for sequential access when the input size *isn't* known ahead of time.

Shuffle Sampling

If you have an array of size n that you would like to take a sample of size k, here is a technique reminiscent of selection sorting. (See Chapter 1.) First, select an element u from the array, chosen randomly, and move it to the first position. The move can be accomplished by exchanging element 0 and *u*. Next select another element, chosen randomly from every element except the first, and move it to the second position. For the third position, choose from all those elements except the first two, and so on. This process continues *k* times. At the end, the first *k* elements of the array will be the desired random sample. Figure 6.1 illustrates the technique, and the following *ShuffleSampler()* function implements it.

```
template<class TYPE>
void ShuffleSampler(TYPE *arr, int n, int k, Ran32 &rg)
// Given an array of size n, this routine generates a random sample
// of k elements. The sample will reside in the first elements.
// NOTE: To shuffle the entire array, use k = n.
{
  if (k > n) k = n; // For safety's sake

  // Starting with the first element, swap it with a random
  // element from the array. Repeat, working toward the
  // right, for k times.

  for (int j = 0; j < k; j++) {
      int u = (int) rg.Next(j, n-1); // j <= u < n
      TYPE t = arr[u];  arr[u] = arr[j];  arr[j] = t;
  }
}
```

 The *ShuffleSampler()* function is given on disk, along with a test program, in the file *shuffle.cpp*.

If you set *k* = *n* when calling *ShuffleSampler()*, you'll cause the entire array to be shuffled. In essence, what you obtain is a *random permutation*. Later you'll see another way to obtain random permutations, generated as sequences.

Figure 6.1 Random sampling via shuffle array.

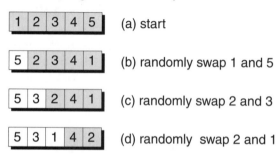

Final sample: 5 3 1

Selection Sampling

Next we present an algorithm that can be used to obtain a random sample when you have only sequential access to the data. This algorithm depends on knowing, ahead of time, the size of the population, n. The basic idea is to read in an item and immediately (and randomly) decide whether to select or discard it. This continues until k items have been chosen, or we run out of input, which isn't supposed to happen, since it's assumed $k \leq n$.

The most obvious approach would be to select each item with a probability of k/n. However, that would only guarantee *on average* that a total of k items would be selected before running out of input. Sometimes you would get less than k items. At other times you would, in effect, choose the k elements too soon and not get a true random sample.

The term k/n is correct as the probability of the first item, but not the others. If the first item is chosen, then there are $k - 1$ items left to choose and $n - 1$ items to choose from. Thus the probability for the second item, having chosen the first, should be $(k - 1)/(n - 1)$. If the first item wasn't chosen, then there are still k items to choose, but out of only $n - 1$ possibilities. Thus the probability for the second item, having discarded the first, should be $k/(n - 1)$. If we let *dk* be the number of items left to choose and *dn* be the number of items left to choose *from*, then the appropriate probability at any given step should be *dk/dn*.

A random choice can be made, then, by obtaining a random number u between 0 and 1, and if u *dk/dn,* the item is discarded, else it's kept. The following *SelectionSampler* class, written generator style, uses this approach to randomly select items from a sequential stream.

```
template<class TYPE>
class SelectionSampler {
// Class template for a selection sampler, where the
// total number of data points is known in advance
private:
  istream &istrm; // Input stream
  fRan32 rg;      // Uniform random number sequencer
  int n, k;       // Number of data points / sample size
  int dn, dk;     // Number of data points left to read / choose
public:
  SelectionSampler(istream &s, int n_, int k_, long seed);
  void Reset();
  int Step(TYPE &d);
  int Eos() const;
};

template<class TYPE>
SelectionSampler<TYPE>
::SelectionSampler(istream &s, int n_, int k_, long seed)
: istrm(s), rg(seed), n(n_), k(k_)
```

```
// Construct a selection sampler, and connect it to the input
// stream s, assumed to be open and in good working order.
{
  Reset();
}

template<class TYPE>
void SelectionSampler<TYPE>::Reset()
// Reset the collection gathering statistics.
{
  dn = n; dk = k;
}

template<class TYPE>
int SelectionSampler<TYPE>::Eos() const
// End of sample can occur due to reaching the desired
// sample size, reaching the end of the data prematurely,
// or due to some other type of stream failure.
// Returns 1 at eos, else 0.
{
  return dk == 0 || !istrm;
}

template<class TYPE>
int SelectionSampler<TYPE>::Step(TYPE &d)
// Gets the next randomly selected data point. Returns 1 if valid,
// or 0 if at eos. Won't modify d unless 1 is returned.
{
  while(1) { // Until we select a data point or reach eos ...
    if (Eos()) return 0;
    if (dn * rg.Next() >= dk) {
      // Discard the next data point.
      TYPE dmy;
      istrm >> dmy;
      if (!istrm) return 0; // Bad read
      dn--;
    }
    else {
      // Choose the next data point for the sample.
      istrm >> d;
      if (!istrm) return 0; // Bad read
      dn--; dk--;
      return 1;
    }
  }
}
```

 Complete code for the *SelectionSampler* class is given on disk in the files
selnsamp.h, *selnsamp.mth*, *samtst1.cpp*, and *samtst.dat*.

As unlikely as it may seem, the *SelectionSampler* class will produce a truly
random sample (assuming the underyling random number generator is good).

Each item is chosen with an overall probability of k/n. But aren't we contradicting ourselves? Didn't we say earlier that using a probability of k/n was bad? The seeming confusion lies in the difference between computing, ahead of time what the probability will be and computing as we go along what it needs to be, depending on the actual choices made during previous steps.

For example, the probability of the second item being chosen can be written as the summation of two components.

$$p_{choose\ 2} = (k/n)\frac{(k-1)}{(n-1)} + (1-k/n)\frac{k}{(n-1)}$$

The first part is the contribution when we choose the first element and also choose the second. The second part is the contribution for the case where we don't choose the first element but do choose the second. The sum is the total probability of choosing the second element. If you simplify the equation, the result becomes.

$$p_{choose\ 2} = \frac{k}{n}$$

A similar analysis can be done for each item, and the result will always be k/n.

As an example of using the *SelectionSampler* class, assume you have the following sequence of integers stored in the input file *samtst.dat*.

1 2 3 4 5 6 7 8 9 10 11 12 13 14 15 16 17

Here's how you could obtain a sample of seven numbers.

```
#include <fstream.h>
#include "selnsamp.h"

main()
{
  fstream fin;
  long seed = 174217;
  int stream_size = 17, sample_size = 7, sample;

  fin.open("samtst.dat", ios::in);

  SelectionSampler<int> ss(fin, stream_size, sample_size, seed);

  while(ss.Step(sample)) cout << sample << ' ';
  cout << '\n';

  fin.close();
  return 0;
}
```

Here's a random sample obtained from the program.

```
4 8 9 11 12 13 17
```

Note that in the sample sequence, the items produced are selected in the order they appear in the input stream. In the next sampling technique we present, this won't be the case.

Reservoir Sampling

Suppose you want to take a random sample from a sequential stream, but don't know ahead of time how long that stream is (assuming, of course, that it's at least as long as the sample you are taking). The technique in *SelectionSampler* isn't the best, because it needs to know n in order to compute the probability for selection. You could make two passes, reading all the items from the stream, computing n, and then rewinding to read the items again using *SelectionSampler*. Surprisingly, there is another method that requires only a single pass through the data.

The idea is to assign a random key to each item in the input stream. Then simply pick the k pairs with the highest keys. The associated items constitute a random sample, as long as each key is unique. You might think you need to store n pairs for the algorithm to work. You actually can get by with storing only k pairs. Here's how: Initialize an array of k pairs, called the *pool* or *reservoir*, and fill it with the first k items from the stream, generating random keys for each. Then read in the rest of the input stream and, for each item, generate a random key. If the key is greater than the smallest key in the pool, replace the corresponding pair in the pool with the new pair. At the end of the process, the pool will contain those pairs with the highest keys and, thus, a random sample of data items.

The following *ReservoirSampler* class implements the technique.

```
template<class TYPE>
class ReservoirSampler {
// Class template for an reservoir sampler, where the total
// number of data points is not known in advance
protected:
  istream &istrm;   // Input stream
  Ran32 rg;         // Random sequencer
  TYPE *pool;       // Pool of data point candidates
  long *keys;       // Corresponding random keys
  int k;            // Number of samples desired
  int cursor;       // Iterator index thru pool
  int IndxToSmallestKey();
public:
  ReservoirSampler(istream &s, int ns, long seed);
  ~ReservoirSampler();
  int Reset();
```

```
    int Step(TYPE &d);
    int Eos() const;
};


template<class TYPE>
ReservoirSampler<TYPE>::ReservoirSampler(istream &s, int ns, long seed)
// This constructor connects the sampler to an input stream,
// presumed to be already opened. ns is the desired sample size,
// and seed is used to initialize the random number generator.
: istrm(s), k(ns), rg(seed)
{
    pool = new TYPE[ns]; // Allocate data pool and keys.
    keys = new long[ns];
    cursor = k; // Thus eos is true till Reset() finishes.
}

template<class TYPE>
ReservoirSampler<TYPE>::~ReservoirSampler()
{
    delete[] pool;
    delete[] keys;
}

template<class TYPE>
int ReservoirSampler<TYPE>::IndxToSmallestKey()
// Returns the index of the smallest key.
{
    long m;
    int i, mi;

    for (i = 0, mi = 0, m = 0x7fffffffL; i<k; i++) {
        if (keys[i] < m) {
            m = keys[i];
            mi = i;
        }
    }
    return mi;
}

template<class TYPE>
int ReservoirSampler<TYPE>::Reset()
// Sets up the open sampler by collecting a pool of candidates.
// At the end of the routine, the pool will contain a random
// sample of the candidations.
// Returns 1 on success, 0 if there is a stream failure or
// premature end of file, or memory allocation error.
{
    cursor = k; // Signals eos, until we're successful.

    if (pool == 0 || keys == 0) return 0;

    // First, read in k records to initially fill the pool.
```

```
     for (int nc = 0; nc < k; nc++) {
        istrm >> pool[nc];
        if (!istrm) return 0; // Abort on stream failure.
        keys[nc] = rg.Next();
     }

     // Now read till end of stream, replacing candidates randomly.

     while(istrm) {
        long u = rg.Next();
        int mi = IndxToSmallestKey();
        if (u < keys[mi]) {
           // u is smaller than the smallest key in the pool,
           // so skip the next data point.
           TYPE dmy;
           istrm >> dmy;
        }
        else {
           // u is larger than or equal to the key value in the
           // pool, so read data and replace smallest candidate.
           istrm >> pool[mi];
           keys[mi] = u;
        }
     }

     if (istrm.eof()) { // Completed normally
        cursor = 0;        // Ready to start iterator portion
        return 1;
     }
     else return 0;      // Stream failure, (note: k = cursor)
}

template<class TYPE>
int ReservoirSampler<TYPE>::Eos() const
// Returns 1 if at end of sample.
{
   return cursor >= k;
}

template<class TYPE>
int ReservoirSampler<TYPE>::Step(TYPE &d)
// Gets the next randomly selected data point.
// Returns 1 if valid, or 0 if at end of data.
{
   if (Eos()) return 0;
   d = pool[cursor++];
   return 1;
}
```

Code for the *ReservoirSampler* class, along with a test program, can be found
on disk in the files *rsvrsamp.h*, *rsvrsamp.mth*, *samtst2.cpp*, and *samtst.dat*.

The function *IndxToSmallestKey* has to do a linear search of *k* items to find the smallest key, requiring $O(k)$ time. If *k* is large, this may not be satisfactory. One efficient alternative is to implement the pool using a priority queue, where the search time can be reduced to $O(k\log k)$. (See Chapter 8.)

The *ReservoirSampler* class is part generator, part iterator. During the *Reset()* function, the data items are generated by being filtered through *pool*, ending up with the random sample upon finishing. Then the iterator phase kicks in, where *Step()* is called repeatedly to retrieve the items from *pool*. Here's an example of using *ReservoirSampler*.

```
#include <fstream.h>
#include "rsvrsamp.h"

main()
{
  fstream fin;
  long seed = 3;
  int sample_size = 7, sample;

  fin.open("test.dat", ios::in);

  ReservoirSampler<int> ss(fin, sample_size, seed);

  if (ss.Reset()) { // Build the pool
    // Iterate through the pool.
    while(ss.Step(sample)) cout << sample << ' ';
    cout << '\n';
  }
  else cout << "Problem occurred building pool\n";

  fin.close();
  return 0;
}
```

Suppose the input stream consists of the following items, shown paired with the random keys assigned to each items.

```
data        : 1  2  3  4  5  6  7  8 9 10 11 12 13 14 15 16 17
random keys:  3 45 10 24 76 23 34 25 4 16 18 22 42  6 12 88 27
```

Here is the pool after being built. (You might want to work through the algorithm by hand.)

```
data:         8  2 13 17  5 16  7
random keys: 25 45 42 27 76 88 34
```

Note that the data does not come out in the order that it was input, as was the case with *SelectionSampler*. That's due to the fact that every time we add a pair to the pool, the old pair with the smallest key is replaced, and the location of that pair changes as the algorithm progresses.

Essentially, the pool serves as a shuffle array, similar to the one in *ShuffleSampler*. Unlike a shuffle array, though, not all of the data is ever present in the pool (unless $k = n$). The idea of using a shuffle array as a filter to randomize input comes up again in the next algorithm.

RANDOM PERMUTATIONS

The three sampling algorithms we've presented all return random combinations. If, however, we set $k = n$, then the *ShuffleSampler* class can be used to generate *random permutations*. Note that *SelectionSampler* can't be used to generate random permutations, since it always presents the data in order. And the *ReservoirSampler* class can't be used either, for if you set $k = n$, then the pool is simply filled with the data, in order, with no chance for shuffling. Of course, you could take the pool and, using the random keys, reorder the data. But that's no better than using the *ShuffleSampler* technique to begin with.

Using a shuffle array to obtain a random permutation is fine, but it does require $O(n)$ storage, and all n data items must be present at once in order to obtain the permutation. When n is large, then the technique becomes less than satisfactory.

Next, we present a random permutation technique, developed by the author, that requires only $O(1)$ storage. This technique allows you to generate permutations of the numbers between 0 and $n - 1$, without storing ahead of time the n numbers being permuted. Instead, the numbers are generated one by one upon request. Algorithms with this kind of behavoir are called *sequential*, or *on-line* algorithms.

With on-line algorithms, you can use *lazy evaluation*: That is, values are produced only when asked for. Little or no work or storage is needed for those values not requested. As you might guess, many of the generators that we've built in the past few chapters implement on-line algorithms.

Where might an on-line algorithm for generating random permutations come in handy? Suppose you have a graphics application where you would like to "dissolve" a 1024×1024 image in such a way that the pixels are erased, seemingly at random. Due to the large number of pixels, you don't want to erase a pixel more than once, yet you must ensure that all pixels are erased eventually.

One way to solve this problem is to treat the image as a one-dimensional array. Then generate a random permutation of the indices to all 1024^2 pixels. This permuted index can be used to erase the pixels randomly. The problem is, the permuted index has over a million elements, too large to be stored practically. The solution is to generate a sequence of 1024^2 indices, one by one, such that a random permutation is generated. Then the pixels can be erased, at random, without requiring a large permutation index. An on-line permutation sequencer is just the tool to use.

Recall from Chapter 5 that, by definition, a full-cycle LCG produces a sequence that is a permutation of *m* numbers, where *m* is the modulus. For example, here is one of the permutations possible for 18 elements, using the parameters shown.

```
m=18, x0=1, a=13, c= 7: 1 2 15 4 5 0 7 8 3 10 11 6 13 14 9 16 17 12
```

It's easy to see that this is a valid permutation, and the sequence of numbers is reasonably random as well. However, the randomness you can obtain with a full-cycle LCG greatly depends on what you use for the parameters. For example, here is one of the permutations possible for $m = 8$.

```
m=8, x0=1, a= 5, c= 3: 1  0  3  2  5  4  7  6
```

While that's certainly a permutation of eight numbers, is it random? If you look closely, you'll notice that the difference between successive numbers follows the alternating sequence −1, 3, −1, 3, and so on. By using the same value for *x0*, but varying the legal values for *a* and *c*, another pattern emerges.

```
m=8, x0=1, a= 5, c= 3: 1  0  3  2  5  4  7  6
m=8, x0=1, a= 5, c= 5: 1  2  7  0  5  6  3  4
m=8, x0=1, a= 5, c= 7: 1  4  3  6  5  0  7  2
m=8, x0=1, a= 9, c= 3: 1  4  7  2  5  0  3  6
m=8, x0=1, a= 9, c= 5: 1  6  3  0  5  2  7  4
m=8, x0=1, a= 9, c= 7: 1  0  7  6  5  4  3  2
```

Notice how the middle value is always the same. This happens any time *m* is even. The problem is more acute when *m* is a prime or prime product, for there the choice of multiplier is limited to $a = 1$. This always leads to sequences where it's easy to see non-randomness. For example, here are the permutations obtainable using a full-cycle LCG, for $m = 10$, fixing $x0 = 1$.

```
m=10, x0=1, a= 1, c= 3: 1  4  7  0  3  6  9  2  5  8
m=10, x0=1, a= 1, c= 7: 1  8  5  2  9  6  3  0  7  4
m=10, x0=1, a= 1, c= 9: 1  0  9  8  7  6  5  4  3  2
```

In each of these cases, the difference between successive numbers is simply equal to the increment *c*, modulo *m*.

There are actually two aspects of randomness to be concerned with here. One is whether you can easily predict the next number in the sequence, as we have just been considering. The other is whether each of the *m*! permutations possible is equally likely. Note that these two aspects can be at odds with one another. For example, the permutation 1 2 3 4 5 of five numbers is just as likely as any other, but we wouldn't normally think of the sequence as random. It all depends on what we are going to do with the sequence.

As you've seen, full-cycle LCGs aren't always capable of producing sequences where the numbers appear random. And only a tiny fraction of the possible permutations can be generated. In the example just given for *m* = 10, you can obtain 30 different permutations, in three sets of ten. Each set is generated by fixing *a* and *c* and then using one of ten different starting values. Generating 30 permutations is a far cry from the 10! = 3,628,800 permutations possible.

While using a full-cycle LCG gives us the desired property of being an on-line algorithm, we must somehow further randomize the sequence produced. One way to do this is to use a filter, similar to the reservoir in the *ReservoirSampler* class. Unlike a reservoir, however, we can't ever discard input data, or we wouldn't end up with a permutation.

The solution is to use a variation of a priority queue. (See Chapter 8.) In a priority queue, items are added to one end of the queue. The next item selected for output from the other end depends on the priority of the item. The element with the highest priority is chosen. For our application, we can give each element a random priority and thus the sequence of elements output will be randomized. Such a data structure is called a *random priority queue*. Since it is a queue, rather than a reservoir, no data will be lost in the process.

Piles: Contiguous Random Priority Queues

We'll call a random priority queue implemented contiguously a *pile* (a term coined by the author). This name was chosen in the spirit of the name given for normal contiguous priority queues, known as *heaps*. A pile works as follows: Items are always added to the end of the queue. When an item is requested, one is selected at random from those available. Because of the random selection process, a pile is similar to a shuffle array. The following *Pile* class shows an implementation.

```
template<class TYPE>
class Pile {
protected:
  Ran32 rg;       // Random generator for extractions
  TYPE *data;     // Pointer to pile data
  long slot;      // Slot to next be extracted from
  int size, len; // Maximum and current lengths
public:
  Pile(int sz, long seed, long starting_slot=-1);
  virtual ~Pile();
  int IsOk() const;
  int IsFull() const;
  int IsEmpty() const;
  int Size() const;
  void Clear(long starting_slot=-1);
  void NewSeed(long seed);
  int Insert(const TYPE &x);
  int Extract(TYPE &x);
```

```
TYPE &Filter(TYPE &outdata, const TYPE &indata);
const TYPE *Peek() const;
const TYPE *Peek(const TYPE &s) const;
};
```

Complete code for the *Pile* class is given on disk in the files *pile.h* and *pile.mth*. A test program resides in *piletst.cpp*.

One critical data member of *Pile* is *slot*. This is a randomly generated number that, when taken modulo *len* or *len + 1*, determines the index of the element to be extracted next. This variable is updated to a random value after every extraction. You can force the starting slot to be a specific value, by setting the *starting_slot* parameter in the constructor to be anything except –1. As you'll see later in the *RandomPermutator* class, this allows you to control what the first output will be, in case that's needed.

Here are the *Insert()* and *Extract()* functions for *Pile*, which are used in the same way as for any queue.

```
template<class TYPE>
int Pile<TYPE>::Insert(const TYPE &x)
// Inserts element x into the queue by adding it
// to the end. If pile is full a 0 is returned.
{
  if (len == size) return 0;
  data[len++] = x;
  return 1;
}
```

```
template<class TYPE>
int Pile<TYPE>::Extract(TYPE &x)
// Extracts, randomly, an element from the pile.
// Returns 0 if pile is empty, else 1.
{
  if (len == 0) return 0;
  int u = int(slot % len);
  x = data[u];
  // Replace hole just made with the element from the end
  if (u != --len) data[u] = data[len];
  slot = rg.Next(); // Needs to be taken module len or len+1
  return 1;
}
```

While inserting and extracting can be done at different times and rates, it's often useful to do both an insertion and an extraction at once. We'll call this the *on-line filter mode*. The following *Filter()* function is used for this mode.

```
template<class TYPE>
TYPE &Pile<TYPE>::Filter(TYPE &outdata, const TYPE &indata)
// Inserts indata into the pile, using it to replace outdata,
```

```
// randomly chosen, which gets extracted. It's possible for
// the indata to be the outdata. That will be a certainty if
// the pile is currently empty.
{
  int u = int(slot % (len+1));
  if (u == len) outdata = indata; // Bypassing pile
  else {
     outdata = data[u]; data[u] = indata;
  }
  slot = rg.Next(); // Needs to be take modulo len or len+1
  return outdata;
}
```

Conceptually (but not physically), the input is inserted before the output is extracted. This makes it possible for the item just inserted to be the one extracted. By doing this, we increase by one the number of items to choose from for the extraction, which means better randomization. If the extracted item is the inserted item, then the item is passed straight through without being stored.

How well a pile randomizes depends on how many items k are currently stored in the pile. If $k = n$, the total number of items to be filtered, the pile becomes essentially a shuffle array and produces fairly random results (unless you force the starting slot to a specific value). When k is smaller than n, the resulting sequence becomes less random, because it's impossible for items near the end of the input sequence to be output first. When the pile is used in the on-line filter mode, it's guaranteed that one of the first $k + 1$ inputs will be output first. If the pile is used with independent insertions and extractions, then one of the first k inputs will be chosen for output.

In general, the randomizing is localized around a moving window of approximately the size k. It's as if the items are randomized loosely in groups. When $k = 0$ (such as when the maximum pile size is 0), the pile becomes an identity filter.

Here's a sample program that uses a pile to help randomize a stream of numbers. Note that the program initially fills the pile to maximum size and keeps it that way as long as possible. This yields better randomization.

```
#include <fstream.h>
#include "pile.h"

main()
{
  fstream fin;
  long seed = 174217;
  int pile_size = 7, indata, outdata;

  Pile<int> pile(pile_size, seed);
```

```
    fin.open("test.dat", ios::in);

    // Best if pile is kept full, so fill it to start with
    for(int i = 0; i<pile_size; i++) {
      fin >> indata;
      if (!fin) break;
      pile.Insert(indata);
    }

    while(1) { // While there is still input, use on-line filter mode.
      fin >> indata;
      if (!fin) break;
      cout << pile.Filter(outdata, indata) << ' ';
    }

    // Now empty the rest of the pile randomly.
    while(pile.Extract(outdata)) cout << outdata << ' ';
    cout << '\n';

    fin.close();
    return 0;
}
```

Here's sample input and output, using a pile of size 4.

```
input:  1  2  3  4  5  6  7  8  9 10 11 12 13 14 15 16 17 18 19 20 21
output: 3  5  2  4  8  7  6  9 12 13  1 10 14 17 18 19 11 15 16 21 20
```

You'll notice that the output is randomized loosely in groups of four numbers, although a number may appear in the output far outside its group, such as the number 1 in the example. Numbers near the beginning of the input sequence may appear toward the end of the output sequence, but not the other way around. In our example, with a pile of size 4, the numbers 6 through 21 could never be output first. A more randomized sequence can be obtained by setting the pile size higher, and the most random sequence is obtained if the pile size equals 21.

A Random Permutation Generator

We can build an on-line random permutation generator by combining a full-cycle LCG with a pile. The pile serves to randomize the output from the LCG further, which by itself won't always yield random sequences (although some LCGs *do* generate random sequences, as you well know). The following *RandomPermutor* class shows an implementation.

```
class RandomPermutor {
protected:
  FullCycler cycler; // Full-cycle LCG
  Pile<long> pile;   // Randomizing priority queue
```

```
    Ran32 rg;           // Generator to configure LCG randomly
    int cursor;         // Cursor into cycle
    void ConfigureCycler(long x0);
public:
    RandomPermutor(long n, long x0, int psize, long pseed=424217L);
    void Reset();
    long Curr() const;
    long Step();
    long Period() const;
    long Multiplier() const;
    long Increment() const;
};
```

Complete code for the *RandomPermutor* class is given on disk in the files *raptor.h*, and *raptor.cpp*, and a test program can be found in *raptst.cpp*.

The constructor sets up the pile, and the full-cycler LCG and also initializes an auxiliary random number generator.

```
RandomPermutor::RandomPermutor(long n, long x0, int psize, long pseed)
// Initializes pile with psize and pseed, the cycler with period n and
// seed x0, and the aux random generator using a combined seed.
// WARNING: ASSUMES 0 <= psize <= n.
: pile(psize, pseed), cycler(n, x0), rg(pseed+x0)
{
    ConfigureCycler(x0);
    Reset();
}
```

Recall that there are different sets of parameters that we can choose to make the LCG full cycle. The *ConfigureCycler()* function randomly chooses one set of parameters, using the reservoir sampling technique in disguise. (The size of the sample is 1.)

```
void RandomPermutor::ConfigureCycler(long x0)
// This routine randomly chooses from the possible sets of
// parms for the full-cycle LCG.
{
    long a = cycler.Multiplier(); // Initial set of parameters
    long c = cycler.Increment();
    long u = rg.Next();           // Random key for initial set

    while(cycler.NextSeq()) {
        // Go through all sets of parms, choose one
        // whose random key is the largest.
        long new_u = rg.Next();
        if (new_u > u) {
            a = cycler.Multiplier();
            c = cycler.Increment();
            u = new_u;
        }
    }
```

```
    cycler.Configure(x0, a, c, cycler.Period());
}
```

After the LCG is configured, the *Reset()* function is called to fill up the pile with input data from the LCG. Again, the idea is to keep the pile as full as possible to aid the randomization process.

```
void RandomPermutor::Reset()
// Clear the pile, and fill it with numbers from the full-cycle LCG.
{
  int i, n;

  cycler.Reset(); // Sets up x0 to be the first number.
  pile.Clear(0);  // The 0 effectively forces x0 to be output first.
  n = pile.Size();
  if (n > cycler.Period()) n = (int) cycler.Period(); // Prevent bad pile.
  for (i = 0, cursor = 0; i<n; i++) {
      pile.Insert(cycler.Curr());
      cycler.Step(); cursor++;
  }
}
```

Note the call to *Pile's Clear()* function, where a 0 is passed. This forces the starting slot to be the 0th slot. During the initial filling of the pile, *x0* will be stored in the 0th slot. Thus it will be *x0* that is output first. Sometimes it's useful to have a random permutation where the first number is a specific value (but all other numbers are random), as you'll see in Chapter 7. If we wanted more randomness here, we could pass −1 to *Clear()* (or pass no parameter at all, since −1 is the default). Another option is to randomize *x0* itself, before it gets passed to the *RandomPermutator* constructor.

Here is the *Step()* function that advances to the next number in the permutation. It filters the LCG output through the pile to randomize it, always keeping the pile as large as possible.

```
long RandomPermutor::Step()
// Advances to the next element in the random permutation.
// Returns -1 at the end of the sequence.
{
  long x;

  if (cursor == cycler.Period()) {
     // We've went a full cycle, so just extract from pile.
     if (pile.Extract(x) == 0) x = -1; // Pile empty too
  }
  else { // Otherwise use on-line filter mode
     pile.Filter(x, cycler.Curr());
     cycler.Step(); cursor++;
  }

  return x;
}
```

The following program shows an example of generating a random permutation of the numbers 0 through 9.

```
#include <iostream.h>
#include "raptor.h"

main()
{
  long n     = 10;       // Size of the permutation
  int psize  = 5;        // Size of the pile
  long x0    = 6;        // Randomly chosen initial value for LCG
  long pseed = 421742L;  // Randomly chosen seed for the pile

  RandomPermutor raptor(n, x0, psize, pseed);

  cout << "Sequence length: " << n;
  cout << "  Pile Size: " << psize;
  cout << "  Pile seed: " << pseed << '\n';
  cout << "m=" << n << ", x0=" << x0;
  cout << ", a=" << setw(2) << raptor.Multiplier();
  cout << ", c=" << setw(2) << raptor.Increment() << ": ";

  while(1) {
    long v = raptor.Step();
    if (v == -1) break;
    cout << v << ' ';
  }
  cout << '\n';

  return 0;
}
```

Here is a sample output from the program.

```
Sequence length: 10  Pile size: 5  Pile seed: 421742
m = 10, x0 = 6, a = 1, c = 7: 6 7 1 5 2 3 9 4 0 8
```

Seeding the Permutation Generator

For any given *n*, you control what permutation is generated by *RandomPermutor* in four indirect ways.

- *The size of the pile.* The larger the pile, the more mixed up the raw, underlying linear congruential sequence will be.

- *The starting value x0.* It's best to choose the starting value of the underlying sequence at random, to guarantee that all the numbers have a chance at being output first. (Recall that with a pile, the numbers at the end of the sequence can never be output first.)

- *The seed for the pile's random number generator.* It's best if this is chosen at random, for it controls the sequence of slots chosen for extraction.
- *The seed for the auxiliary generator used to configure the LCG's parms randomly.* It's best if this is chosen at random as well.

All of these controls work indirectly in choosing what permutation is produced. You have no direct control over the process, other than specifiying what you would like *x0* to be (which, in general, also should be chosen at random). Ideally, it would be nice if all possible *n!* permutations were equally likely. Then *RandomPermutor* would produce permutations truly at random. But because of the underlying non-randomness of the LCG for certain values of *n* (particularly when *n* is a prime or a prime product), and the fact that a pile can't always randomize completely due to its limited size, it's unlikely that all of the *n!* permutations are possible, let alone equally probable.

The only time you are guaranteed a truly random permutation is when you set the pile size to *n* and randomly choose *x0*. But then you might as well use a shuffle array, since the code is much simpler. Using the pile gives you more flexibility, for you can trade space for randomness. In some applications, such as the image dissolver of the next section, the sequence doesn't have to be that random but only needs to appear random to the casual observer. You can use a small pile of 10 or 20 items, for example, and randomize large permutations quite effectively, even for *n* in the millions.

IMAGE DISSOLVING

In this section, we come back to our original problem of dissolving images and show how an on-line random permutation generator can be used. The following program, written for BC++ 3.1, illustrates the idea.The program draws a "man" on one side of the screen and "transports" the man to the other side of the screen, moving his "atoms" one by one. Figure 6.2 shows the program in progress.

```
// scotty.cpp: A test program for the random permutor class
#include <conio.h>
#include <dos.h>
#include <time.h>
#include "raptor.h"

const int nrows = 12;
const int ncols = 13;

char atoms[nrows][ncols+1] = {
    "    @@@@@@@   ",
    " @@@ #|# @@@ ",
    " @@@  v  @@@ ",
    "    @@@@@@@   ",
    "       @@@       ",
```

```
   "@@@@@@@@@@@@",
   "@@    @@@    @@",
   "@@    @@@    @@",
   "@@    @@@    @@",
   "      @@@      ",
   "      @@@      ",
   "    @@@@@@    "
};

main()
{
  long xseed, pseed;
  int i;
  clrscr();

  for (i = 1; i<23; i++)   { gotoxy(40, i);   putch('|'); }
  for (i = 0; i<nrows; i++){ gotoxy(10, i+5); cputs(atoms[i]); }

  gotoxy(2, 20); cputs("Beam me over Scotty! <Press return>");
  getch();
  gotoxy(2, 20); cputs("                                   ");

  xseed = (long)time(0);
  pseed = xseed + 421742L + (long)time(0);

  RandomPermutor raptor(nrows*ncols, xseed, 10, pseed);
```

Figure 6.2 Image dissolving using a random permutation.

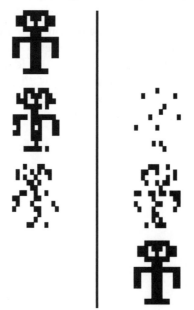

```
for (i = 0; i<nrows*ncols; i++) {   // For all atoms
    int atom = (int) raptor.Curr(); // Get the current random atom.
    int x = atom % ncols;
    int y = atom / ncols;
    gotoxy(x+10, y+5); putch(' ');              // Erase at old posn.
    gotoxy(x+45, y+5); putch(bodybits[y][x]); // Draw at new posn.
    delay(25);
    raptor.Step(); // Randomly select next atom
}

gotoxy(42, 20); cputs("Transport complete!");

return 0;
}
```

This program can be found on disk in the file *scotty.cpp*.

In case you think image dissolving is a frivolous use of random permutation sequences that doesn't justify their existence, in Chapter 7 you'll see where these sequences are useful, at least in theory, in implementing hash tables.

Hashing Algorithms

I n this chapter we take a look at *hashing*, a technique used for building highly efficient search tables. The hashing algorithms are a logical continuation of the last few chapters, since the concepts of permutations, modulo arithmetic, and randomization are used.

In the companion volume we showed several methods of searching that were based on binary search trees. These data structures led to $O(\log n)$ searching algorithms. However, it is possible to devise data structures called *hash tables* that have $O(1)$ search times. In a well-designed hash table, finding an entry may take only one or two comparisons, regardless of the number of entries present.

The trade-off made with hashing is that the entries are not ordered in the table. You can retrieve the associated data for an entry, but you can't tell what the next or previous entry is. In contrast, with binary search trees, you can move through the entries in sorted order.

Trading searching functionality for speed makes sense in many applications. The prime example is a symbol table used in a compiler. During a compilation, the compiler must look up keywords and variable names and obtain their associated attributes quickly. Fast search times are paramount, since during compilation thousands of searches might be made. However, the symbols are not needed in sorted order, except possibly at the end of the compilation when a cross-reference listing is to be printed. Here a fast sort might be used to put the symbols in order.

HASHING—RANDOM SAMPLING IN REVERSE

If you've studied the search tree data structures in the companion volume, you may have been struck by the complexity of it all. Why does searching seem to

149

be hard? Why not simply use the key as an index into an array and be done with it? Then no searching would be needed at all! There are two fundamental problems with this idea, of course. Very often the keys are character strings rather than numbers. Even if the keys are numeric, we still have a problem, as shown in the following example.

Suppose you have a "reverse" telephone directory for a small town of 100 people, where you can search the directory based on a seven-digit phone number and retrieve the name and address of the person having that number. To achieve the fastest possible lookup, with no searching at all, you could use the phone number as an index into an array of customer records. The problem is, to cover all possible seven-digit numbers, you would need an array of 10 million records!

We're never going to use all 10 million possible phone numbers, however, since only 100 people live in the town. Suppose we have some function $h(k)$ that takes a 7-digit number k as the key and maps it into a number in the range 0 to 99. The number can then be used to index the array elements in the telephone directory. In hashing terminology, each array element is called a *bucket,* and the array is called a *hash table.*

One possible mapping uses the two rightmost digits of the phone number as the bucket number. For example, here is a list of four phone numbers and the buckets they map to:

```
Key        Bucket

6741088    88
6741212    12
6746711    11
6743288    88   Oops! Same index as first
```

As the example shows, our mapping strategy works fine—as long as no two phone numbers have the same rightmost digits. When this occurs, the two numbers map to the same bucket, a situation known as a *collision.* Methods for handling collisions are known as *collision resolution strategies,* and we'll discuss them later.

Mapping functions like the one just given are called *hash functions.* In Webster's Dictionary, to *hash* means to "chop into small pieces" or to "confuse and muddle." A hash function must chop up a key, which comes from a large range of values, and squeeze it into the smaller range of buckets. When mapping from many to few, collisions are bound to happen. One way to minimize collisions is to mix up the keys in such a way that the output of the hash function is essentially random. Functions with this property are known as *uniform hash functions.* Ideally, we want each bucket to have a $1/m$ chance of being chosen, given a random set of keys, and where m is the number of buckets in the table.

It's possible to have really bad hash functions. For example, had we chosen to use the two leftmost digits for the mapping of our phone numbers, then all keys would have mapped to the same bucket. In such worst case scenarios, hashing performs no better than simply doing a linear search of the buckets.

Surprisingly, hashing is related to random sampling. Recall from Chapter 6 that in a random sample, k items are chosen at random from a set of r elements. Suppose you assign a unique number in the range 0 to $m - 1$ to each sample in a set of m unique random samples. These numbers are equivalent to bucket indices, and the items in each sample are the entries whose keys map to the same bucket.

We can't use our sampling algorithms for hashing though. The problem is a difference in directionality between the two techniques. Hashing is essentially random sampling in reverse. With random sampling, we are free to choose from a large set of keys, the small set we want for our single sample. With hashing, we have many samples, chosen a priori, and we must determine from among the samples which sample a given key belongs to.

HASH FUNCTIONS

We must design for two basic features of hash tables: the hash function and the collision resolution strategy. In this section we'll focus on hash function design. A hash function must satisfy two requirements, which are usually at odds with each other. We would like the hash function to be fast, taking just a few computations, yet somehow the function must be clever enough to minimize collisions.

Ideally, we would like a hash function that produces a unique index for each key. That way there would be no collisions at all. Hash functions that cause no collisions are called *perfect hash functions*. Finding perfect hash functions isn't easy. There are algorithms that make the attempt (see [Frakes 92], for example), but they are all plagued by either being very expensive (many have exponential time complexities) or aren't guaranteed to work all the time. The worst problem is that finding perfect hash functions requires knowing, up front, all of the keys that are expected. This precludes the use of hashing for more dynamic situations, where the keys aren't known ahead of time.

In general, then, we must settle with finding hash functions that minimize collisions while still being quick to compute. We'll look at a few such functions next, first for hashing numerical keys and then for hashing character strings.

Numerical Hashing

One effective technique for hashing numerical keys is simply to take the key modulo m, where m is the number of buckets in the table. The following equation is used.

$$h(k) = k \bmod m$$

This is known as the *division method* of hashing. There is a counterpart method involving multiplication instead (see [Knuth 73]), but for whatever reason, it doesn't appear to have widespread use today. One reason the division method is popular is that by taking the key modulo m, we automatically transform the key into the desired range of bucket indices, from 0 to $m - 1$.

As a simple example of the division method, Figure 7.1 shows the result of hashing the following eight phone numbers, in order, into 11 buckets: 6741088, 6741212, 6746711, 6743288, 6745117, 6749211, 6745412, 6742517. Colliding entries are indicated by having buckets with multiple entries.

Instead of the rightmost digits determining where the keys are mapped, the mapping is fairly randomized. There are only two collisions, which isn't too bad.

It's crucial that we choose a good value for m, or the results may not be very good. If we had made m even (12, for example), then the even keys would have mapped to even buckets and odd keys to odd buckets. If all of the phone numbers had been even, then half of the buckets would never have been used, and we would expect more collisions.

If the keys are truly random, the choice of m isn't as critical. In fact, for truly random keys, we merely can take a few bits out of each key as the hash value. But in the real world, keys are hardly ever random. In our small-town phone directory, the phone numbers all have the same three-digit prefix. That means there is a lot of bias in the keys.

The most effective hash functions use all of the information in a key, so that similar keys can be scattered as evenly as possible. Ideally, a hash function

Figure 7.1 A hash table using the division method.

0	
1	6742517
2	6741088, 6743288
3	6745412
4	6746711
5	6741212, 6745117
6	
7	6749211
8	
9	
10	

should work as well for non-random keys as it does for random keys. That gives us a degree of robustness. We won't have to spend a lot of time experimenting with different hash functions for each application.

It turns out that the division method of hashing seems quite robust, if we make *m* prime. That is, it's best to use a prime number of buckets whenever possible. For instance, if we determine that we would like around 100 buckets, then tables with sizes of 103, 107, 109, or 113 buckets would be suitable.

The division method isn't infallible, but there's probably no hash function that is. You'll always be able to concoct a pathological set of keys that will cause a given hash function to perform poorly. In the worst case, all keys map to the same bucket. For well-designed hash functions, however, the chance of this happening in a real application is very small, but it should be noted that you *are* relying on the laws of probability. If you absolutely must guarantee that a worst-case scenario won't happen, then don't use hashing.

There seems to be a little "black magic" involved in designing hash tables. For example, some prime numbers seem "cleaner" than others and are better choices to use as the hash table size. Experience has shown that *m* = 211 works quite well, for instance. Many symbol tables have been implemented with 211 buckets for just this reason. But curiously, no one knows exactly why the prime 211 is better than, let's say, its nearest prime neighbors of 199 and 223.

Hashing Character Strings

While you may have applications where the keys are numeric, it's quite common for the keys to be character strings. Examples are names of people, street addresses, or identifiers and keywords in a program. What should we use for a hash function in these cases?

One possibility is to sum up the integer values of each character in the string and use the result as the bucket index. An example is the following *StrHash()* function.

```
int StrHash(char *key)
// A not-so-good hash function
{
  int h = 0;
  while(*key) h += *key++;
  return h;
}
```

Since it's possible for the resulting index to be greater than the number of buckets, a common technique is to take the result modulo *m*. For example

```
int h = StrHash("Lillehammer");
int b = h % 211; // Assuming m = 211 buckets
```

In other words, once a character string is converted into numerical form, we use the division method both to further hash the result and to keep the value in range. As before, it's best if *m* is prime.

Note The computation by modulo *m* is properly a part of the hash function for character strings. However, you'll find that it's more convenient to leave the modulo computation out of the function. By convention, we'll show hash functions for character strings without the modulo arithmetic, and assume the calculation is performed afterward as needed.

The *StrHash()* function is not a good hash function. It does poorly with keys that are similar, for instance. That's unfortunate, for similar keys are quite common in hashing applications such as symbol tables. Programs tend to have variable names such as *foo1*, *foo2*, *newfoo*, *oldfoo*, and so on.

To show you how poorly *StrHash()* performs, Figure 7.2 plots the bucket distribution for the three hundred keys *v100*, *v101*, ... v399, when using 211 buckets. The buckets not shown have zero hits.

You'll notice that *StrHash()* tends to bunch the keys around the bucket 62, which itself receives 28 hits. Interestingly enough, a familiar bell-shaped curve is produced. Only a few of the 211 buckets are used, which means there are a lot of collisions. A uniform hash function would give each bucket the same probability *n/m* of being chosen, so the curve would be a straight line. In our example, each bucket would have approximately 300/211 (that is, 1 to 2) entries.

It's hard to find a general-purpose function that hashes uniformly for all types of character strings, but the following *ElfHash()* function comes very close to the ideal. This function is the epitome of hashing, using a magic formula that chops, mixes, slices, and dices the input character string, using just a few statements.

Figure 7.2 A poor hashing distribution.

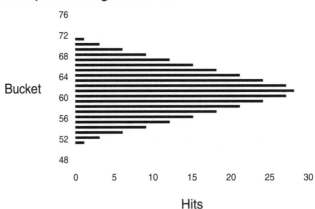

```
unsigned long ElfHash(const unsigned char *key)
// A very good hash function. The return value should be
// taken modulo m, where m is the prime number of buckets.
{
  unsigned long h = 0;
  while (*key) {
    h = (h << 4) + *key++;
    unsigned long g = h & 0xF0000000L;
    if (g) h ^= g >> 24;
    h &= ~g;
  }
  return h;
}
```

You may wonder why this function is called *ElfHash()*. Perhaps it's due to the magic sprinkled in its incantations! Actually, ELF is an acronym for "Executable and Linking Format," which is a format used for executable and object files in Unix System V Release 4. The *ElfHash()* function is used to hash character strings for ELF files.

When used on the keys *v100, v101, ..., v399, ElfHash()* distributes the keys quite evenly for *m* = 211, with most buckets getting one or two hits. No bucket gets more than three hits, and only a few get none. In other words, the distribution curve is quite flat, exactly as you want. If you need an excellent general-purpose hash function for character strings, then use *ElfHash()*.

COLLISION RESOLUTION STRATEGIES

Three basic strategies are used to handle collisions.

- *Open addressing.* When a collision occurs, the table is probed, looking for an empty bucket to place the new entry.
- *Separate chaining.* Each bucket becomes the head of a list. When a key is mapped to a bucket, the entry is simply added to the bucket's list.
- *Coalesced chaining.* This is a hybrid of the first two techniques. The bucket array holds entries, which are threaded into multiple lists as collisions occur. Unlike separate chaining, the lists are allowed to coalesce. Each list may contain entries with keys having different hash values.

Since the collision resolution strategies are tied directly to how a hash table is implemented, we'll show the strategies in detail as we give the different implementations.

SEPARATE CHAINING

The simplest way to resolve collisions is to keep the entries on lists, one list for each bucket. Figure 7.3 shows this arrangement for the phone number example

Figure 7.3 Hash table with separate chaining.

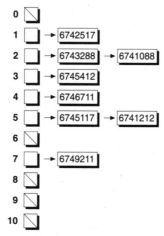

given earlier. The following *SepHashTable* class and the auxiliary *Chain* class (used for linked list nodes) implements this type of hash table.

```
// Some function return value codes
const int ALLOC_FAILURE   = -1;
const int NOT_FOUND       =  0;
const int SUCCESS         =  1;
const int DUPLICATE_KEY   =  2;
const int TABLE_FULL      =  3;
const int NOT_SUPPORTED   =  4;

template<class TYPE>
class Chain {
// Hash bucket chain link class.
public:
  TYPE info;     // Entry data
  Chain *next;   // Next colliding entry
  Chain(const TYPE &s, Chain<TYPE> *n);
};

template<class TYPE, class KEY>
class SepHashTable {
protected:
  Chain<TYPE> **buckets;
  long num_entries, num_collisions, num_comparisons;
  unsigned num_buckets;
  char ok;
public:
  SepHashTable(unsigned tsize);
  virtual ~SepHashTable();
  void Clear();
```

```
int Insert(const TYPE &x);
int Update(const TYPE &x);
TYPE *Search(const KEY &key);
const TYPE *Search(const KEY &key) const;
int Delete(const KEY &key);
  ...
};
```

Complete code for the *SepHashTable* and *Chain* classes can be found on disk
in the files *shash.h* and *shash.mth*. Test programs can be found in *shtst1.cpp*
and *shtst2.cpp*. Also, a common set of functions used by many of the test
programs in this chapter can be found in *hashtst.cpp*.

You'll notice *SepHashTable* has two template parameters. The *TYPE* parameter indicates the entry type. The *KEY* parameter is the type of key used
during searching. It's assumed that each *TYPE* object has a *KEY* as one of its
members. Also, *TYPE* must have at least three member functions: (1) a *Key()*
function that returns the key in the *TYPE* object, (2) a *Match()* function used in
determining matching keys, and (3) a *Hash()* function used to hash keys of
type *KEY*. In addition, *TYPE* should have the usual complement of constructors
and destructors as needed.

We chose this method for specifying the entry type in order to make the
SepHashTable class as general as possible. You'll notice in the hash table
classes to follow that the tables need to know only the size of the entries (for
which the *TYPE* parameter will suffice) and the type of key stored in the entry.
The hash tables never need to know what other kind of associated data the
entry has. The test programs found on disk use the following prototypical
Entry class as an example of how to correctly specify a class for hash table
entries. (We show only the most relevant member functions.)

```
// Sample class for hash table entries

struct Entry {
  char key[32]; // Key used in hashing
  int data;     // The associated data
  Entry();
  Entry(const Entry &e);
  Entry(const char *k, int d);
  static unsigned long Hash(const char *quay);
  int Match(const char *quay);
  const char *Key() const;
};

  ...

int Entry::Match(const char *quay)
{
  return strcmp(key, quay) == 0;
}
```

```
const char *Entry::Key() const
{
  return key;
}

unsigned long Entry::Hash(const char *quay)
// We'll use elf-hashing.
{
  unsigned long h = 0;
  while (*quay) { // Elf magic follows ...
    h = (h << 4) + *((unsigned char *)quay)++;
    unsigned long g = h & 0xF0000000L;
    if (g) h ^= g >> 24;
    h &= ~g;
  }
  return h;
}
```

Complete code for the *Entry* class is given on disk in the files *entry.h* and *entry.cpp*.

The *Hash()* function is made a static member, since it doesn't need a *this* pointer. It works on a key of the same type as that stored in an *Entry*. Note that we chose to use ELF hashing for the key string. If *KEY* were a numerical type instead, *Hash()* could simply return the key. It's assumed that the result of *Hash()* will be taken modulo *m* to determine the actual bucket index. Even though our *Hash()* function returns an *unsigned long*, the *Hash()* function for your entry class doesn't have to. All that's required is that it return a numerical type for which modulo arithmetic is defined.

Let's now discuss some of the functions of *SepHashTable*. The constructor and destructor (not shown) are fairly straightforward. The constructor allocates a bucket array of *Chain<TYPE>* pointers, which serve as list heads for the buckets. The pointers are set to zero so that the table starts out empty. The destructor deletes all nodes of all bucket lists and then deletes the buckets themselves.

A more interesting function is *Insert()*. It takes a *TYPE* object and inserts it in the table.

```
template<class TYPE, class KEY>
int SepHashTable<TYPE, KEY>::Insert(const TYPE &x)
// Inserts new entry into table having specified key and data.
// Returns DUPLICATE_KEY if matching entry already exists, else
// returns SUCCESS or ALLOC_FAILURE.
{
  unsigned h = unsigned(TYPE::Hash(x.Key()) % num_buckets);
  Chain<TYPE> *p = buckets[h];

  while(1) {
    if (p == 0) { // Bucket list currently empty
      // Insert new node at front
      Chain<TYPE> *q = new Chain<TYPE>(x, buckets[h]);
```

```
        if (q == 0) return ALLOC_FAILURE;
        buckets[h] = q;
        return SUCCESS;
      }
      if (p->info.Match(x.Key())) return DUPLICATE_KEY;
      p = p->next;
    }
  }
}
```

▼ **Note**   The member functions of the hash table classes we give on disk are slightly more complicated than what we show herein. Code is added to collect statistics, such as the number of collisions, for testing purposes. We don't show the full code here so that you can see the relevant techniques of hashing more easily.

Note how *TYPE*s *Hash()* function is called, and the result taken modulo *num_buckets*. This gives us a bucket index. If the bucket stores a null pointer, we know that no collision has occurred, so a new *Chain<TYPE>* node is created and the entry added to the bucket's list. Otherwise the list is scanned for a matching entry. If one is found, we have a duplicate key, so an error code is returned. If the entry isn't found in the list, we should add it.

In our example we always add the entry to the front of the list. The optimum way to add entries is application dependent, but keep in mind that the lists are supposed to be relatively short. For example, if a well-designed hash function is used, each list will have approximately n/m nodes. Because the lists are short, it probably won't matter too much how the list is organized.

In the companion volume, we spend an entire chapter discussing various methods for implementing linked lists. We could have used one of the list classes given there for our hash table, such as the *Slist<TYPE>* class. However, this list class was written with generality in mind and has more power and functionality than what we need here. It is just as simple to embed list functionality directly into the *SepHashTable* class.

The following *Search()* function is used to search the table for existing entries. Again, we hash the entry's key to find the right bucket and then scan the bucket's list, searching for a match.

```
template<class TYPE, class KEY>
TYPE *SepHashTable<TYPE, KEY>::Search(const KEY &key)
// Searches for a node having specified key. Returns pointer
// to the data in the node if found, else returns 0.
{
  unsigned h = unsigned(TYPE::Hash(key) % num_buckets);
  Chain<TYPE> *p = buckets[h];
  while(p) {
    if (p->info.Match(key)) return &p->info;
    p = p->next;
  }
  return 0; // No match found
}
```

Two other useful functions for our hash table are the following *Update()* and *Delete()* functions. They allow us to search for a matching entry and then either update the associated data or remove the entry from the table.

```
template<class TYPE, class KEY>
int SepHashTable<TYPE, KEY>::Update(const TYPE &x)
// Updates the entry having the key in x with the new data
// in x. Returns either SUCCESS or NOT_FOUND.
{
  TYPE *p = Search(KEY(x.Key())); // KEY typecast necessary in BC++
  if (p) {
    *p = x;
    return SUCCESS;
  }
  return NOT_FOUND;
}

template<class TYPE, class KEY>
int SepHashTable<TYPE, KEY>::Delete(const KEY &key)
// Search for node having the specified key and deletes
// it from the table. Returns SUCCESS or NOT_FOUND.
{
  unsigned h = unsigned(TYPE::Hash(key) % num_buckets);
  Chain<TYPE> *p = buckets[h], *q = 0;

  while(p) { // Scan the list looking for matching entry.
    if (p->info.Match(key)) {
      // Be sure to test for deleting the first node of the list.
      if (q == 0) buckets[h] = p->next; else q->next = p->next;
      delete p;
      return SUCCESS;
    }
    q = p; p = p->next;
  }

  return NOT_FOUND;
}
```

Of the collision resolution methods you'll find in this chapter, separate chaining is the most intuitive method and the easiest to implement. Unlike the other hash table classes to come, there is no fixed limit on how many entries can be added to the table. The bucket lists can grow to arbitrary lengths.

Of course, the longer the lists, the slower the searching will be. Ideally, you want the lists to have only a few nodes each. Since the average length of the bucket lists will be n/m, separate chaining essentially reduces the searching that would be required for a brute force linear search by a factor of m. The larger you make m, the faster the searching will be.

Of course, m must be kept within reason, or the overhead for the array of bucket pointers will become significant. Also, don't forget that there is overhead in the list nodes themselves, both in time spent allocating and deallocating the nodes and in memory for the *next* pointers.

OPEN ADDRESSING

If you know ahead of time how many entries to expect, then the technique presented next, called *open addressing*, may give you the most compact table. Here each bucket stores one entry. As before, a hash function determines a bucket index for a given key. If that bucket is empty, the entry is simply copied into the bucket. If the bucket isn't empty, then a collision occurs. The term *open addressing* stems from the fact that after a collision, we probe the table, looking for either a match or an open (that is, empty) bucket.

When there is a collision, the rest of the buckets in the table are probed in some predetermined sequence, until an empty bucket is found to insert the entry into. If no empty bucket is found, then the table is full. When doing a search as opposed to an insert, the same probe sequence is followed to look for a match. If an empty bucket is encountered, it means the entry isn't in the table.

There are two problems to handle with open addressing. The easiest problem is how to determine when a bucket is empty. At least one bit needs to be set aside for this purpose, either in the entries themselves or stored as an auxiliary flag in each bucket.

A harder problem is what to use for the bucket probe sequence. In general, you want the probe sequence to have the following properties: It should be able to scan all buckets in the table if necessary, and the probes should be scattered as much as possible, to minimize the number of collisions that occur. The first requirement implies that the probe sequence, which we can write as

$$p_0, p_1, ..., p_{m-1}$$

is actually a permutation of the numbers from 0 to $m - 1$, with p_0 being the bucket returned by the hash function. Next we'll take a look at four different types of probe sequences.

Linear Probing

The simplest probe sequence you can use is p_0, $p_0 + 1$, ... $p_0 + m - 1$, taken modulo m. (The numbers wrap around to the beginning of the table when the end is reached.) This type of probing is known as *linear probing,* and the ith probe is given by the equations

$$p_0 = Hash\left(key\right)$$
$$p_i = \left(p_0 + i\right) \bmod m$$

Figure 7.4 shows the 11-bucket hash table that results when inserting the same eight phone numbers used earlier. In the figure, the numbers in parentheses are the original hash values of the keys.

Figure 7.4 Hash table with linear probing.

0	
1	6742517 (1)
2	6741088 (2)
3	6743288 (2)
4	6746711 (4)
5	6741212 (5)
6	6745117 (5)
7	6749211 (7)
8	6745412 (3)
9	
10	

Let's look what happens when the key 6743288 is added. This number maps to bucket 2, which is already occupied. Bucket 3 is checked and found empty, so the entry is added there. The key 6745117 also causes a collision, this time at bucket 5, and the corresponding entry is added in bucket 6.

A bucket filled via probing tends to cause further collisions to occur. With linear probing, this situation gets worse as entries are added. In fact, the occupied buckets tend to coalesce into clusters. The larger the cluster, the more chance a collision will occur, and, unfortunately, the longer the probe sequence is likely to be. For example, after adding the key 6743288 to bucket 3, a key that would normally hash to bucket 3, such as 6745412, causes a collision. In this case, we have to scan all the way down to bucket 8 to find an empty bucket.

The average number of probes needed in linear probing for a successful search (where a matching entry exists) and an unsuccessful search (such as during insertions) is given by the equations

$$S(\alpha) = \frac{1}{2} + \frac{1}{2}\left(\frac{1}{1-\alpha}\right), \quad U(\alpha) = \frac{1}{2} + \frac{1}{2}\left(\frac{1}{1-\alpha}\right)^2$$

Here, $\alpha = n/m$, the load factor of the table. For example, when the table is half full, it takes on average 1.5 probes to search for an entry already in the table and 2.5 probes to insert a new entry. When the load factor is 75 percent, the probes average 2.5 and 8.5, respectively. The numbers go up rapidly from there, until the table is full. (At that point, the formulas break down.) In a full table, it takes on average $m/2$ probes for a successful search and m probes for an unsuccessful one. This is no better than if we used a simple linear search through the table, without hashing.

Note that the equations for $S()$ and $U()$ depend only on the load factor and not directly on the number of entries stored in the table. In other words, in a half-full table of 40,000 buckets, it still takes only 1.5 probes on average to find an existing entry. This behavior is at the very heart of why hashing is such a popular searching technique. Only when the table gets close to becoming full does the number of entries start to matter.

The type of clustering caused by linear probing is known as *primary clustering*. There are other probe sequences that don't cause primary clustering, as you'll see next.

Quadratic Probing

One probe sequence that eliminates primary clustering is known as *quadratic probing*, where the ith probe is given by the equations

$$p_0 = Hash\left(key\right)$$
$$p_i = \left(p_0 + i^2\right) \bmod m$$

With quadratic probing, the increment between each probe starts out at 1 and then goes to 4, 9, 16, and so on. (As before, wraparound occurs at the end of the table.) Note that you don't actually have to square i to compute the next bucket in the sequence. The following recurrence equations show equivalent computations.

$$c_0 = 1$$
$$c_i = c_{i-1} + 2$$
$$p_0 = Hash\left(key\right)$$
$$p_i = \left(p_{i-1} + c_i\right) \bmod m$$

These equations are based on the fact that

$$1 + 3 + 5 + ... + \left(2i - 1\right) = i^2$$

Because quadratic probing skips around more than linear probing does, it causes less clustering to take place, so the probe sequences tend to be shorter. However, all keys that map to the same bucket will follow the same probe sequence. This is known as *secondary clustering*. In practice, secondary clustering causes slightly more searching than the ideal. In the next section we'll take a look at the theoretically best probe sequence.

While quadratic probing is easy to compute and performs fairly well, it has one significant drawback. Only half the buckets in the table can be probed before the sequence begins to repeat. Because of this, the number of probes usually is limited in the code to $(m + 1)/2$. Thus during insertions, it's possible for empty buckets to be missed, since not all buckets are scanned. The table behaves as if it were full, when, in fact, half of the buckets may still be empty.

It's possible to get around this problem, and have all buckets searched by using a more complicated form of quadratic probing, given by the equation

$$p_0 = Hash\left(key\right)$$
$$p_i = \left(p_0 \pm i^2\right) \bmod m$$

In other words, the increment i^2 is used to go both forward and backward through the table. This sequence is more costly to implement and may not be worth the trouble, since, as you'll see shortly, there are other probe sequences that are better anyway.

Uniform Probing

A probe sequence that causes no clustering, primary or secondary, is known as a *uniform probe sequence* (not to be confused with a uniform hash function). In uniform probing, the buckets chosen for the sequence are random, with each bucket having an equal probability of being chosen. It can be shown that with uniform probing, the average number of probes for searching can be given by the following two equations.

$$S(\alpha) = \frac{1}{\alpha}\ln\left(\frac{1}{1-\alpha}\right), \quad U(\alpha) = \frac{1}{1-\alpha}$$

For example, when the table is half full ($\alpha = 0.5$), it takes an average of 1.38 probes to search for an existing entry and 2 probes for an unsuccessful search. For $\alpha = 0.75$, the numbers become 1.8 and 4 probes, respectively. This compares with 2.5 and 8.5 probes used in linear probing on a table three-quarters full.

Double Hash Probing

Uniform probing is the best you can do, but it's a mythical form of probing, for no implementation is known to achieve it exactly. However, a probe sequence called *double hash probing* comes close.

The idea behind double hash probing is that, if a collision occurs, the key is hashed again and the hash value is used as the offset between buckets during successive probes. In other words, the following equations are used.

$$p_0 = Hash\left(key\right)$$
$$c = Hash\,2\!\left(key\right)$$
$$p_k = \left(p_{k-1} + c\right) \bmod m$$

In order for all buckets to be in the probe sequence, c must be relatively prime to m. This is easy to see if you recognize that the equation for p is nothing more than an LCG, where the multiplier is 1 and the increment is c. By making m prime, we can guarantee a full-cycle LCG, for any $0 < c < m$. This gives us another reason for making m prime.

Double hash probing is known to produce results very close to uniform probing, as long as the second hash function is independent of the first. Even though several keys may map to the same bucket, as given by *Hash()*, these keys shouldn't cause the same probe increment to be used, as given by *Hash2()*. Only identical keys should probe the same buckets. By doing this, secondary clustering is eliminated.

One set of functions, suggested in Knuth [Knuth 73], causes virtually no secondary clustering.

$$Hash(k) = k \bmod m$$
$$Hash2(k) = k \bmod (m - 2)$$

This is really the division method of hashing applied twice, and we are implying that the keys are numeric. Recall for character string keys that we first use a hash function to convert to a number and then take the result modulo m. Thus if we were to use ELF hashing, here's how the code should look.

```
unsigned long h = ElfHash(key); // Convert string to a number.
unsigned long b = h % m;        // Compute first bucket b.
unsigned long c = h % (m-2);    // Compute probe increment.
```

By using two different moduli, keys that map to the same bucket may use a different probe increment, and thus, no secondary clustering will occur. Also, Knuth suggests that you pick m so that m and $m - 2$ are twin primes, such as 1021 and 1019, to help the randomization. In the experiments we ran while researching this book, not much difference was noticed whether you used twin primes or not. Figure 7.5 shows the result of using double hash probing on our sample set of phone numbers. In the figure, the numbers in parentheses are the original hash values of the keys.

When adding the key 6743288, we have a collision at bucket 2. The increment needed to find the next bucket to probe is computed as 6743288 mod 9, which equals 2. Thus the next bucket probed is 4, which has an entry, so we try the next bucket in the sequence at bucket 6, where the entry is

---

Figure 7.5 Hash table with double hash probing.

0	
1	6742517 (1)
2	6741088 (2)
3	6745412 (3)
4	6746711 (4)
5	6741212 (5)
6	6743288 (2)
7	6749211 (7)
8	
9	6745117 (5)
10	

---

added. Similarly, there is a collision at bucket 5 when adding 6745117. Here the increment is computed as 6745117 mod 9, which equals 4. Thus bucket 9 is tried, and the entry is added there. No other collisions occur.

Random Probing

For both linear and double hash probing, the probe sequences are actually permutations of the numbers between 0 and $m - 1$. Why not instead use a random permutation, as in Chapter 6, as the probe sequence? Such *random probing* is intuitively appealing, but as you'll see, it doesn't work so well in practice.

To implement random probing, you could use a shuffle array of size m, and iterate through the array to obtain bucket indices for probing. For any given key, you must use the same permutation each time. This means the seed of the random number generator used for shuffling must depend in some way on the key. One way of doing this is to use the output of the hash function (before the modulo arithmetic) as the seed. Then the first element in the array after the shuffle will be the same bucket index that is obtained by taking the hash value modulo m.

There are several drawbacks to the random probing approach. Extra $O(m)$ overhead is needed for the shuffle array, and the array must be reshuffled at the beginning of each probe sequence, which takes $O(m)$ time. However, since only a few buckets may need to be probed, you can reduce the time somewhat by shuffling each element only as needed. That is, you could use a shuffle generator. Unfortunately, at the beginning of each sequence you still need to put the shuffle array in order, which takes $O(m)$ time, although it's faster than actually shuffling the elements.

Alternatively, you could use the *RandomPermutor* class of Chapter 6. However, experiments show that this class produces mediocre probe sequences. There are two reasons for that. First, the underlying full-cycle LCGs aren't always very random, especially when m is prime, as is usually the case for hash tables. Second, it seems that the piles used to randomize the sequence must have a size close to m in order to minimize collisions effectively. So you might as well use the simpler shuffle array approach.

If you restrict what values you use for m, you can get around using a shuffle array or pile. If m is a power of two, on-line algorithms exist to generate random permutations (besides the algorithm used in the *RandomPermutor* class). See Aho [Aho 83]. However, setting m to a power of two is exactly what we don't want when the division method of hashing is used.

Regardless of how you implement random probing, it turns out that the algorithm isn't as good as you might think. That's because, like quadratic probing, random probing has secondary clustering. All the keys that map to the same bucket use the same permutation and thus the same probe sequence. Because of this secondary clustering, random probing performs somewhere in between double hash probing and quadratic probing. With some data, random probing is as good as double hash probing; with other data, it's no better than quadratic probing. In any case, the shuffle array is costly to implement, both in space and time, so you are better off with one of the other techniques.

THE OPENHASHTABLE CLASS

Now that you've seen various strategies for handling collisions, let's put them to work in a hash table class designed for open addressing. Like the *SepHashTable* class, we've made the following *OpenHashTable* class a template, requiring *TYPE* and *KEY* parameters, with the same semantics as before. Some sophisticated C++ code follows, so hold on to your hats!

```
enum HashProbe {
  linear_probe, quadratic_probe, double_probe, random_probe
};

template<class TYPE, class KEY>
class OpenHashTable {
protected:
  Ran32 rg;            // For random probing
  unsigned *shuffle;   // For random probing
  unsigned cursor;     // For random probing
  TYPE *buckets;       // The data in the buckets
  char *inuse;         // In-use flags for the buckets
  long h;              // Used during probing
  unsigned b, c, n;    // Used during probing
```

```
      unsigned num_buckets;
      long num_entries, num_collisions, num_comparisons;
      char ok;
      void SetupLinear();
      void SetupQuadratic();
      void SetupDoubleHash();
      void SetupRandomHash();
      void NextLinear();
      void NextQuadratic();
      void NextDoubleHash();
      void NextRandomHash();
      void (OpenHashTable<TYPE, KEY>::*SetupProbe)();
      void (OpenHashTable<TYPE, KEY>::*NextProbe)();
public:
      OpenHashTable(unsigned tsize, HashProbe hp=double_probe);
      virtual ~OpenHashTable();
      void ChooseProbe(HashProbe hp);
      void Clear();
      int Insert(const TYPE &x);
      int Update(const TYPE &x);
      TYPE *Search(const KEY &key);
      const TYPE *Search(const KEY &key) const;
      int Delete(const KEY &key); // Not actually supported
      ...
};
```

 Complete code for the *OpenHashTable* class is given on disk in the files *ohash.h* and *ohash.mth*, with test programs in *ohtst1.cpp* and *ohtst2.cpp*.

In this class, two main arrays are used. The *buckets* array stores the entry data, and the *inuse* array stores flags indicating which buckets are in use. Unless you have read the companion book, you might be surprised to find an unusual method for constructing the *buckets* array. Unlike the default behavior for arrays in C++, the elements of *buckets* are not constructed until actually used. As we explain in the companion book, this seems to be a more semantically correct way of handling the elements.

In order to prevent the default construction of elements, the constructor allocates *buckets* as an array of characters. The destructor deallocates the array the same way.

```
template<class TYPE, class KEY>
OpenHashTable<TYPE, KEY>::OpenHashTable(unsigned tsize, HashProbe hp)
// Sets up a hash table with tsize buckets and probe sequence hp.
// The buckets are left unconstructed until used.
: rg(555) // For random probing. Seed not really used here.
{
  ok = 0;
  buckets = (TYPE *) new char[tsize * sizeof(TYPE)];
  inuse   = new char[tsize];
  shuffle = new unsigned[tsize];
```

```
    if (buckets && inuse && shuffle) {
      ok = 1;
      ChooseProbe(hp);
      num_buckets = tsize;
      for (unsigned i = 0; i < num_buckets; i++) inuse[i] = 0;
      num_entries = 0; num_collisions = 0; num_comparisons = 0;
    }
    else { // Allocation failure. Deallocate those we did allocate
      delete[] shuffle;
      delete[] inuse;
      delete[] (char *)buckets; // As allocated, so deallocated
    }
}

template<class TYPE, class KEY>
OpenHashTable<TYPE, KEY>::~OpenHashTable()
{
  if (ok) {
    Clear();
    delete[] shuffle;
    delete[] inuse;
    delete[] (char *)buckets; // As allocated, so deallocated
  }
}
```

Soon you'll see how the elements are actually constructed as *TYPE* objects, but the following *Clear()* function, called by the destructor, shows how each constructed element is destroyed with an explicit *TYPE* destructor call.

```
template<class TYPE, class KEY>
void OpenHashTable<TYPE, KEY>::Clear()
// Clears the table.
{
  for (unsigned i = 0; i<num_buckets; i++) {
      // If element is constructed, explicitly destruct it.
      if (inuse[i]) (buckets+i)->TYPE::~TYPE();
      inuse[i] = 0;
  }
  num_entries = 0; num_collisions = 0; num_comparisons = 0;
}
```

The *OpenHashTable* class was designed with testing in mind. We've incorporated all the types of probing discussed in this chapter. Two member function pointers, *(::*SetupProbe)()* and *(::*NextProbe)()*, are used to set up the probe sequence and step along it. Of course, in a real implementation, you would choose just one type of probing, taking out the function pointers, and place the probing code in line. The following code shows how the function pointers are initialized, followed by the probing functions themselves. If you've never had an occasion to use member function pointers, now is your chance to see them in action.

```
template<class TYPE, class KEY>
void OpenHashTable<TYPE, KEY>::ChooseProbe(HashProbe hp)
// Choose the probe sequence. NOTE: You should call Clear()
// afterward if there are buckets currently in use.
{
  switch(hp) {
    case linear_probe:
      SetupProbe = &OpenHashTable<TYPE, KEY>::SetupLinear;
      NextProbe  = &OpenHashTable<TYPE, KEY>::NextLinear;
    break;
    case quadratic_probe:
      SetupProbe = &OpenHashTable<TYPE, KEY>::SetupQuadratic;
      NextProbe  = &OpenHashTable<TYPE, KEY>::NextQuadratic;
    break;
    case double_probe:
      SetupProbe = &OpenHashTable<TYPE, KEY>::SetupDoubleHash;
      NextProbe  = &OpenHashTable<TYPE, KEY>::NextDoubleHash;
    break;
    case random_probe:
      SetupProbe = &OpenHashTable<TYPE, KEY>::SetupRandomHash;
      NextProbe  = &OpenHashTable<TYPE, KEY>::NextRandomHash;
    break;
  }
}

template<class TYPE, class KEY>
void OpenHashTable<TYPE, KEY>::SetupLinear()
// Set up a linear probing sequence.
{
  c = 1;
  n = num_buckets;
}

template<class TYPE, class KEY>
void OpenHashTable<TYPE, KEY>::NextLinear()
// Go to next bucket in a linear probing sequence.
{
  b++;
  if (b >= num_buckets) b -= num_buckets;
}

template<class TYPE, class KEY>
void OpenHashTable<TYPE, KEY>::SetupQuadratic()
// Set up a quadratic probing sequence.
{
  c = 1;
  n = (num_buckets + 1) / 2;
}

template<class TYPE, class KEY>
void OpenHashTable<TYPE, KEY>::NextQuadratic()
// Go to next bucket in a quadratic probing sequence.
{
```

```
  b += c;
  while (b >= num_buckets) b -= num_buckets;
  c += 2;
}

template<class TYPE, class KEY>
void OpenHashTable<TYPE, KEY>::SetupDoubleHash()
// Set up a double hashing probing sequence.
{
  c = 1 + unsigned(h % (num_buckets-2));
  n = num_buckets;
}

template<class TYPE, class KEY>
void OpenHashTable<TYPE, KEY>::NextDoubleHash()
// Go to next bucket in a double hash probing sequence.
{
  b += c;
  if (b >= num_buckets) b -= num_buckets;
}

template<class TYPE, class KEY>
void OpenHashTable<TYPE, KEY>::SetupRandomHash()
// Set up a random probing w/secondary clustering sequence.
{
  for (unsigned j = 0; j<num_buckets; j++) shuffle[j] = j;
  n = num_buckets;
  cursor = 0;
  rg.Reset(h);       // So that first bucket always first in shuffle
  NextRandomHash(); // To prevent first bucket being used twice
}

template<class TYPE, class KEY>
void OpenHashTable<TYPE, KEY>::NextRandomHash()
// Go to next bucket in a random probing.
// w/secondary clustering sequence.
{
  // Compute b such that cursor <= b < num_buckets
  b = (unsigned) rg.Curr(cursor, num_buckets-1);
  unsigned t = shuffle[b];
  shuffle[b] = shuffle[cursor];
  shuffle[cursor++] = t;
  rg.Next();
}
```

In the probing functions, the variable b is used to store the result of the hash function, b is the index to the current bucket, c is the probe increment, and n is the number of buckets left in the probe sequence. Note how all the probes set n = *num_buckets* initially, except for quadratic probing, where (*num_buckets* + 1)/2 is used. For random probing, we use a shuffle generator embedded directly into the *OpenHashTable* class.

Now we can show the meat of the *OpenHashTable* class, in the following *Insert()* and *Search()* functions.

```
template<class TYPE, class KEY>
int OpenHashTable<TYPE, KEY>::Insert(const TYPE &x)
// Inserts a new entry x. If a matching entry is found, DUPLICATE_KEY
// is returned, else SUCCESS or TABLE_FULL is returned.
{
  // Compute the first bucket to try.

  h = TYPE::Hash(x.Key());
  b = unsigned(h % num_buckets);

  // Check for collision.

  if (inuse[b] == 0) goto build_bkt; // We got lucky
  if (buckets[b].Match(x.Key())) return DUPLICATE_KEY;

  // We have a collision, so start the probe sequence.

  (this->*SetupProbe)();

  while(1) {
    if (--n == 0) return TABLE_FULL;
    (this->*NextProbe)();
    if (inuse[b] == 0) break;
    if (buckets[b].Match(x.Key())) return DUPLICATE_KEY;
  }

  build_bkt:

  new (buckets + b) TYPE(x); // Copy construct the data in-place.
  inuse[b] = 1;
  num_entries++;
  return SUCCESS;
}

template<class TYPE, class KEY>
TYPE *OpenHashTable<TYPE, KEY>::Search(const KEY &key)
// Searches for a bucket having specified key. Returns pointer
// to the data in the bucket if found, else returns 0.
{
  h = TYPE::Hash(key);
  b = unsigned(h % num_buckets);

  if (inuse[b] == 0) return 0;
  if (buckets[b].Match(key)) return buckets + b;

  (this->*SetupProbe)();

  while(--n) {
    (this->*NextProbe)();
    if (inuse[b] == 0) break;
```

```
    if (buckets[b].Match(key)) return buckets + b;
  }

  return 0; // No match found
}
```

Note how *SetupProbe()* and *NextProbe()* are called to initiate the probing of the table. We've optimized the *Insert()* and *Search()* functions so that the probe sequence is not set up unless necessary. When there is no collision, the only overhead will be the initial hashing. This is an important optimization, for when the hash function is properly designed and the table isn't close to being full, there will be few collisions.

When a new entry is added in the *Insert()* routine, you'll notice what may seem to be a strange line of code:

```
new (buckets + b) TYPE(x); // Copy construct the data in-place.
```

This code causes bucket *b* to be constructed, using *TYPE*'s copy constructor. (Remember that the buckets are unconstructed until needed). A special form of the overloaded new operator is used, sometimes called the *placement new operator.*

```
inline void *operator new(size_t, void *p)
// Placement new operator
{
  return p;
}
```

Here the hidden *size_t* parameter gives the size of the object being constructed (not used here since it's assumed the memory is already allocated for the object), and the pointer *p* is where the object is to reside in memory. This address is simply passed on to the constructor as the *this* pointer.

Deletions with Open Addressing

With the exception of clearing *all* entries in the table, deletions are not supported in *OpenHashTable*. There are too many problems in allowing deletions in conjunction with open addressing. Suppose that we allow an entry to be deleted. We could call the destructor for the entry and then set the *inuse* flag for that bucket to 0. But remember that a probe sequence stops when either a match or an empty bucket is found (or we've probed all the buckets). Thus, if the entry we delete is in the middle of a probe sequence, then the entries that follow in the sequence will never be found during a search.

The problem can be partially solved by using a second type of flag to indicate that a bucket has been deleted, as opposed to being empty because it was never filled. Then, during searching, when a deleted bucket is found, it is treated like a full bucket so that the probe sequence can continue. For insertions, the first empty or deleted bucket is used for the new entry.

This trick is workable only when just a few deletions are going to occur. Once a bucket is occupied, it is never flagged as empty again. (It can only be flagged deleted.) If enough deletions occur, then all the "empty" buckets will disappear, and an unsuccessful search will take *m* probes. There is a technique that attempts to rearrange the buckets to collapse the probe sequences when an entry is deleted (see [Knuth 73]), but the process works only for linear probing.

Because of all these reasons, if you want to support deletions, use separate chaining. Alternatively, the type of hash table to be discussed next can support deletions if need be.

COALESCED CHAINING

Coalesced chaining is a hybrid between separate chaining and open addressing. Like separate chaining, collisions are resolved by keeping the colliding entries on linked lists. However, no separate nodes are used for the lists. Instead, an array of buckets storing entries is used, as in open addressing, with an additional array that stores *next* pointers, initially set to null. (The *inuse* array also is used, as before.) In addition, a cursor is kept to help find empty buckets to insert into a given list. Initially this cursor points to the first bucket and is incremented sequentially through the table, as needed.

Figure 7.6 shows the state of a coalesced hash table after adding our familiar eight phone numbers. For reasons to be explained later, we start the table indexing at 1 instead of 0. To compensate, we add one to the hashed values. The numbers 6741088, 6741212, and 6746711 are inserted without collisions. When we attempt to add 6743288, a collision at bucket 3 occurs. The corresponding *next* pointer is null, so an empty bucket is found by scanning sequentially down the table, using the cursor, which starts out at bucket 1. Since bucket 1 is empty, the entry is added there, and bucket 3's *next* pointer is set to point to bucket 1.

The same type of collision occurs when 6745117 is added; its hash value collides at the bucket 6. Bucket 2 is the next available empty bucket and is used to store the new entry. Bucket 6's *next* pointer is set to link to bucket 2.

The numbers 6749211 and 6745412 are added without incident. However, when the last number, 6742517, is added, another collision occurs. This number maps to bucket 1, which unfortunately already has an entry from the chain for bucket 3. At this juncture, the cursor also points to bucket 3 and must be incremented all the way to bucket 7 before an empty bucket is found. The new entry is added there, and bucket 1's *next* pointer is set to link to bucket 7.

We end up with two probe chains that are now sharing the same list. One chain is for all keys that hash to bucket 3. The other is for all keys that hash to

Figure 7.6 Hash table with coalesced chaining.

Next Entries

	Next	Entries
1	7	6743288 (3)
2	0	6745117 (6)
3	1	6741088 (3)
4	0	6745412 (4)
5	0	6746711 (5)
6	2	6741212 (6)
7	0	6742517 (1)
8	0	6749211 (8) ← cursor
9	0	
10	0	
11	0	

bucket 1, and this chain starts in the middle of the first chain. In other words, we've allowed the lists to *coalesce*, something that doesn't occur with separate chaining.

Because of the coalesced lists, a certain amount of clustering can occur. However, coalesced chaining leads to much less clustering than with any open addressing scheme, although you may not believe it with the small example we gave. In practice, the amount of clustering will be somewhere in between that of separate chaining, where no clustering occurs, and the open addressing schemes, where a great deal of clustering can occur.

Coalesced hashing is an excellent method to use when you want a fixed-size table. The only drawback is that the *next* pointer array represents additional $O(m)$ space over that needed in open addressing. However, if the entries are large to start with, this overhead may be masked. Or other auxiliary pointers might be needed anyway, such as in applications like a compiler symbol table.

The following *CoalHashTable* class shows an example of implementing coalesced chaining. The class is very similar to all the other hash table classes that we've shown.

```
template<class TYPE, class KEY>
class CoalHashTable {
protected:
  TYPE *buckets;  // Data for the buckets
  unsigned *next; // Next pointers for the buckets
  char *inuse;    // In-use flags for the buckets
  long num_entries, num_collisions, num_comparisons;
```

```
  unsigned num_buckets, cursor;
  int ok;
public:
  CoalHashTable(unsigned tsize);
  virtual ~CoalHashTable();
  void Clear();
  int Insert(const TYPE &x);
  int Update(const TYPE &x);
  TYPE *Search(const KEY &key);
  const TYPE *Search(const KEY &key) const;
  int Delete(const KEY &key);
  ...
};
```

 Complete code for the *CoalHashTable* class is given on disk in the files *chash.h* and *chash.mth*. Test programs can be found in *chtst1.cpp* and *chtst2.cpp*.

Like the *OpenHashTable* class, we don't construct the buckets until they actually are used. The *next* pointers are implemented as ordinary array indices. Also, we start the indexing from 1 instead of 0, so that we can use *next == 0* to indicate a null pointer. The modified indexing is accomplished by decrementing the array pointers by one after allocation. The following constructor and destructor, along with the *Clear()* function, show how this is done.

```
template<class TYPE, class KEY>
CoalHashTable<TYPE, KEY>::CoalHashTable(unsigned tsize)
// Sets up a hash table with tsize buckets. The data portion
// of the buckets is left unconstructed until used.
// NOTE: We decrement the pointers from the actual start
// of the arrays, so that we can index from 1.
{
  ok = 0;
  buckets = (TYPE *) new char[tsize * sizeof(TYPE)];
  next = new unsigned[tsize];
  inuse = new char[tsize];
  if (buckets && next && inuse) {
    ok = 1;
    buckets--; next--; inuse--;
    num_buckets = tsize;
    for (unsigned i = 1; i <= num_buckets; i++) inuse[i] = 0;
    Clear();
  }
  else { // Memory allocation failure, so clean up everything.
    delete[] inuse;
    delete[] next;
    delete[] (char *)buckets; // As allocated, so deallocated
  }
}

template<class TYPE, class KEY>
CoalHashTable<TYPE, KEY>::~CoalHashTable()
```

```
{
  if (ok) {
    Clear();
    // Must adjust pointers back to real start of arrays.
    delete[] ++inuse;
    delete[] ++next;
    delete[] (char *)(++buckets); // As allocated, so deallocated
  }
}

template<class TYPE, class KEY>
void CoalHashTable<TYPE, KEY>::Clear()
// Clears the table, destroying the data for all buckets in use.
{
  unsigned i;
  for (i = 1; i<=num_buckets; i++) {
      if (inuse[i]) (buckets+i)->TYPE::~TYPE();
  }
  for (i = 1; i<=num_buckets; i++) inuse[i] = 0;
  for (i = 1; i<=num_buckets; i++) next[i] = 0;
  cursor = 0;
  num_entries = 0; num_collisions = 0; num_comparisons = 0;
}
```

Instead of starting the indexing from 1, we also could have used *signed ints* instead of *unsigned ints* for the array indices and used *next == −1* to flag null pointers. Or we simply could leave the first bucket empty and never use it. This would waste space, however, especially if the entries are large.

Next we give the *Insert()* and *Search()* functions to illustrate how the coalesced chaining is implemented.

```
template<class TYPE, class KEY>
int CoalHashTable<TYPE, KEY>::Insert(const TYPE &x)
// Inserts a new entry having the specified key. If a matching
// entry is found, DUPLICATE_KEY is returned, else SUCCESS or
// TABLE_FULL is returned.
{
  // Note: indexing from 1
  unsigned h = unsigned(TYPE::Hash(x.Key()) % num_buckets) + 1;

  while (inuse[h]) { // Okay, we've got a collision.
    if (buckets[h].Match(x.Key())) return DUPLICATE_KEY;
    unsigned p = h;
    h = next[h];
    if (h == 0) { // End of list, search for place to add node.
      while(1) {
        if (cursor == num_buckets) return TABLE_FULL;
        if (inuse[cursor] == 0) break;
        ++cursor;
      }
      h = cursor;          // This is the new node for list.
      next[p] = cursor++;
```

```
      break;
    }
  }

  new(buckets+h) TYPE(x); // Copy construct the data in place.
  inuse[h] = 1;
  next[h] = 0;
  return SUCCESS;
}

template<class TYPE, class KEY>
TYPE *CoalHashTable<TYPE, KEY>::Search(const KEY &key)
// Searches for a bucket having specified key. Returns pointer
// to the data in the bucket if found, else returns 0.
{
  // Note indexing from 1.
  unsigned h = unsigned(TYPE::Hash(key) % num_buckets) + 1;

  while(1) {
    if (inuse[h] == 0) break;
    if (buckets[h].Match(key)) return buckets + h;
    h = next[h];
    if (h == 0) break;
  }

  return 0; // No match found
}
```

As implemented, *CoalHashTable* doesn't support deletions. The singly linked chains are too awkward for the task. However, if you make the chains doubly linked, by adding an extra array of *prev* pointers, deletion can be supported. You'll have to handle the fact that the entry being deleted might be the first in a chain. One of the entries that follow it in the chain will have to be moved to the deleted entry's bucket.

FILE-BASED HASHING

So far we have been considering only hash tables that reside in memory. Often it's useful to store the tables in files, to give them a bit more permanence. There are two basic approaches. If the table is small enough to fit into memory, you can load the whole table into memory, use it, and then store it back in the file when finished. Or you can load only portions of the table in memory.

For the first approach, using a fixed-size table is the best. Either the open addressing or coalesced chaining method works fine. Recall that we implemented the *next* pointers used in coalesced chaining as simple array indices. That makes the table easily relocatable as you move it in and out of memory. Using separate chaining is harder, since the *next* pointers are true pointers. Storing true pointers in a file is problematic, and the table is not as easy to relocate.

If the table is fairly large, you'll want to keep only portions of it in memory. Although all of the hashing methods will work, separate chaining seems the most appropriate, for now the table can grow to any size, (as long as the chains don't get too long). Here you must change the *next* pointers to long integers, which represent offsets into the file. It's useful to have a small cache in memory that stores chain nodes to help speed things up. Also, you can expect to double the performance if you cache the bucket elements (the long integers that represent the heads of the chains). Better yet, you can keep a complete copy of the bucket elements in memory. That will result in another doubling in performance, according to our experiments.

Another optimization you might want to do is to allow each bucket (in the case of fixed-size tables) or each chain node (in the case of separate chaining) to store more than one entry. That way one access to the file will retrieve multiple entries, which will be ones that have the same hash values. While in theory this should speed things up, in our experiments that used a cached table implemented with separate chaining, we found very little difference in performance.

To give you a working example of file-based hashing, we have provided a *FileHashTable* class on disk in the files *fhash.h* and *fhash.mth*, with test programs in *fhtst1.cpp* and *fhtst2.cpp*. The *FileHashTable* class uses the separate chaining method, with caching as provided by classes developed in the companion book. A text file on disk, *fhash.txt*, explains how these additional files work.

REHASHING

With all the techniques we've presented, a problem occurs if you try to add too many entries, especially for the fixed-size tables. Even with separate chaining, which allows any number of entries to be added, eventually you will run into problems, as the chains will become too long to be practical. You must either know ahead of time how many entries to expect and design your tables accordingly, or you must provide some way for the table to grow.

One technique that's used when a table becomes full is to create another, larger table, perhaps twice the size, and insert the entries from the old table into it. This is called *rehashing*, for each key must be rehashed, since the number of buckets is different. Depending on the size of the table, this can be rather expensive, so you don't want to do it very often.

There are techniques that allow the tables to grow on the fly (for example, see [Larson 88]), but they are somewhat complicated, and almost all the techniques suffer slight performance problems even if the table doesn't need to grow. Another popular technique is called *extendible hashing*, which sets up a hierarchical directory structure based on the bit patterns of the hashed keys. This technique is suitable for file-based hashing and is somewhat reminiscent of using binary trees. (In some ways, extendible hashing is a hybrid between tree-based searching and hashing.) See Horowitz [Horowitz 90] for an example.

COMPARISON OF TECHNIQUES

In this chapter you've seen a variety of techniques for building hash tables. We've shown two important hashing functions that are quite effective: the division method for numerical keys and ELF hashing for string-based keys. You won't go wrong using these two hashing functions.

Of the collision resolution strategies, separate chaining is the easiest to implement. It results in no clustering and allows an arbitrary number of entries to be added. Of course, if too many entries are added, the bucket lists become too long and performance suffers. Open addressing offers compact albeit fixed-sized hash tables, but more probing is necessary than with separate chaining, since there will be some clustering of the probe sequences. Finally, coalesced hashing is an attractive hybrid of the first two, offering fast probing yet relatively small and compact tables.

Of the probe sequences that can be used for open addressing, double hash probing is the best general-purpose approach. Its performance is very close to uniform probing, which theoretically is as good as you can get. However, setting up the probe sequence is slightly slower than with, say, quadratic probing, since an extra division is required. Realistically, though, that extra division is likely to be masked by the time spent in comparing keys, especially when the keys are long character strings. Also, double hash probing allows all buckets to be scanned. The simple form of quadratic probing can only scan half the buckets, causing the table to appear full when it really isn't.

To give you an idea of the performance of the different types of hashing, Table 7.1 shows the number of probes (i.e; comparisons) required to insert 1000 words, extracted from a customer database, into a hash table with 1021 buckets. The table shows statistics at very stages, to illustrate how the various strategies perform as the table fills up. When the table is less than half full, there isn't much difference between any of the strategies, but as the table becomes full, the separate chaining method wins easily, with coalesced chaining second. Double hash probing and random probing are the best of the open addressing schemes.

Keep in mind that hashing performance depends on the patterns present in the keys, so you might get different results with your own data. For example, even though random probing performed well in our test data, with other data it will do only slightly better than quadratic probing. And remember that random probing is expensive to implement, so the runtime is likely to be worse than either quadratic probing or double hash probing.

Table 7.1 Hashing Performance (Number of Comparisons)

No. of Entries	L	Q	D	R	C	S
250	51	55	44	45	38	35
500	846	370	194	184	185	135
750	1,925	1,488	626	599	506	301
1000	61,162	4,858	2,784	2,776	989	527

Legend:

L = Linear probing   D = Double hash probing   C = Coalesced chaining

Q = Quadratic probing   R = Random probing   S = Separate chaining

E I G H T

Heaps

I n this chapter we move away from our discussion of generators, modulo arithmetic, and randomization, and start focusing on sorting algorithms. One type of sort, *heap sort*, is based on a data structure called a *heap*, which also is useful for implementing contiguous priority queues. (In the companion book we looked at a way to implement non-contiguous priority queues using *splay trees*.) To start our discussion, we'll examine the heap data structure and its derivatives in depth and see how they can be used as priority queues.

PRIORITY QUEUES

A priority queue is a data structure where you can insert items at one end of the queue and, at the other end, extract items based on their *priority*. Those items with a higher priority are extracted first. Figure 8.1 shows the basic model of a priority queue.

The priority of an item is often a number or key stored in the item. Usually, a higher number or key means a higher priority. But sometimes it's the other way around, depending on the application.

Figure 8.1 Abstract model of a priority queue.

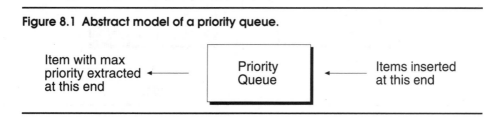

An example of a priority queue is the line outside the building hosting the Oscar awards, where movie stars have a higher priority and can move directly to the front of the line. In the programming world, priority queues are used for applications such as the scheduling of tasks in a multitasking operating system, where some tasks get a higher priority than others. Another example is an event-driven simulation, where each event has a time that it needs to be completed by. Those with the earliest times are executed first.

There are a couple of simple ways to implement a priority queue. One is to use an unordered array. Items are inserted at the end of the array. Upon extraction, the array is searched, looking for the highest-priority item. This item is extracted, and the hole left behind is replaced with the last item in the array. The time complexity is $O(n)$ comparisons. Figure 8.2 shows an example.

Using the last item in the array to fill the hole left behind is a useful trick to remember. It avoids a lot of data movement. You saw this same trick in the *pile* data structure in Chapter 6, and you'll see it again shortly with heaps.

Another alternative is to use an unordered linked list. The technique is similar to using an array, except the extraction is accomplished by pointer manipulation, rather than by moving the last item. Again, the extraction requires $O(n)$ comparisons.

We can improve things somewhat by keeping the array or linked list sorted in order of priority. Then the time for extraction is $O(1)$. However, during insertion, the new item must be placed in order. On average, we would have to scan down at least half the array or list to find the appropriate place for the new item, implying $O(n)$ time.

Another idea is to use a binary search tree, as given in Chapters 10, 11, and 12 in the companion book. Now insertion is reduced to $O(\log n)$ complexity, and extraction is $O(\log n)$. Indeed, Chapter 12 explained how splay trees make good priority queues (for non-contiguous implementations, that is). However, search trees are in some ways overkill. We don't need to know the complete ordering of the items. All we need is to know, at any given time, which item has the highest priority. Can we take advantage of the reduced ordering requirements and come up with a more efficient structure than binary trees for priority queues? The answer is yes, in the heap structure you are about to see.

Figure 8.2 Extracting the highest-priority item from an unordered array.

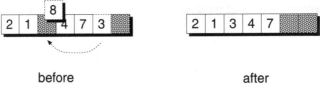

before   after

HEAPS—COMPLETE BINARY TREES

Heaps are binary trees, but they have different characteristics from binary search trees. A heap is always structured as a *complete binary tree*. A tree is complete if all levels of the tree are filled, except for possibly the last level. Figure 8.3 shows an example.

A complete binary tree is as balanced as possible. This will mean optimum time complexities in the heap algorithms to come. Because a complete binary tree has no "holes" in it, you can represent it compactly in an array, with no need for left and right pointers. For example, Figure 8.4 shows the array representation of the tree in Figure 8.3.

The mapping from tree nodes to array items is as follows. The root node contains the first item. The two children of the root contain the second and third items. The children of the left child of the root contain the fourth and fifth items, and so on. If you've studied the companion book, you'll recognize this ordering as *level ordering*. Given a parent node at position p in the array, its children will be at positions $2p$ and $2p + 1$. Conversely, the parent of a child at position c will be at $c/2$ (using integer division).

Just as light can be thought of as both particles and waves, a heap can be thought of as both a tree and array. The compactness of the array form is what makes heaps useful as contiguous implementations of priority queues.

Heaps have a special ordering known as the *heap property*: A parent node always has a higher priority than its children. When a higher priority means a higher key, the heap is known as a *max-heap*. When a higher priority means a lower key, the heap is known as a *min-heap*. The root node key will be the largest of the keys for max-heaps and smallest for min-heaps. As an example, Figure 8.3 and Figure 8.4 show the different forms of the same min-heap.

Figure 8.3 A complete binary tree.

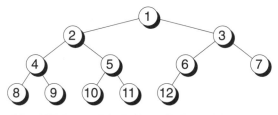

Figure 8.4 Array implementation of a complete binary tree.

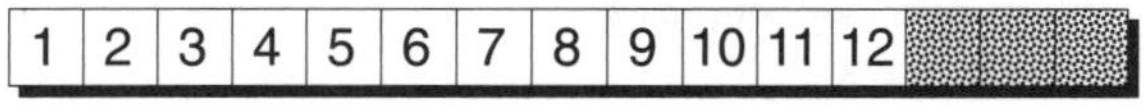

| 1 | 2 | 3 | 4 | 5 | 6 | 7 | 8 | 9 | 10 | 11 | 12 | |

Figure 8.5 An example of a max-heap.

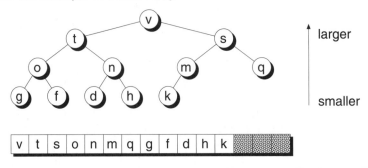

> **Note** In the examples to follow, we will be using max-heaps, and unless otherwise noted, when we say "heap," we mean "max-heap." Also, we'll use the term *large item* to mean an item with the high priority, and *small item* to mean an item with a low priority. In most of the illustrations, we'll represent priorities using letters.

Using the array form, the heap property means that items on the left will be larger than those on the right. However, it's not necessary for the items to be in strict sorted order. For example, Figure 8.5 shows a max-heap in both tree and array form, where the items are not in strict descending order. All that's required to satisfy the heap property is that a parent is larger than its children.

THE HEAP CLASS

Before we study operations on heaps, we'll introduce a *Heap* class to give us a framework for discussion.

```
const int MIN_HEAP = 0;
const int MAX_HEAP = 1;

template<class TYPE>
class Heap {
protected:
  TYPE *items;
  int maxitems, nx, aliased_data, doing_max_heap;
  static int Higher(const TYPE &a, const TYPE &b);
  static int Lower(const TYPE &a, const TYPE &b);
  void TrickleUp(int k, const TYPE &x);
  void TrickleDown(int k, const TYPE &x);
public:
  Heap(int n, int sense=MAX_HEAP);
  Heap(int n, TYPE *m, int sense=MAX_HEAP);
  Heap(const Heap<TYPE> &h);
  virtual ~Heap();
  int Copy(const Heap<TYPE> &s);
```

```
    void operator=(const Heap<TYPE> &s);
    int Insert(const TYPE &x);
    int Extract();
    int Extract(TYPE &x);
    int Replace(int p, const TYPE &x);
    int Delete(int p);
    void Build();
    void Arrange();
    void Sort();
    ...
};
```

Complete code for the *Heap* class is given on disk in the files *heap.h* and *heap.mth*. Test programs are given in *tstheap.cpp* and *tstheap2.cpp*.

The *Heap* class is template-based, where *TYPE* is the type of item stored on the heap. It's assumed TYPE has the comparison operators > and < defined, as well as the standard complement of constructors, destructors, and assignment operators. While many housekeeping functions are listed in the class declaration, we're going to be examining only the relevant ones in the manuscript.

By default, the *Heap* class uses max-heap semantics, which you can change during the constructor call. The following static member functions *Lower()* and *Higher()* define the semantics.

```
template<class TYPE>
int Heap<TYPE>::Lower(const TYPE &a, const TYPE &b)
// Returns 1 if a has a lower priority than b.
// Note: Use a < b for a max-heap and a > b for a min-heap.
{
  return doing_max_heap ? (a < b) : (a > b);
}

template<class TYPE>
int Heap<TYPE>::Higher(const TYPE &a, const TYPE &b)
// Returns 1 if a has a higher priority than b.
// Note: Use a > b for a max-heap and a < b for a min-heap.
{
  return doing_max_heap ? (a > b) : (a < b);
}
```

Note that adding this flexibility costs us some runtime speed. If you want ultimate speed, you might "hard-code" classes for min- or max-heaps.

The *Heap* class dynamically allocates room for a fixed number of items, as shown in the following constructor and destructor:

```
template<class TYPE>
Heap<TYPE>::Heap(int n, int sense)
// Creates a heap holding n items.
{
  doing_max_heap = sense;
```

```
    items = new TYPE[n]; // Default TYPE constructor called.
    items--; // So that indices start with 1
    maxitems = n;  nx = 1; aliased_data = 0;
}

template<class TYPE>
Heap<TYPE>::~Heap()
{
    if (!aliased_data) delete[] (items+1); // Be sure to adjust pointer!
}
```

There is a little bit of trickery going on here. Notice how, after allocating room for the items, we decrement the *items* pointer by one. By doing this, the first item is given by the subscript *items[1]* instead of *items[0]*. There is a good reason for doing this. Here are the equations we gave earlier for computing the parent of a child and the children of a parent.

```
// For 1-based subscripting
p = c/2;  // To compute parent index given a child index
c = 2p;   // To compute left child index, given a parent index
c = 2p+1; // To compute right child index, given a parent index
```

These equations work only if subscripting starts at 1. If you start subscripting at 0, the equations must be changed to:

```
// For 0-based subscripting
p = (c-1)/2; // To compute parent index given a child index
c = 2p+1;    // To compute left child index, given a parent index
c = 2p+2;    // To compute right child index, given a parent index
```

It's simpler and faster to use the 1-based form. However, C++ uses 0-based subscripting, so we compensate by adjusting the pointer to the *items* array. Another alternative is to go ahead and used 0-based subscripting, but to allocate an extra item. The first item, (*items[0]*), then is used as a sentinel, having the highest priority of any item stored on the heap. We'll talk more about the sentinel approach in a little bit.

The *items* array is a variable length array, meaning that initially, no items are in use. Items are inserted at the end of the array. The *nx* index is used to point the next available location for insertion. When the heap is empty, *nx* = 1. When it is full, *nx* > *maxitems*. The current length of the array (that is, the number of items in use) is given by *nx* − 1.

HEAP OPERATIONS

Next we take a look at the basic operations used on heaps. Recall that in the simple, unordered array representation of a priority queue, we always add items to the end of the array, and when the largest item is extracted, the hole left behind is replaced with the last item. The same technique is used with heaps, except that after adding or removing an item, we must ensure that the heap property still holds.

Heap Insertion—Bubbling Up

Inserting an item into a heap involves adding the item to the end of the array and then comparing the item with its parent. If the item is larger than its parent, the two are swapped. Then, the item is compared with its new parent and swapped if necessary. This bubbling up process continues until either the item is smaller than its parent or the root is reached. Figure 8.6 illustrates this process.

In can be shown that, in the worst case, the bubble-up operation will peform $O(\log n)$ comparisons of items. This happens when the item added is the largest and must be moved all the way to the root. Of course, it's likely that the bubble-up process will finish much sooner than that. In any case, the insertion time is quite fast.

The insertion operation is coded in the following *Insert()* and *BubbleUp()* functions.

```
template<class TYPE>
int Heap<TYPE>::Insert(const TYPE &x)
// Inserts x into heap by grabbing the next available slot
// and then bubbling up the tree until the heap property
// is restored. Returns 1 if successful, 0 if heap is full.
{
  if (IsFull()) return 0;
  BubbleUp(nx++, x); // Heap grows by one item.
  return 1;
}
```

Figure 8.6 The bubble-up operation.

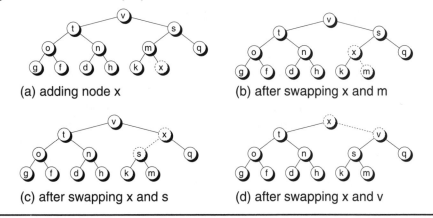

(a) adding node x

(b) after swapping x and m

(c) after swapping x and s

(d) after swapping x and v

```
template<class TYPE>
void Heap<TYPE>::BubbleUp(int k, const TYPE &x)
// Bubble the item at position k, which we pretend has x
// stored in it, up the tree until the heap property is
// restored. During this process, items with a lower
// priority than x will be shifted down toward the bottom
// of the tree. ASSUMES the heap is not empty.
{
  while(k > 1 && Higher(x, items[k/2])) {
    // While not at root and x has a higher priority,
    // shift parent to child position.
    items[k] = items[k/2];
    k /= 2; // Repeat, using parent.
  }

  items[k] = x; // Copy x into the final resting place.
}
```

While there is an inherent swapping process taking place in the inner *while* loop of *BubbleUp()*, we have carefully arranged the assignments so that a full swap doesn't take place until the very end.

You'll notice that we perform two tests at the beginning of the *while* loop, a check to see if we've reached the root and the comparison of the items. It's possible to get rid of the first test by storing a sentinel at *items[0]* (using 0-based subscripting, and allocating one more than the number of items needed). This sentinel should have a higher priority than any item in the heap.

Using a sentinel will speed up the heap operations somewhat. The trouble is, you have to remember to put in the sentinel, and you must arrange for it to have the highest priority. It's easy to forget the sentinel, resulting in hard-to-find bugs. And it can be hard to determine what the sentinel value should be, for any given type. The problem is particularly difficult when you are using templates. We chose not to go the sentinel route, sacrificing some speed for safety and robustness.

Heap Extraction—Trickling Down

The largest item always is the one extracted from the heap. That item always will be at the root (or in array form, the first item in the array). Once the item is extracted, the hole left behind is filled with the rightmost leaf in the tree (or in array form, the last item in the array). This keeps the tree balanced and complete. However, the tree may now be out of order, so the order is restored by using the inverse of the bubble-up operation, in an operation called *trickle-down*. This process is coded in the following *Extract()* and *TrickleDown()* functions.

```
template<class TYPE>
int Heap<TYPE>::Extract(TYPE &x)
```

```
// Copies the data in the front item of the heap into x,
// and then removes that item from the heap.
// Returns 0 if heap is empty, else 1. If heap is empty,
// x remains untouched.
{
  if (IsEmpty()) return 0;
  x = items[1]; // Copy root into x
  // Pretend as if the rightmost leaf has been moved
  // from where it is and placed in the root, then trickle
  // the item down where it belongs. Note that the item isn't
  // really moved until its final resting place is known.
  TrickleDown(1, items[--nx]);
  return 1;
}


template<class TYPE>
void Heap<TYPE>::TrickleDown(int k, const TYPE &x)
// Trickle the item at position k, which we pretend has x
// stored in it, down the tree until the heap property is
// restored. Note that items below k are shifted up as
// needed. ASSUMES heap isn't empty.
{
  int j, last;

  last = nx - 1; // Last occupied position in the heap
  while(k <= last/2) { // item at k has at least one child
    // Set j to index child with highest priority.
    j = k*2;
    if (j < last && Higher(items[j+1], items[j])) j++;
    if (Higher(x, items[j])) break; // x larger than children
    // Otherwise, move higher priority child up, and repeat.
    items[k] = items[j];
    k = j;
  }

  items[k] = x; // Copy x into the final resting place.
}
```

The trickle-down operation works as follows. If the item at position k is smaller than either child, the item is swapped with the largest child. At the new position, the item is again compared with its children and swapped if necessary. We always use the largest child when swapping, to guarantee that the parent will be larger than its children. This process repeats until either the item is larger than both children, or we're at the bottom of the tree. As we did with *BubbleUp()*, we carefully arranged the assignments to avoid doing a full swap until the very end. Figure 8.7 shows the trickle-down operation.

In the worst case, the trickle-down operation takes $O(\log n)$ comparisons. That's also the same complexity in the average case, for remember that we are trickling down the tree what used to be the rightmost leaf. It's likely that we'll have to trickle the item all the way back down to the bottom. Note, though,

Figure 8.7 The trickle-down operation.

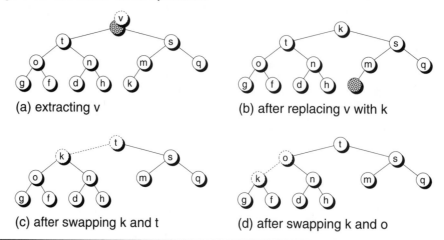

(a) extracting v

(b) after replacing v with k

(c) after swapping k and t

(d) after swapping k and o

that the item won't be moved back to its original position. That's partially because there will be one less node in the tree. But the item also may be moved to the other side of the tree, as illustrated in Figure 8.7.

Replacing Items

In the strictest terms, the only operations defined for a heap are insertion and extraction. However, there are applications where you may want to replace one of the existing items on the heap. For example, if the heap is used as a priority queue for task scheduling, at some point you may want to give one task a higher priority.

Next we'll show you how to do replacement. It's assumed that you magically have knowledge of where an item resides on the heap. An example of this is given in Chapter 15, where we'll be using a fully indirect priority queue. Later in this chapter you'll see a simple form of an indirect queue.

If you replace an item with new data, but don't change the priority of the item, only an assignment is needed. However, if the priority is to change, you must restore the heap order after the replacement. Basically, if the item is larger than its parent, a bubble-up operation must be employed. If the item is smaller than either of its children, a trickle-down operation must be employed. Figure 8.8 shows both cases.

As with most of the heap operations, replacement takes $O(\log n)$ time for worst and average cases. The following function implements replacement.

```
template<class TYPE>
int Heap<TYPE>::Replace(int p, const TYPE &x)
```

Figure 8.8 Replacing items on the heap.

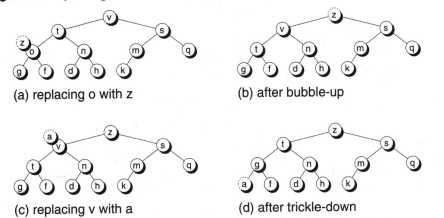

(a) replacing o with z

(b) after bubble-up

(c) replacing v with a

(d) after trickle-down

```
// Replace the item at position p with new data,
// which might change its priority.
// Returns 1 if successful, 0 if p is out of range.
{
  if (p < 1 || p >= nx) return 0;

  if (Higher(x, items[p])) BubbleUp(p, x);
  else if (Lower(x, items[p])) TrickleDown(p, x);
  else items[p] = x;

  return 1;
}
```

Deleting Items

Another operation you might want to have is the ability to delete an item from
a heap, before it is extracted. An example would be killing a task waiting in a
multitasking queue. Deletion is actually quite simple. Just replace the item to
be deleted with the data stored in the rightmost leaf node, using the *Replace()*
function. Figure 8.9 shows an example.

Since the deletion is really using the replacement operation, $O(\log n)$ time
is required, for both worst and average cases. The following function implements the delete operation.

```
template<class TYPE>
int Heap<TYPE>::Delete(int p)
// Delete the item at position p from the tree.
// Returns 1 if successful, else 0 if p is out of range.
{
  if (p < 1 || p >= nx) return 0;   // Out of range
  nx--; // Heap's going to shrink.
```

Figure 8.9 Deleting an item from a heap.

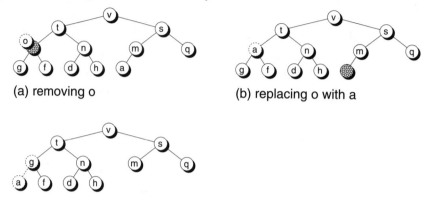

(a) removing o

(b) replacing o with a

(c) after swapping a with g

```
        if (p != nx) Replace(p, items[nx]);
        return 1;
}
```

Building a Heap Bottom Up

If you have an existing array of data that you would like to turn into a heap, the following two functions will allow you to do so. The first function is a constructor that lets you alias an existing array. It's assumed this array is already filled with data.

```
template<class TYPE>
Heap<TYPE>::Heap(TYPE *m, int n, int sense)
// Sets up a heap that aliases the array m, which presumably
// is already loaded with n items.
{
    doing_max_heap = sense;
    items = m-1; // So that indices start with 1
    maxitems = n;  nx = n+1;  aliased_data = 1;
    MakeHeap();
}
```

Since the array may not be in heap order, the following *Arrange()* function is called, which rearranges the items to be in heap order.

```
template<class TYPE>
void Heap<TYPE>::Arrange()
// Assumes the items are already loaded, but unordered.
// This routine orders the items to have the heap
// property. The algorithm works from the bottom up.
```

Figure 8.10 Bottom-up heap construction.

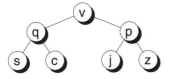

(a) initial, unordered heap

(b) after swapping p with z

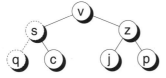

(c) after swapping s with q

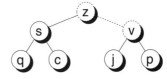

(d) after swapping z with v

```
{
  if (IsEmpty()) return;

  TYPE temp; // Default constructor called

  for (int i = (nx-1) / 2; i >= 1; i--) {
      // The source data, in items[i], is on the trickle path,
      // so we must save a copy of it before starting.
      temp = items[i];
      TrickleDown(i, temp);
  }
}
```

The *Arrange()* function uses a bottom-up algorithm known as *Floyd's algorithm*, which happens to be $O(n)$ in complexity. It works on the fact that we can treat each subtree in the heap as a sub-heap, and restore the heap property for each sub-heap, working our way up the tree from right to left until the root is reached. We start with the parent of the rightmost child, which always can be found at index $(nx-1)/2$. (Recall that nx is always one past the last item in the heap.) Figure 8.10 illustrates the process of turning an unordered tree into a heap.

Using a Heap to Sort

You can use a heap to sort data. First, place the unordered data in the heap, then call *Arrange()* to ensure the data is in heap order. Then extract the items one at a time, until the heap is empty. If the heap is a max-heap, the data comes out in descending order. If the heap is a min-heap, the data comes out in ascending order. For example, in the following code,

```
char sorted_arr[7], arr[7] = {'v', 'q', 'p', 's', 'c', 'j', 'z'};

Heap<int> heap(arr, 7); // Puts arr in max-heap order automatically.

for (int i = 0; i<7; i++) heap.Extract(sorted_arr[i]);
```

the array *sorted_arr* will have the contents *[z, v, s, q, p, j, c]*.

The process just described is known as the *heap sort algorithm*. This algorithm is an important sorting algorithm, for both its worst-case and average complexities are $O(n \log n)$, the best that can be done with a general-purpose sort that uses comparisons.

While our example used another array to hold the results of the sort, it's possible to sort the array in place. Instead of extracting the root item, you can swap it with the rightmost leaf item. That will become the root's final resting place. Decrement the length of the heap by one. Then trickle the new root (which has the rightmost leaf's original value) down the tree to restore the heap property. Again, swap the rightmost leaf and the root, and trickle down the tree. Repeat this process until the length of the heap is zero. Figure 8.11 illustrates this process, using the heap we built in the last section.

Figure 8.11 In-place heap sorting.

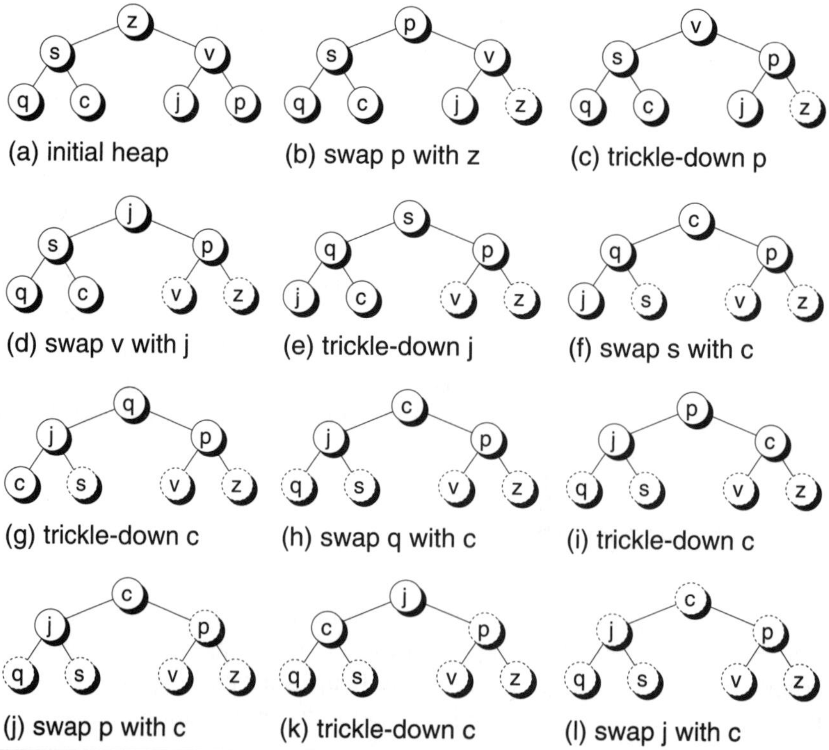

(a) initial heap

(b) swap p with z

(c) trickle-down p

(d) swap v with j

(e) trickle-down j

(f) swap s with c

(g) trickle-down c

(h) swap q with c

(i) trickle-down c

(j) swap p with c

(k) trickle-down c

(l) swap j with c

Contrary to the auxiliary array approach, sorting a max-heap in place will put the array in ascending order, while sorting a min-heap in place results in a descending order. For example, the array produced in Figure 8.11 would have the contents *[c, j, p, q, s, v, z]*.

The following *Sort()* function implements the heap sort algorithm.

```
template<class TYPE>
void Heap<TYPE>::Sort()
// Assuming the data in the array is already constructed, and
// in heap order, this routine does an in-place heap sort.
// NOTE: The data is sorted in reverse order!
{
  if (nx <= 2) return; // Nothing to sort.

  TYPE leaf; // Default constructor called.
  int old_nx = nx; // So we can restore length afterward

  while(1) {
    // First, store the rightmost leaf in temporary storage
    // and the root in the rightmost leaf. Shrink the heap too.
    leaf = items[--nx];
    items[nx] = items[1];
    // Now, pretending that the saved data from the rightmost
    // leaf is in the root, restore the heap property by
    // trickling down.
    if (nx > 2) { // Really something to do
       TrickleDown(1, leaf);
    }
    else { // Only two items left, so finish the swap, and quit.
       items[1] = leaf;
       break;
    }
  }

  nx = old_nx; // Restore length of heap.
}
```

The *Sort()* function is given mainly for illustrative purposes. When doing heap sorts for real, you would condense the code contained in *TrickleDown()*, *Arrange()*, and *Sort()* into a single function. In the next chapter we'll do just that.

DEAPS—DOUBLE ENDED HEAPS

A heap allows you to extract either the minimum or maximum item efficiently, depending on whether you are using a min-heap or a max-heap. However, you can't efficiently extract *both* the minimum and maximum. There are times when it's convenient to do so, as in the external quicksort algorithm in Chapter 10. A data structure that allows you to extract both the smallest and largest items is known as a *double-ended priority queue*. As you might guess, if you implement the queue using a heaplike data structure, you end up with a *double-ended heap*.

Several different data structures can be used to implement double-ended heaps. We'll show one that is probably the simplest and most efficient of the lot. This data structure is called a *deap*.

Like a normal heap, a deap is a complete binary tree and can be stored in array form. In a deap, the root node is always empty. The subtree to the left holds a min-heap. The subtree to the right holds a max-heap. One additional property of a deap is that an item *i* in the min-heap will always be less than or equal the corresponding item *j* of the max-heap. One consequence of this rule is that the root of the min-heap holds the minimum item of the deap and the root of the max-heap holds the maximum item. It's no coincidence that these two items are right at the top of the tree. Figure 8.12 shows an example of a valid deap.

In the figure, note how each node in the min-heap is less than the corresponding node in the max-heap. Examples are $k < n$, $f < p$, and $c < x$.

It's possible that the max-heap will have fewer nodes than the min-heap, as shown in the figure. If there is no corresponding node in the max-heap, then item *i* must be less than the item p_j in the max-heap that corresponds to *i*'s parent. For example, the corresponding node for both *o* and *h* is *p*. Again, we have the proper relation that $o < p$ and $h < p$. Of course, that has to be the case since *f*, the parent of *o* and *h*, is less than *p*.

To give you more experience with deap properties, Figure 8.13 shows an invalid deap, where node *y* is greater than its corresponding node *n*, and node

Figure 8.12 A valid deap.

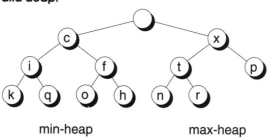

min-heap   max-heap

Figure 8.13 An invalid deap.

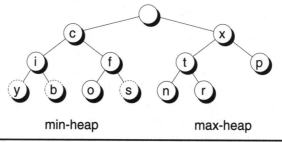

min-heap   max-heap

s is greater than the corresponding *p*. Also, *b* is less than its parent *i* in the min-heap, making the min-heap portion invalid.

A Deap Class

Here is the declaration of a *Deap* class that we'll be using as the basis for discussion. It's similar to the *Heap* class, and as always, we'll be examining only the most relevant functions in detail.

```
template<class TYPE>
class Deap {
protected:
  TYPE *items;
  int maxitems, nx;
  int DeapPartner(int k, int &p);
  void BubbleUpMin(int k, const TYPE &x);
  void TrickleDownMin(int k, const TYPE &x);
  void BubbleUpMax(int k, const TYPE &x);
  void TrickleDownMax(int k, const TYPE &x);
public:
  Deap(int n);
  Deap(const Deap<TYPE> &s);
  virtual ~Deap();
  int Copy(const Deap<TYPE> &s);
  void operator=(const Deap<TYPE> &s);
  int Insert(const TYPE &x);
  int ExtractMin();
  int ExtractMin(TYPE &x);
  int ExtractMax();
  int ExtractMax(TYPE &x);
  int Replace(int p, const TYPE &x);
  int Delete(int p);
  ...
};
```

Complete code for the *Deap* class is given on disk in the files *deap.h* and *deap.mth*. A test program is provided in *tstdeap.cpp*.

As with heaps, the subscripting for a deap should start at 1, to make the parent-child index calculations simpler. However, since the root element is always empty, we don't have to store it. We can start the subscripting at 2 instead. The following constructor and destructor reflect this 2-based subscripting.

```
template<class TYPE>
Deap<TYPE>::Deap(int n)
// Constructs a deap of size n.
{
  maxitems = n; nx = 2;
  items = new TYPE[maxitems];
```

```
    items -= 2; // For 2-based indexing
}


template<class TYPE>
Deap<TYPE>::~Deap()
{
    delete[] (items+2); // Be sure to adjust pointer.
}
```

With this subscripting arrangement, the root of the min-heap will be at index 2. The root of the max-heap will be at index 3.

The variable *nx* is used to index the next available slot for storing an item. An empty heap has *nx == 2*, a full heap has *nx > maxitems + 1*, and the current length of the heap is given by *nx – 2*.

Deap Insertions

When an item is inserted into a deap, it is first added in the rightmost leaf node (at the end of the array). This means the item is always added to the max-heap (unless it's the first item to be added). The item must then be bubbled-up the max-heap to its appropriate location. Figure 8.14 shows an example of adding node *z* to the deap originally shown in Figure 8.12.

Sometimes the newly added item doesn't belong in the max-heap at all. For example, suppose we add *a* instead of *z*. Now, *a* is less than its corresponding node *o* in the min-heap, violating the deap property. To fix the problem, *a* is swapped with *o* and then bubbled up the min-heap as appropriate. Figure 8.15 illustrates this process.

Regardless of which heap we bubble up, it's important to note that the deap property—the corresponding nodes having the proper relations—still holds, even after all the shuffling that goes on.

Insertion into a deap is implemented in the following *Insert()* function. The *DeapPartner()* function is used to find the corresponding partner of the item we're about to add.

Figure 8.14 Bubbling up the max-heap

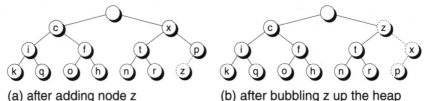

(a) after adding node z   (b) after bubbling z up the heap

Figure 8.15 Bubbling up the min-heap.

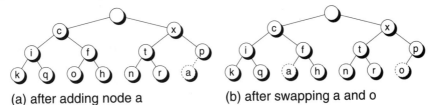

(a) after adding node a   (b) after swapping a and o

(b) after bubbling a up the heap

```
template<class TYPE>
int Deap<TYPE>::Insert(const TYPE &x)
// Inserts x into the deap. Returns 1 if successful, 0 if deap full.
{
  if (IsFull()) return 0;

  int k = nx; // k is the slot to add item.

  if (k == 2) { // x is first item added.
     items[2] = x; nx++;
     return 1;
  }

  // Find the partner to slot k.

  int partner_k;

  if (DeapPartner(k, partner_k)) {
     // k is in the max-heap, so x's priority should
     // be higher than its partner's. If not, we need
     // to swap before bubbling up.
     if (x < items[partner_k]) {      // Need to move x to min-heap
        items[k] = items[partner_k]; // by swapping with partner.
        BubbleUpMin(partner_k, x);
     }
     else BubbleUpMax(k, x);
  }
  else {
     // k is in the min-heap, so x's priority should
     // be lower than its partner's. If not, we need
     // to swap before bubbling up.
     if (x > items[partner_k]) {      // Need to move x to max-heap
        items[k] = items[partner_k]; // by swapping with partner.
```

```
        BubbleUpMax(partner_k, x);
    }
    else BubbleUpMin(k, x);
  }

  nx++; // Record growing deap. (Don't do this before now!)

  return 1;
}


template<class TYPE>
int Deap<TYPE>::DeapPartner(int k, int &p)
// Finds the corresponding node to the node at index k,
// Passes back the corresponding node index in p.
// Returns 1 if k is in the max-heap portion (thus p
// is in the min-heap portion), or 0 if vice-versa.
{
  int m = 1, j = k;

  while(1) { // Compute m = 2^log(k)-1.
    j /= 2;
    if (j < 2) break;
    m *= 2;
  }

  p = k - m; // Assume k in max-heap portion.

  if (p < m * 2) {
    // Nope, k is in the min-heap portion, so p must be in
    // the max-heap portion, and will be the parent of the
    // actual corresponding node if the corresponding node
    // doesn't exist.
    p = k + m;
    if (p >= nx) p /= 2;
    return 0;
  }
  else return 1;
}
```

The *DeapPartner()* function not only determines the corresponding node for node *k*, it also returns which heap *k* is on. Using this information, *InsertAt()* then determines which heap the new item *x* *really* belongs to, swapping partners if necessary. Note that *x* actually isn't copied into the deap until its final resting place is found. This avoids some assignments as the bubble-up operation takes place.

There are two bubble-up functions, one for the max-heap and one for the min-heap. Here is the *BubbleUpMax()* version. It's basically the same as that used in the *Heap* class, but notice that the root node for the max-heap is at index 3.

```
template<class TYPE>
void Deap<TYPE>::BubbleUpMax(int k, const TYPE &x)
```

```
// Bubble the item at position k, which we pretend
// has x stored in it, up the max-heap until the heap
// property is restored. During this process, items
// with a lower priority than x will be shifted down
// toward the bottom of the tree.
{
  while(k > 3 && x > items[k/2]) {
    // While not at root and x has a higher priority,
    // shift parent to child position.
    items[k] = items[k/2];
    k /= 2; // Repeat, using parent.
  }

  items[k] = x; // Copy x to final destination.
}
```

The *BubbleUpMin()* function is virtually identical, except the *while* loop test is replaced with the following code.

```
while(k > 2 && x < items[k/2]) // Use this instead for BubbleUpMin().
```

For min-heaps, the root is at index 2, hence the change. Also, we're checking for *x* being at a lower priority instead of a higher priority.

Extracting the Minimum Item from a Deap

The minimum item of the deap is in the root of the min-heap, at index 2. To extract this item, it is copied, and then the rightmost leaf of the deap is put in its place and trickled down as needed. This operation is basically the same as what we used in the *Heap* class, except that the item replacing the old min-heap root might be borrowed from the max-heap portion of the deap. Figure 8.16 illustrates the extract-minimum process.

Figure 8.16 Extracting the minimum item from a deap.

(a) after extracting c

(b) after replacing c with r

(c) after trickling down r

(d) after swapping r and n

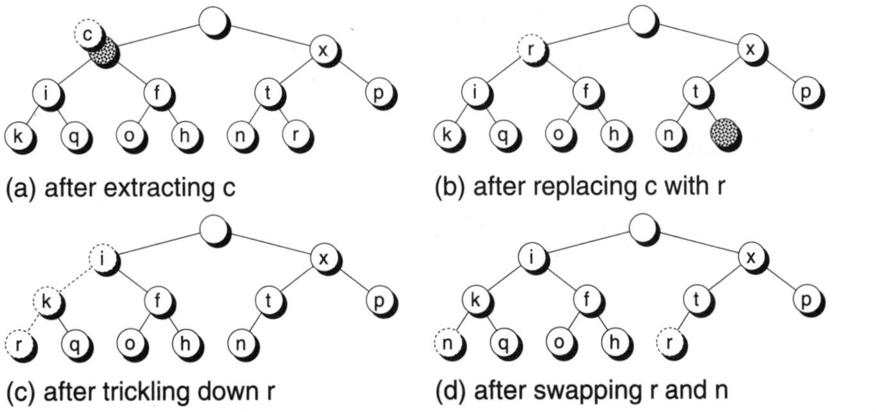

The *ExtractMin()* function is called to do the extraction, which in turns calls *TrickleDownMin()* to do most of the work.

```
template<class TYPE>
int Deap<TYPE>::ExtractMin(TYPE &x)
// Copies the data from the minimum item of the deap
// into x, and then calls the other ExtractMin() to do
// the actual removal.
{
  if (IsEmpty()) return 0;
  x = items[2]; // Copy root of min-heap into x.
  // Shrink deap by one. If we still have items, pretend
  // last item is at min-heap root, and trickle it down.
  if (--nx > 2) TrickleDownMin(2, items[nx]);
  return 1;
}


template<class TYPE>
void Deap<TYPE>::TrickleDownMin(int k, const TYPE &x)
// Trickle the item at position k, which we pretend has
// x stored in it, down the min-heap until the deap
// property is restored. Note that items below k are
// shifted up as needed. ASSUMES heap isn't empty.
{
  int j, last;

  last = nx - 1;  // Last occupied position in the heap
  while(k <= last/2) { // item at k has at least one child.
    // Set j to index child with lowest priority.
    j = k*2;
    if (j < last && items[j+1] < items[j]) j++;
    if (x < items[j]) break; // x smaller than children.
    // Move lower priority child up, and repeat.
    items[k] = items[j];
    k = j;
  }

  // x is ready to go into slot k. However, it may be larger
  // than its partner in the max-heap, if so, we need to swap
  // and then bubble x up the max-heap. Otherwise, simply
  // copy into slot k. Be sure to handle the case when only
  // one item is in the heap (partner_k < 2).

  int partner_k;
  DeapPartner(k, partner_k);
  if (partner_k >= 2 && x > items[partner_k]) {
    items[k] = items[partner_k];
    BubbleUpMax(partner_k, x);
  }
  else items[k] = x;
}
```

TrickleDownMin() is similar to the *TrickleDown()* function of the *Heap* class, except we must be careful to ensure that the deap property holds. The last few lines of *TrickleDownMin()* handles this.

Extracting the Maximum Item from a Deap

The maximum item in a deap will reside in the root of the max-heap, at index 3. To extract the maximum item, it is copied and the rightmost leaf of the deap put in its place. The new root is then trickled down to its proper place. Note that if there is only one item in the heap, then the maximum is the same as the minimum and is stored in the root of the min-heap. This special case must be trapped. Figure 8.17 illustrates the process of extracting the maximum item.

The *ExtractMax()* function is called to do the extraction, which in turns calls *TrickleDownMax()* to do most of the work.

```
template<class TYPE>
int Deap<TYPE>::ExtractMax(TYPE &x)
// Copies the data from the maximum item of the deap
// into x, and then removes the item.
{
  if (IsEmpty()) return 0;

  x = items[(nx == 3) ? 2 : 3]; // Copy max item into x.

  // Shrink deap by one. If there is a child on a lower level
  // to place in the max-heap root, pretend you place it there
  // and trickle it down.

  if (--nx > 3) TrickleDownMax(3, items[nx]);
  return 1;
}
```

Figure 8.17 Extracting the maximum item from a deap.

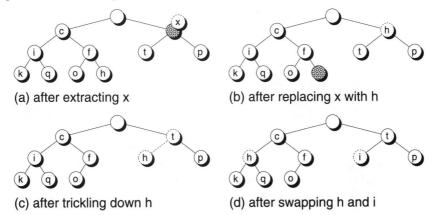

(a) after extracting x

(b) after replacing x with h

(c) after trickling down h

(d) after swapping h and i

```
template<class TYPE>
void Deap<TYPE>::TrickleDownMax(int k, const TYPE &x)
// Trickle the item at position k, which we pretend has
// x stored in it, down the max-heap until the deap
// property is restored. Note that items below  k are
// shifted up as needed. ASSUMES heap isn't empty.
{
  int j, last;

  last = nx - 1;  // Last occupied position in the deap
  while(k <= last/2) { // item at k has at least one child.
    // Set j to index child with highest priority.
    j = k*2;
    if (j < last && items[j+1] > items[j]) j++;
    if (x > items[j]) break; // x larger than children.
    // Move higher priority child up, and repeat.
    items[k] = items[j];
    k = j;
  }

  // x is ready to go into slot k. However, it may be smaller
  // than its partner in the min-heap; if so, we need to swap
  // and then bubble x up the min-heap. Otherwise, simply
  // copy into slot k.

  int partner_k;
  DeapPartner(k, partner_k);

  if (2*k >= nx) { // k has no children
    j = 2 * partner_k;
    if (j < nx) { // but partner does, so find largest.
      if (j < last && items[j+1] > items[j]) j++;
      partner_k = j;
    }
  }

  if (x < items[partner_k]) {
    items[k] = items[partner_k];
    BubbleUpMin(partner_k, x);
  }
  else items[k] = x;
}
```

As was the case with *TrickleDownMin()*, we must compare node *k* with its partner before copying *x* into to its final resting place. One special case in *TrickleDownMax()*, though, occurs when *k* has no children. When this occurs, we must compare with the larger of the partner's children (if it has any) rather than with the partner itself. Figure 8.18 illustrates this special case.

In the figure, notice the configuration after step (b). Node *k* has *c* as its partner, and *k* > *c*, so you would think the deap property is maintained. However, *c*'s child, *m*, is really the maximum node, so we should swap *k* and

Figure 8.18 Special case: Extracted maximum node has no children.

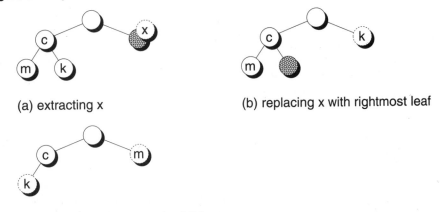

(a) extracting x

(b) replacing x with rightmost leaf

(c) swapping with partner's child

m. After the swap, the *BubbleUpMin()* function is called, although in this case, *k* doesn't move any farther.

Replacing and Deleting Items in a Deap

Replacing items in a deap works much the same way it does for heaps. If the replacement has a different priority, we may have to bubble up or trickle down to compensate. We also must determine which heap we are on (min-heap or max-heap) and thus call the appropriate functions. The following *Replace()* function shows how.

```
template<class TYPE>
int Deap<TYPE>::Replace(int p, const TYPE &x)
// Replace the item at position p with new data,
// which might change its priority. Returns 1 if
// successful, 0 if p is out of range.
{
  if (p < 2 || p >= nx) return 0;

  if (x == items[p]) { // Priority the same as before
    items[p] = x;
  }
  else {
    int unused_parm;
    if (DeapPartner(p, unused_parm)) { // p is in max heap.
      if (x > items[p]) { // New priority is higher.
        BubbleUpMax(p, x);
      }
      else { // New priority is lower.
```

```
                TrickleDownMax(p, x);
            }
        }
        else { // p is in min_heap
            if (x < items[p]) { // New priority is lower.
                BubbleUpMin(p, x);
            }
            else { // New priority is higher.
                TrickleDownMin(p, x);
            }
        }
    }

    return 1;
}
```

To delete an item, you merely replace it with the rightmost leaf and let *Replace()* do its magic to clean things up, as shown in the following *Delete()* function.

```
template<class TYPE>
int Deap<TYPE>::Delete(int p)
// Delete the item at position p from the tree.
// Returns 1 if successful, else 0 if p is out of range.
{
    if (p < 2 || p >= nx) return 0;  // Out of range
    nx--; // Deap's going to shrink.
    if (p != nx) Replace(p, items[nx]);
    return 1;
}
```

Deap Performance

Like heaps, deaps are $O(\log n)$ for all the main operations. However, the constant factors are higher than for heaps, so you will have slower performance. That's to be expected, since deaps are maintaining more information so that you can extract efficiently both the minimum and maximum elements. As double-ended heaps go, deaps are one of the most efficient types.

DELAYED CONSTRUCTION OF HEAP ELEMENTS

Both heaps and deaps are implemented using what constitute variable-length arrays. If you've studied the companion volume, you'll know that one of the issues with variable-length arrays is the construction and destruction of the elements.

By default, C++ will construct all elements of an array when the array itself is constructed. Also, the elements are all destructed when the array itself is destructed. This default behavior makes sense for fixed-length arrays. But it doesn't necessarily make sense for variable-length arrays. Why should elements be constructed that aren't ever used? That's analogous to constructing nodes for a linked list that are never added.

The companion book shows how to circumvent the default construction of array elements: You declare the array as an array of characters and then construct the elements individually in place, using the placement new operator. We gave examples of this technique in the *OpenHashTable* and *CoalHashTable* classes of Chapter 7 in this book.

The same technique can be used on the *Heap* and *Deap* classes. However, we didn't show the technique in the manuscript because it's rather tricky and messy, and it's harder to see the underlying heap operations. However, we've provided sample code on disk that shows you how to define heaps this way. Note that if *TYPE* is a simple structure or built-in type, then which method you use is moot, so you might as well use the simpler default way.

Alternative code for the *Heap* class and *Deap* class that uses delayed construction of elements is given on disk in the files *altheap.mth* and *altdeap.mth*. These files are to be substituted for the files *heap.mth* and *deap.mth*.

INDIRECT HEAPS

The bubble-up and trickle-down operations of a heap involve a lot of data movement. If the items are large, this can take a significant portion of the heap running time. It's possible to never actually move the items themselves. Instead, an auxiliary array can be used, called a *map*, which maps the *logical index* of an item (where the item is on the heap) to the *physical index* of the item (where the item actually resides in the *items* array). During all heap operations, it's the map elements that are shuffled around. A heap with an auxiliary map is called an *indirect heap*. Figure 8.19 gives an example.

Figure 8.19 An indirect heap

```
                    v
          t                   s
      o       n           m       q
    g   f   d   h       k
```
(a) conceptual heap

9	1	11	5	10	7	4	12	2	6	8	3		

(b) map

t	f	k	q	o	d	m	h	v	n	s	g		

(c) actual data

To give you a feel for how an indirect heap can be coded, here is the *Insert()* function from a hypothetical *iHeap* class. (The lowercase *i* in front of the class name indicates an indirect heap.)

```
template<class TYPE>
int iHeap<TYPE>::Insert(int q)
// Inserts the data in items[q] on the heap indirectly.
// WARNING: items[q] should not already be on the heap!
// Returns 1 if successful, 0 if heap is full.
{
  if (IsFull()) return 0;
  // Trickle up the tree until items[q]'s proper spot is found.
  map[nx] = q; // Initial mapping
  BubbleUp(nx, q);
  nx++;
  return 1;
}
```

Passed to the *Insert()* function is the physical index *q* of an item we wish to insert on the heap. It's assumed that some outside code has added the item to the *items* array and given us the location of that item in *q*. To add the item, what we need to do is link a *map* element to *q*. Using the next available *map* element, given in *map[nx]*, we set up this initial mapping. Then the *BubbleUp()* function is called, which indirectly restores the heap order.

```
template<class TYPE>
void iHeap<TYPE>::BubbleUp(int k, int q)
// Bubble items[q] up the tree until the heap property
// is restored. map[k] is the initial mapping of items[q].
// ASSUMES heap is not empty.
{
  while(k > 1 && Higher(items[q], items[map[k/2]])) {
    // While not at root and items[q] has a higher priority,
    // shift parent to child position, indirectly, that is.
    map[k] = map[k/2];
    k /= 2; // Repeat, using parent.
  }

  map[k] = q; // We found the final mapping for the item.
}
```

The *BubbleUp()* function is passed a map index *k* and a physical index *q*, such that *map[k]* is the initial mapping of the data in *items[q]*. During comparisons, we compare *items[q]* to the parent of the item at location *k* in the heap, which is given by *items[map[k/2]]*. If we need to swap items, it's done indirectly, involving only the *map* elements. Finally, at the end of the *while* loop, we set up the final mapping for the item added.

While the actual items are used in the comparison operations, the only data ever moved are the map elements. If the items are large and have expensive assignment operations, manipulating the map instead can save a significant amount of time.

Complete code for an indirect heap class, as well as an indirect deap class, is given on disk in the files *iheap.h*, *iheap.mth*, *tstiheap.cpp*, *ideap.h*, *ideap.mth*, and *tstideap.cpp*.

Basic Sorting Algorithms

In this chapter we'll take a look at some of the basic algorithms used in sorting. While there are many algorithms available, the ones we've chosen are perhaps the most popular. We'll work with algorithms that sort arrays in memory and that use key comparison as the fundamental sorting operation. The latter requirement leads to the most general sorting algorithms. At the end of the chapter, we'll touch briefly on other sorting algorithms that use "tricks" taking advantage of special types of keys.

We'll use template-based C++ functions for the sorting algorithms. It's assumed in these functions that the item type being sorted, called *TYPE*, has the necessary operations defined to support the functions. These include having a default constructor, an assignment operator, and the appropriate comparison operators, such as > and <. Of course, if *TYPE* is a built-in type, then all of these operations are already defined.

In the text we give each algorithm in a form that does the sorting in ascending order. Descending-order sorts can be accomplished by reversing comparisons at opportune places in the algorithms. It's not always readily apparent where these comparisons are. For that reason, you'll find code supplied on disk to implement the algorithms both for ascending and descending order.

Selection Sort

The selection sort algorithm was described in Chapter 1. Here is a template function to implement the algorithm, for ascending order.

```
template<class TYPE>
void SelectionSort(TYPE arr[], int n)
// Ascending selection sort
```

213

```
{
  TYPE tmp;
  int i, j, m;

  // Select the smallest item, swap it with the first
  // item, select the second smallest item, swap it
  // with the second item, and so on.

  n--;

  for (i = 0; i<n; i++) {
      m = i;
      for (j = i+1; j<=n; j++) { // Select
          if (arr[j] < arr[m]) m = j;
      }
      if (m != i) { // Swap if necessary.
          tmp = arr[i]; arr[i] = arr[m]; arr[m] = tmp;
      }
  }
}
```

 Complete template-based code for the selection sort algorithm, including both ascending sort and descending sort versions, is given on disk in the file *slctsort.h*. Test programs are provided in *slcttst.cpp*, *slcttst2.cpp*, and *slcttst3.cpp*.

Figure 9.1 shows a snapshot of an array being sorted by selection, at each pass through the outer loop of the algorithm.

The selection sort algorithm requires $O(n^2)$ comparisons in all cases, but it is only $O(n)$ in terms of the number of swaps or exchanges that occur.

Figure 9.1 Selection sorting.

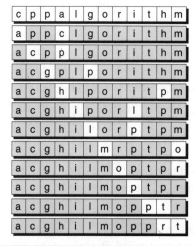

c	p	p	a	l	g	o	r	i	t	h	m	(a) initial order

(a) initial order
(b) after swapping c and a
(c) after swapping p and c
(d) after swapping p and g
(e) after swapping p and h
(f) after swapping l and i
(g) after swapping p and l
(h) after swapping o and m
(i) after swapping r and o
(j) p already in proper place
(k) after swapping t and p
(l) after swapping t and r

Selection sort therefore is useful when the items are very large and expensive to move around. Later you'll see another technique called *indirect sorting* that also can be used to handle cases when the array items are large.

One nice property of the selection sort algorithm is that it is *stable*. This means that items with equal keys maintain their relative positions after the sort. For example, suppose you have an array of items consisting of an integer key and an integer data field, where keys 5 and 9 are duplicated.

```
key  : 5 9 2 3 9 4 7 5 6
data : 1 2 3 4 5 6 7 8 9
```

Here's how the array would look after being sorted with selection sort.

```
key  : 2 3 4 5 5 6 7 9 9
data : 3 4 6 1 8 9 7 2 5
```

Note that the two items with keys of 5 are sorted together but that their relative ordering is the same as in the original array. Ditto for the items with keys of 9. While this seems natural and intuitive, only a few of the sorting algorithms have this property.

INSERTION SORT

The insertion sort algorithm was also described in Chapter 1. Here is a template function to implement the algorithm, for ascending order.

```
template<class TYPE>
void InsertionSort(TYPE arr[], int n)
// Ascending insertion sort
{
  TYPE item_to_sort;
  int i, j;

  // Starting with the second item, take each item,
  // scan to the left, and insert the item where it
  // belongs, i.e., stopping when you find an item smaller
  // or the beginning of the array is reached.

  for (i = 1; i < n; i++) {
      item_to_sort = arr[i];
      j = i;
      while(j >= 1 && item_to_sort < arr[j-1]) {
        arr[j] = arr[j-1];
        j -= 1;
      }
      arr[j] = item_to_sort;
  }
}
```

Figure 9.2 Insertion sorting.

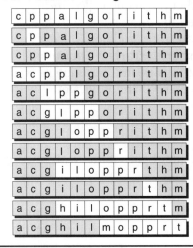

c	p	p	a	l	g	o	r	i	t	h	m	(a) initial order
c	p	p	a	l	g	o	r	i	t	h	m	(b) after inserting p (no movement needed)
c	p	p	a	l	g	o	r	i	t	h	m	(c) after inserting p (no movement needed)
a	c	p	p	l	g	o	r	i	t	h	m	(d) after inserting a
a	c	l	p	p	g	o	r	i	t	h	m	(e) after inserting l
a	c	g	l	p	p	o	r	i	t	h	m	(f) after inserting g
a	c	g	l	o	p	p	r	i	t	h	m	(g) after inserting o
a	c	g	l	o	p	p	r	i	t	h	m	(h) after inserting r (no movement needed)
a	c	g	i	l	o	p	p	r	t	h	m	(i) after inserting i
a	c	g	i	l	o	p	p	r	t	h	m	(j) after inserting t (no movement needed)
a	c	g	h	i	l	o	p	p	r	t	m	(k) after inserting h
a	c	g	h	i	l	m	o	p	p	r	t	(l) after inserting m

Complete template-based code for the insertion sort algorithm, including both ascending sort and descending sort versions, is given on disk in the file *isort.h*. Test programs are provided in *istst.cpp*, *istst2.cpp*, and *istst3.cpp*.

Figure 9.2 shows a snap shot of an array being sorted by insertion at each pass through the outer loop of the algorithm.

Insertion sort requires $O(n^2)$ comparisons in the average and worst cases and $O(n)$ in the best case (when the array is sorted or nearly sorted coming in). Insertion sort is useful for small arrays (10 to 50 elements, for instance) and for arrays that are already close to being sorted. Insertion sort is also stable.

SHELL SORT

The problem with insertion sort is that only adjacent items are compared and moved. Sorting algorithms with this behavior always leads to $O(n^2)$ complexities. Although selection sort moves items in big jumps (leading to $O(n)$ movements), it too compares items consecutively and thus is $O(n^2)$ for comparisons. What if we were to use a more global comparison technique that compares items far apart in the array? The *shell sort* algorithm does just that.

Shell sort is a generalized version of insertion sort. In each pass, an insertion sort is done on array items spaced *h* elements apart. For example, Figure 9.3 shows an example where *h* = 8.

On each successive pass, *h* is made smaller, until *h* = 1. Thus the overall shell sort routine in pseudo-code is the following.

Figure 9.3 One pass of shell sort with h = 8.

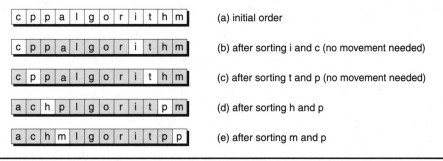

| | | | | | | | | | | | | |
| c | p | p | a | l | g | o | r | i | t | h | m | (a) initial order |

| c | p | p | a | l | g | o | r | i | t | h | m | (b) after sorting i and c (no movement needed) |

| c | p | p | a | l | g | o | r | i | t | h | m | (c) after sorting t and p (no movement needed) |

| a | c | h | p | l | g | o | r | i | t | p | m | (d) after sorting h and p |

| a | c | h | m | l | g | o | r | i | t | p | p | (e) after sorting m and p |

```
// Shell sort pseudo-code
for each h until h = 1 do begin
  insertion sort using h as the increment
end
```

Each pass of shell sort can be coded by taking the insertion sort routine given in the last section and replacing the constant 1 everywhere with h.

```
// Insertion sort with increment of h
for (i = h; i < n; i++) {
    item_to_sort = arr[i];
    j = i;
    while(j >= h && item_to_sort < arr[j-h]) {
      arr[j] = arr[j-h];
      j -= h;
    }
    arr[j] = item_to_sort;
}
```

Because h decreases each pass, shell sort is also known as *diminishing increment sort*. Suppose we are sorting 12 elements. We could use an increment sequence of h = 8, 4, 2, 1, as given in Figure 9.4.

If an array is sorted with an increment of h, the array is said to be *h-sorted*. An important property of shell sort is that rearrangements made in previous passes are not "undone" in later passes. That is, if an array is sorted with h = 8 and then later sorted with h = 4, the array will be 4-sorted as well as 8-sorted. Any diminishing sequence of values for h will work, as long as the last value of h is 1. This is a rather surprising, but fortuitous result, the proof of which we leave to the references (for example, Knuth [Knuth 73]). The result is fortuitous because by the time we use h = 1, the array will be nearly sorted, and the last pass will execute quickly.

Shell sort uses insertion sort to the best advantage. When h is large, the number of items sorted is small, a case that insertion sort is well suited for. As

Figure 9.4 Diminishing increment (shell) sorting.

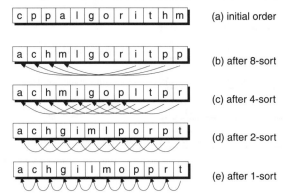

(a) initial order

(b) after 8-sort

(c) after 4-sort

(d) after 2-sort

(e) after 1-sort

h gets smaller, the number of items involved increases. But these items tend to be almost sorted, again a situation that is highly suited to insertion sort.

Even though any diminishing sequence of *h* ending in 1 will work, some sequences work better than others, leading to fewer comparisons. The sequence obtained by dividing *h* by 2 each pass turns out to be a relatively bad sequence. There is some theory about why some sequences are better than others. (See Knuth [Knuth 73].) However, nobody knows what the perfect sequence is. Here is one sequence, given in Gonnet and Baeza-Yates [Gonnet 91], that appears to work well.

$$h_0 = \alpha n$$
$$h_k = \alpha h_{k-1}$$

such that $h \geq 1$, and where $\alpha = 0.45454...$

While this formula seems to imply using floating point arithmetic, you can compute *h* using integer arithmetic with the following C++ statements.

```
h = n;
...
if (h < 5) h = 1; else h = (5*h-1)/11;
```

For some values of *n*, *h* will not converge to 1. That's the reason for the *if* statement. Given this increment sequence, and using our generalized insertion sort routine, the following *ShellSort()* routine can be derived.

```
template<class TYPE>
void ShellSort(TYPE arr[], int n)
// Ascending shell sort
{
```

```
TYPE item_to_sort;
int i, j, h;

for (h = n; h > 1;) {
  if (h < 5) h = 1; else h = (5*h-1)/11;
  // Perform insertion sort with increment h.
  for(i = h; i < n; i++) {
    item_to_sort = arr[i];
    j = i;
    while(j >= h && item_to_sort < arr[j-h]) {
      arr[j] = arr[j-h];
      j = j-h;
    }
    arr[j] = item_to_sort;
  }
}
}
```

 Complete code for shell sort, for both ascending and descending order, is given on disk in the file *ssort.h*. Test programs can be found in *sstst.cpp*, *sstst2.cpp*, and *sstst3.cpp*.

One particular problem with the increment sequence given is that you must worry about overflow. For example, if you're using an implementation where *int*s are 16 bits, then the computation for *h* will overflow for *n* > 6553. To avoid this, you can force *long* arithmetic to be used, as follows.

```
if (h < 5) h = 1; else h = int((5L*long(h)-1L)/11L);
```

Here, we're assuming that *n* has a small enough value to fit into an *int*, and a *long* is 32 bits, and that *int* is 16 bits.

This particular increment sequence has been shown experimentally to be consistently better than many of the others. Another popular sequence is 1, 4, 13, 40, ..., in reverse order, where the equation *h* = 3**h* + 1 is used to find the largest *h* ≤ *n* at which to start the sequence. Then at each pass, you can divide *h* by 3. In our tests, this sequence proved 10 to 20 percent slower than the 0.45454 sequence we used.

Although there are better sequences than both of those discussed, the cost of computing the other sequences often outweighs the savings made by having fewer comparisons. Where the breakpoint occurs depends on the type of data, but you should know that most of the good sequences are within 20 percent of each other anyway, in terms of number of comparisons performed. Any good method you use won't be far from optimum.

While no one knows analytically what the time complexity of shell sort is, for most good increment sequences, it appears to be around $O(n^{1.5})$, in the average and worst cases. Shell sort does less work when the array is partially sorted. However, you won't see the same $O(n)$ performance, as with insertion sorting, since many passes of insertion sort take place in a shell sort.

In practice, shell sort is a popular technique because it performs well and is easy to implement. Shell sort has no discouraging worst-case behavior, and it requires no additional storage during operation. One drawback of shell sort is that it is not stable. Also, the next two algorithms appear to work faster on average, at least in our tests.

HEAP SORT

In Chapter 8 you learned how the heap data structure can be used to sort arrays with an algorithm called *heap sort*. Like shell sort, heap sort is fast because it compares items in a non-localized fashion (in fact, a logarithmic one), so that the sorting tends to converge quickly on a solution. Next we present a *HeapSort()* function that combines the *TrickleDown()*, *Arrange()*, and *Sort()* functions of the *Heap* class into a single, condensed function.

```
template<class TYPE>
void HeapSort(TYPE arr[], int n)
// Sorts array of n elements in ascending order.
{
  TYPE x;
  int i, j, k;

  if (n < 2) return; // Trap for nothing to do.

arr--; // Child/parent calculations rely on 1-based indexing!

  // Bottom-up heap construction phase:

  for (i = n/2; i >= 1; i--) {
      k = i;
      x = arr[k]; // Start first swap
      while(k <= n/2) { // Trickle x down the heap.
        j = k*2;                            // Find left child.
        if (j < n && arr[j+1] > arr[j]) j++; // Right child bigger
        if (x >= arr[j]) break;            // x >= largest child
        arr[k] = arr[j];
        k = j;
      }
      arr[k] = x; // Finish final swap.
  }

  // Sorting phase:

  if (n == 2) return; // Heap construction sorted us.

  while(n > 1) {
    // Do the first two steps in swapping the rightmost
    // leaf and the root. Shrink the heap too.
```

```
    x = arr[n];
    arr[n--] = arr[1];
    k = 1;
    while(k <= n/2) { // Trickle x down the heap.
        j = k*2;                              // Find left child.
        if (j < n && arr[j+1] > arr[j]) j++;  // Right child bigger
        if (x >= arr[j]) break;               // x >= largest child.
        arr[k] = arr[j];
        k = j;
    }
    arr[k] = x; // Finish final swap.
  }
}
```

Complete code for heap sort, for both ascending and descending order, is given on disk in the file *hsort.h*. Test programs can be found in *hstst.cpp*, *hstst2.cpp*, and *hstst3.cpp*.

Figure 9.5 illustrates some of the steps used in a heap sort.

Heap sort's claim to fame is that it is one of the few $O(n\log n)$ sorting algorithms in both worst and average cases, and it requires no additional storage. Heap sort is a safe and efficient sorting routine, no matter how large n is. In practice, heap sort is faster on average than shell sort for most non-trivial item types and for all but small values of n. However, heap sort usually is slower than quick sort.

Figure 9.5 Heap sorting.

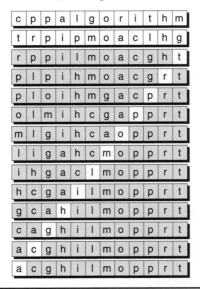

(a) initial order

(b) after heap construction phase

(c) moving the root to the end, and so on...

Like shell sort, heap sort can sort the array in place, requiring no extra storage. It isn't a stable sorting algorithm, but neither is shell sort. Heap sort is twice as large in code size as shell sort. Note that the two trickle-down loops can be combined into one, with the help of some auxiliary flags. This cuts the code size roughly in half, to just a little bit larger than shell sort. However, with this change you will notice a slight slowdown in speed, and the code is more obscure.

QUICK SORT

Next we look at yet another sorting technique called *quick sort*. This algorithm gets its name because, on average, it is the fastest sort known (for those sorts that use comparisons anyway). Quick sort works on the principle of *divide and conquer*, an effective technique that shows up quite frequently in algorithm design. You saw divide and conquer with the binary search algorithm in Chapter 2.

The idea behind quick sort is to partition the array into two parts, one part whose items are less than or equal to some chosen item called the *pivot* and one part whose items are larger than or equal to the pivot. For example, Figure 9.6 shows an array that's been partitioned.

Note that the partitions themselves are not in sorted order. Instead, they are sorted by being split in half, recursively, until the partitions are of size one. (The partitions are not always of equal size, as you'll see later.) Here is a high-level sketch of the algorithm.

```
template<class TYPE>
void QuickSort(TYPE arr[], int a, int b)
// Sort the portion of the array from a to b.
{
  int i, j;
  if (b > a) { // While there's still something to sort
      // Split array in half. Upon return, item i will be in
      // its final resting place. Items a -> j will be less than
      // or equal to item i, and items i+1 -> b will be greater
      // than or equal to item i.
      Partition(arr, a, b, i, j);
      QuickSort(arr, a, j);      // Sort left half.
      QuickSort(arr, i+1, b);    // Sort right half.
  }
}
```

Figure 9.6 Partitioned array with pivot point.

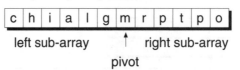

As you can see, at a high level, quick sort is quite simple and elegant. Of course, we haven't defined what the *Partition()* function does. That's where all the dirty work is.

Partitioning

Choosing a pivot for partitioning turns out to be the most crucial design issue. Pretend for now that you pick as the pivot the last element in the portion of the array being sorted (that is, at index *b*). The partitioning works as follows. Set index *i* to *a* − 1. Set index *j* to *b*. Now scan in both directions, incrementing *i* to the right and decrementing *j* to the left. For *i*, scan until you find an item that is larger than the pivot. For *j*, scan until you find an item that is smaller than the pivot. These locations represent items that are out of place, so swap them. Then continue scanning in the same fashion, swapping at the next stopping point of the two indices.

This process goes on until the indices cross. At that point, the array has been fully partitioned. The pivot element is then swapped with the *i*th item. This puts the pivot in its final resting place. Figure 9.7 illustrates the technique.

Code for the partitioning procedure is as follows.

Figure 9.7 Array partioning with the pivot at the end.

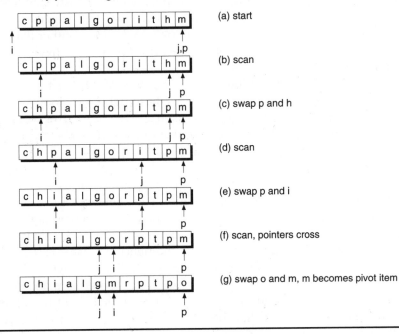

```
template<class TYPE>
void Partition(TYPE arr[], int a, int b, int &i, int &j)
{
  TYPE t, pivot = a[b];
  i = a-1; j = b;

  while(1) {
    while(arr[++i] < pivot); // Pivot is sentinel.
    while(arr[--j] > pivot); // Infinite loop if pivot is smallest.
    if (j < i) break; // Pointers crossed.
    t = arr[i]; arr[i] = arr[j]; arr[j] = t; // Swap items.
  }

  // Put pivot in its final resting place.
  pivot = arr[i]; arr[i] = arr[b]; arr[b] = pivot;
}
```

There are several things to note about this algorithm. First, in order for *Partition()* to work properly, there must be sentinel on the left side that's smaller than any other item in the array. Otherwise the loop for j may run indefinitely. This would happen if the pivot happens to be the smallest item in the array. No sentinel is needed for i, since the pivot itself serves as a sentinel. Later you'll see how to get rid of the sentinel for j.

Second, some issues arise concerning when to stop the scans. What should you do if an item is equal to the pivot? And should you stop partitioning only when $j < i$, or would $j = i$ work as well? Extensive studies show that it's best to have an index scan stop when an item equal to the pivot is encountered. Also, stopping the partitioning when $j < i$ rather than $j = i$ allows two items to be swapped in some cases. Both of these behaviors are coded in the *Partition()* function given.

Notice how tight the inner loops are. Besides the fact that quick sort is using a divide-and-conquer strategy, it's these tight inner loops that give quick sort all its speed. On average, quick sort is $O(n\log n)$. The constant factor is smaller than other sorts—so much so that quick sort can be anywhere from 1.5 to 2 times faster than its nearest competitor!

But all is not well with quick sort as we've given it here. Although it is $O(n\log n)$ on the average, in the worst case, it's $O(n^2)$. In fact, in the worst case, quick sort is much slower than insertion sort. This worst case happens if the items are ordered in such a way that the partitions become highly unbalanced and that the unbalanced partitions occur consistently as the algorithm recurses. In the pivot location we've chosen (on the right), the worst case happens when the array is in reverse sorted order. Figure 9.8 illustrates what goes wrong.

From the figure, you can see that each pass sorts only one element. After the partitioning, one side is empty and the other has all the remaining elements. Besides the fact that this causes $O(n^2)$ time complexity, you can see where the recursion would take place roughly n times. This implies $O(n)$ stack space, a very undesirable circumstance.

Figure 9.8 Degenerate partitioning.

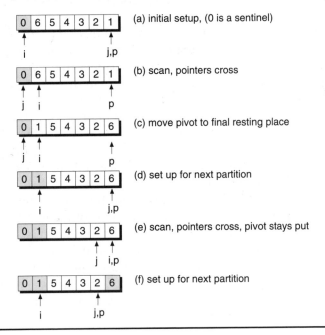

(a) initial setup, (0 is a sentinel)

(b) scan, pointers cross

(c) move pivot to final resting place

(d) set up for next partition

(e) scan, pointers cross, pivot stays put

(f) set up for next partition

Had we chosen the first item in the array as the pivot, the worst-case scenario would occur if the array came in already sorted. That's unfortunate, for someone using the algorithm could easily pass a sorted array in by mistake. Obviously, we must make a more intelligent choice for the pivot.

Choosing a Pivot

In order to avoid the worst case from happening with quick sort, we must choose the pivot wisely. In the ideal case, we would like the partitions to be as balanced as possible. This is analogous to the case with binary search trees, where you want the trees to be balanced for optimum performance. In theory, you could analyze the data and find optimum pivots for each partitioning pass. However, in all likelihood that would take more time than actually doing the sort.

One idea is to choose the pivot at random for each partitioning pass. While this would reduce the chances of the worst case happening, it doesn't eliminate it entirely. There would still be a chance, however small, that the initial array ordering would be just "right" (or wrong if you prefer) to cause the partitions to degenerate. Another problem with choosing a random pivot is that generating a random number takes time. For example, if you were to use the Minimal Standard Generator (probably overkill), you're looking at a substantial number of calculations. Whether the calculations cause a noticeable slowdown depends on what type of keys are being used during comparisons. (The time for comparisons may mask everything else.) In any event, you still have the problem of providing a sentinel for *j.*

It turns out there is a partitioning method that tends to keep the partitions balanced and eliminates the sentinel problem as well. This method is called *median-of-three partitioning.* The idea is to take the left, middle, and right items of the current partition and sort them with respect to each other. Then the middle item is used as the pivot. Figure 9.9 illustrates the process.

It's important that the three items are sorted before choosing the middle one as the pivot. That guarantees us that the pivot is neither the smallest nor the largest item, and thus no sentinels are needed. It also gives us a reasonable chance that the partitions will end up balanced. In effect, we're taking a sample of three items from the array and, after sorting them, using the middle item as an educated guess for the median of the array. At times it will be close to being right. At other times it will be way off, but that will only affect performance, not correctness.

In order to start the partitioning, we must tuck the pivot out of the way. This is done by swapping it with the item just to the left of the largest item of the three chosen. After the partitioning, the pivot will be swapped back and will land in its final resting place. Here is a C++ function implementing median-of-three partitioning.

```
template<class TYPE>
void Partition(TYPE arr[], int a, int b, int &i, int &j)
// Median-of-three partitioning
{
  TYPE pivot, t;
```

Figure 9.9 Median-of-three partitioning.

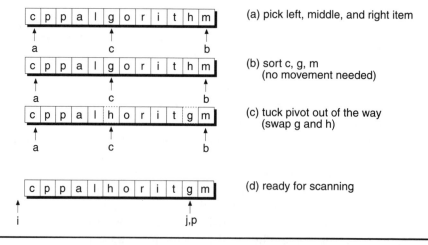

(a) pick left, middle, and right item

(b) sort c, g, m
(no movement needed)

(c) tuck pivot out of the way
(swap g and h)

(d) ready for scanning

```
// Sort the left, middle, and right items.
int c = (unsigned(a) + unsigned(b)) / 2;
if (arr[a] > arr[c]) {
   t = arr[a]; arr[a] = arr[c]; arr[c] = t;
}
if (arr[a] > arr[b]) {
   t = arr[a]; arr[a] = arr[b]; arr[b] = t;
}
if (arr[c] > arr[b]) {
   t = arr[c]; arr[c] = arr[b]; arr[b] = t;
}

// Tuck the pivot, at arr[c], away.
pivot = arr[c]; arr[c] = arr[b-1]; arr[b-1] = pivot;
i = a; j = b-1;

while(1) {
  while(arr[++i] < pivot);
  while(arr[--j] > pivot);
  if (j < i) break;
  t = arr[i]; arr[i] = arr[j]; arr[j] = t;
}

// Put pivot into its final resting place.
  pivot = arr[i]; arr[i] = arr[b-1]; arr[b-1] = pivot;
}
```

This routine sorts the three items chosen "by hand." One common implementation bug is to not handle the possibility of overflow when calculating the index *c*. We've handled it by using *unsigned* arithmetic.

Like choosing a random pivot, the median-of-three partitioning method does not guarantee that you won't encounter a worst-case scenario that's $O(n^2)$. But like the random pivot method, the chance of that occurring is exceedingly small. Basically two out of the three items chosen must be among the smallest or largest items in the array, and this must happen consistently throughout the partitioning. Keep in mind that you are relying on the laws of probability to work for you. If it's critical that the sort complete quickly *every* time (perhaps in a real-time environment where a slowdown is highly undesirable), then don't use quick sort. Instead, shell sort or heap sort makes a better choice.

Optimizing Quick Sort

While we can't entirely avoid having a worst case that performs more poorly than the average $O(n\log n)$ case, there are steps you can take to minimize the impact of the worst-case scenario, so that it will be less than $O(n^2)$. There are three steps involved.

1. Remove the recursion.
2. Order the simulated recursive calls "intelligently."

3. Use some other sorting algorithm for small partitions.

You can remove the recursion from quick sort as follows. The last call is tail recursive, so you can use the tail recursion removal technique from Chapter 2. For example:

```
template<class TYPE>
void QuickSort(TYPE arr[], int a, int b)
// Tail recursion optimized.
{
  int i, j;
  while(b > a) {
    Partition(arr, a, b, i, j);
    QuickSort(arr, a, j);
    a = i+1; // Simulate QuickSort(arr, i+1, b);
  }
}
```

To remove the other recursive call, we must use an explicit stack.

```
template<class TYPE>
void QuickSort(TYPE arr[], int a, int b)
// Iterative version
{
  int stk[MAX_STACK_SIZE];
  int i, j, top;

  top = 0;
  while(1) {
    while(b > a) {
      Partition(arr, a, b, i, j);
      // Simulate QuickSort(arr, a, j) fby QuickSort(arr, i+1, b);
      stk[top++] = i+1; stk[top++] = b; // Defer second call on stack.
      b = j; // Do first "call"
    }
    if (top == 0) break; // End of recursion
    b = stk[--top]; a = stk[--top]; // Do deferred call.
  }
}
```

Here we've embedded a hard-coded integer-based stack right into the routine. Recall that in the degenerate case of partitioning, n recursive calls are made, meaning the $O(n)$ stack space is used up. It turns out, however, that if we always make the simulated recursive call for the smaller partition first, (thus never placing it on the stack), it's guaranteed that only about $2\log n$ stack elements are needed. For example, if n can be as large as 2^{32} items (probably larger than any array you are going to sort), then a stack size around 64 integers should suffice. Here is the quick sort routine modified for this optimization.

```
template<class TYPE>
void QuickSort(TYPE arr[], int a, int b)
```

```
// Optimized iterative version
{
  int stk[64]; // Should be more than you'll ever need.
  int i, j, top;

  top = 0;
  while(1) {
    while(b > a) {
      Partition(arr, a, b, i, j);
      if (j-a > b-i) {
        // Do QuickSort(arr, i+1, b) fby QuickSort(arr, a, j);
        stk[top++] = a; stk[top]++ = j; // Defer second call.
        a = i+1; // Do first call.
      }
      else {
        // Do QuickSort(arr, a, j) fby QuickSort(arr, i+1, b);
        stk[top++] = i+1; stk[top++] = b; // Defer second call.
        b = j; // Do first "call".
      }
    }
    if (top == 0) break; // End of recursion
    b = stk[--top]; a = stk[--top]; // Do deferred call.
  }
}
```

The final optimization is to prevent small partitions from being passed through quick sort. For partitions of around 10 or 20 items, it's faster to use a simpler sorting routine, because quick sort does have some overhead. A popular choice is to use insertion sort. It's ideal, since the partitions are small and almost sorted. A minor modication will prevent quick sort from sorting small partitions. Change the second *while* loop test to the following:

```
while(b > a+CUTOFF) {
```

where *CUTOFF* is the smallest partition size you wish to quick sort. While the optimum value for *CUTOFF* depends on the actual data, you won't go wrong making it somewhere between 10 and 20 elements. You want to keep it fairly small so that insertion sort works fast, but not so small that the higher-overhead quick sort algorithm is employed when it shouldn't be.

By putting in this change, when *QuickSort()* finishes, the array will be sorted loosely in partitions of size *CUTOFF*. All you need to do then is call *InsertionSort()* to do the final sorting pass. Since the array will be almost sorted, this final insertion sort pass will go quickly.

One other advantage to sorting small partitions with *InsertionSort()* is that you reduce the bad behavior of the worst-case scenario dramatically. While you can't guarantee an $O(n \log n)$ algorithm, the worst case won't be $O(n^2)$, but something in between. And keep in mind that the median-of-three partitioning reduces the chances of the worst case from occurring to a very small probability.

Tread Carefully over the Quick Sort Code

There is one thing to caution you about when using the hybrid sorting scheme. The final insertion sort pass will sort the array regardless of whether quick sort worked correctly or not. You may not notice anything wrong unless the sort takes a long $O(n^2)$ time. The best thing to do is get quick sort working first without the follow-up insertion sort.

Getting the quick sort code correct is tricky. Seemingly minor changes can break the algorithm. Suppose you decided to start the value of i so that it points directly at the first item in the items to be partitioned, instead of one back, and did the same for j (instead of j pointing one forward). You could modify the inner loops of the partition algorithm to the following:

```
i = a; j = b-2;
while(1) {
  while(arr[i] < pivot) { i++; } // Modified. Will it work?
  while(arr[j] > pivot) { j--; } // Modified. Will it work?
  if (j < i) break;
  t = arr[i]; arr[i] = arr[j]; arr[j] = t;
}
```

This change looks innocuous enough. However, an infinite loop occurs if *arr[i] = arr[j] = pivot*.

Using the Built-in Quick Sort Routine

Rather than try to code quick sort yourself, you might be tempted to use the built-in *qsort()* routine provided in the standard C library. The *qsort()* function is written generically to handle any type of data, through the use of function pointers for the comparison operations. However, because function pointers are being used rather than direct comparison operations, *qsort()* may run slower than the optimized *QuickSort()* function to be given next. For example, in sorting integers, *qsort()* is twice as slow.

A more significant concern is that *qsort()* doesn't know anything about the type of data it's sorting. All it knows is the size of the data and what comparison function to use. This presents a problem if you are sorting objects that have embedded pointers to other data allocated separately. In *qsort()*, exchanges are done by using bitwise copying. As a consequence, with embedded pointers, only the pointers are copied, not the data they point to. This may or may not cause a problem, depending on the nature of the objects.

With the upcoming *QuickSort()* template, the assignment operator for the object is called, which presumably guarantees that the copying is done correctly, most likely in a memberwise fashion. However, this may lead to a slower sort, since in all likelihood a memberwise copy will be slower than a bitwise copy. You could arrange the assignment operator to do a bitwise copy, if you think that's okay. Or a special version of *QuickSort()* can be written that avoids assignments and does bitwise copying instead. As you can imagine, you must be very careful if you do this. We did something similar with the alternate heap classes that use delayed construction of elements, given on disk for Chapter 8.

The Final Optimized Version of Quick Sort

The following *QuickSort()* routine puts all the optimizations we've shown to use.

```
const int CUTOFF = 20; // A good number to start with

template<class TYPE>
void QuickSort(TYPE arr[], int n)
// Sorts the array of size n.
{
  QuickSort(arr, 0, n-1); // Does most of the sorting.
  InsertionSort(arr, n);  // Does the final pass.
}

template<class TYPE>
void QuickSort(TYPE arr[], int a, int b)
// Sorts the partition of the array from a to b in
// ascending order.
{
  int stk[64]; // Should be more than you'll ever need.
  TYPE pivot, t;
  int i, j, c, top;

  top = 0;

  while(1) {
    while(b > a+CUTOFF) {

      // Start: i = median-of-3-partition(a, b)
      c = (unsigned(a) + unsigned(b)) / 2;
      if (arr[a] > arr[c]) {
        t = arr[a]; arr[a] = arr[c]; arr[c] = t;
      }
      if (arr[a] > arr[b]) {
        t = arr[a]; arr[a] = arr[b]; arr[b] = t;
      }
      if (arr[c] > arr[b]) {
        t = arr[c]; arr[c] = arr[b]; arr[b] = t;
      }

      // Could be median-of-3 sort has sorted us completely.
      if (b-a <= 2) goto recur_return; // You MUST do this!

      pivot = arr[c]; arr[c] = arr[b-1]; arr[b-1] = pivot;
```

```
    i = a; j = b-1;

    while(1) {
      while(arr[++i] < pivot);
      while(arr[--j] > pivot);
      if (j < i) break;
      t = arr[i]; arr[i] = arr[j]; arr[j] = t;
    }
    pivot = arr[i]; arr[i] = arr[b-1]; arr[b-1] = pivot;
    // End: i = median_of_3_partition(a, b)

    if (j-a > b-i) {
      // Do QuickSort(arr, i+1, b) fby QuickSort(arr, a, j);
      stk[top++] = a; stk[top++] = j; // Defer on stack.
      a = i+1;
    }
    else {
      // Do QuickSort(arr, a, j) fby QuickSort(arr, i+1, b);
      stk[top++] = i+1; stk[top++] = b; // Defer on stack.
      b = j;
    }
  }

  recur_return:

  if (top == 0) break;
  b = stk[--top]; a = stk[--top];
  }

}
```

Code for the optimized quick sort routine, for both ascending and descending order, can be found on disk in the file *qsort.h*. Test programs are given in *qstst.cpp*, *qstst2.cpp*, and *qstst3.cpp*.

COMPARISON OF THE SORTING ALGORITHMS

You might wonder why we showed you so many different sorting algorithms. Wouldn't just one do? Unfortunately, none of the algorithms is perfect. They all have drawbacks that prevent them from being recommended all the time. However, you'll see in this section that if you were to pick just one sorting algorithm, heap sort would probably be the best choice. Table 9.1 lists the advantages and disadvantages of each algorithm.

To give you an idea of the performance of the sorting algorithms, Tables 9.2 through 9.5 compare the execution times of the different algorithms under various conditions. The times for quick sort were obtained using the optimized version given in the last section.

Choosing a sorting algorithm is a bit like playing the stock market. If you want a high rate of return and are willing to take some risks, quick sort is the choice for you. If you prefer absolutely no risk at the expense of some performance, heap sort is a safe bet. If you want a good all-around performer with minimum overhead, shell sort is a reasonable choice. Finally, selection sort and insertion sort are useful in "speciality portfolios."

Table 9.1 Feature Comparison of the Sorting Algorithms

Algorithm	Advantages	Disadvantages
Selection sort	Only $O(n)$ movements used. Thus good for very large items. Code is simple. Array sorted in place. Sorting is stable.	Requires $O(n^2)$ comparisons. Not good unless items are very large.
Insertion sort	Roughly $O(n)$ comparisons when array partially sorted coming in. Code is simple. Array sorted in place. Sorting is stable.	Requires $O(n^2)$ comparisons on the average.
Shell sort	Roughly $O(n^{1.5})$ with no bad worst-case behavior. Code is simple. Array sorted in place.	Quick sort is twice as fast on average. Heap sort is faster for all but trivial item types and as n becomes large. Not a stable sort.
Heap sort	Always $O(n\log n)$. The safest of the bunch, and reasonably fast. Array sorted in place.	Code is moderately complex. Not a stable sort.
Quick sort	Twice as fast on average than shell sort and about 1.5 times faster than heap sort.	Has probability (although very small) of poor performance. Code tricky to implement correctly. Requires $O(\log n)$ stack space. (Practically speaking, 64 integers is enough.) Not a stable sort.

Table 9.2 Average Relative Performance Sorting 16-bit Integers

Algorithm	n=1,000	n=5,000	n=20,000
Selection sort	57.9	166.0	559.0
Insertion sort	18.0	52.0	185.0
Shell sort	2.0	2.0	2.1
Heap sort	3.0	2.4	2.4
Quick sort	1.0	1.0	1.0

Table 9.3 Relative Performance Sorting 8-byte Records with Integer Keys

Algorithm	n=1,000	n=2,000	n=5,000
Selection sort	14.3	26.2	64.2
Insertion sort	21.1	40.2	96.2
Shell sort	2.0	1.9	2.1
Heap sort	1.5	1.5	1.6
Quick sort	1.0	1.0	1.0

Table 9.4 Relative Performance Sorting 1024-byte Records with Integer Keys

Algorithm	n=200	n=400
Selection sort	1.0	1.0
Insertion sort	13.5	20.1
Shell sort	3.21	3.0
Heap sort	2.20	2.1
Quick sort	1.83	1.6

Table 9.5 Relative Performance Sorting 64-byte Character Strings

Algorithm	n=1,000	n=2,000
Selection sort	14.3	41.7
Insertion sort	23.3	43.8
Shell sort	1.7	1.8
Heap sort	1.4	1.4
Quick sort	1.0	1.0

INDIRECT SORTING

When the items being sorted are large, a lot of data movement takes place, which can significantly slow down the sorting algorithms. You've seen where selection sort is useful in situations like these. In this section we'll show another way to sort large items. The idea is to use an auxiliary array, called a *map*. This array is a set of indices that maps the logical ordering of the items to their physical ordering. This is the same strategy that was used with the indirect heaps in Chapter 8.

To convert one of the sort algorithms to an indirect sort, first add the *map* array and initialize it to be sequentially ordered. Then almost everywhere the expression *arr[i]* is found, replace it with *map[i]*, where *i* is some subscript. The two exceptions are as follows. First, when items are being compared, use *arr[map[i]]*. Second, when a temporary item is being used, replace it with an integer index instead, and use it to index the proper item when necessary. This way, the actual items are used only in comparisons. The only exchanges that take place are with indices in the *map* array. As an example, here is the shell sort algorithm converted to an indirect sort.

```
template<class TYPE>
void IndirectShellSort(TYPE arr[], int map[], int n)
// Ascending indirect shell sort. Assumes map[] comes in
// initialized to be the integers 0,1,...n-1, or whatever
// the initial mapping should be.
{
  int item_to_sort_indx;
  int i, j, h;

  for (h = n; h > 1;) {
    if (h < 5) h = 1; else h = (5*h-1)/11;
    // Perform insertion sort with increment h.
    for(i = h; i < n; i++) {
      item_to_sort_indx = map[i];
      j = i;
      while(j >= h && arr[item_to_sort_indx] < arr[map[j-h]]) {
        map[j] = map[j-h];
        j = j-h;
      }
      map[j] = item_to_sort_indx;
    }
  }
}
```

Complete code for the indirect shell sorting method, for both ascending and descending order, and including the map sort method to be discussed next, is given on disk in the file *imsort.h,* with test programs in *imtst.cpp, imtst2.cpp,* and *imtst3.cpp.*

Figure 9.10 shows an unsorted array and corresponding map created by using indirect sort. (Part b of this figure is explained in the next section.)

Using indirect sorting, especially with the underlying shell sort algorithm we show, will work faster than using selection sort. However, indirect sorting does require the $O(n)$ map, and the array elements aren't actually physically sorted.

Figure 9.10 Indirect sorting.

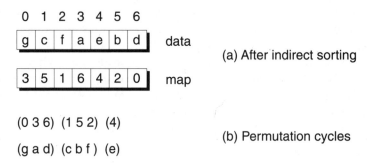

(a) After indirect sorting

(b) Permutation cycles

Map Sorting

If after indirect sorting you wish to have the items physically ordered, you can use another auxiliary array that becomes the newly sorted array. A loop like the following will finish the sort.

```
for(int i = 0; i<n; i++) new_arr[i] = arr[map[i]];
```

However, there is a way to sort the original array in place, without requiring another new array. This second method is a rather interesting use of permutations. (See Chapter 4.) Notice that the *map* array represents a permutation of the original array.

One theory from mathematics states that every permutation is made up of one or more disjoint cycles. To find a cycle, start with an index *i*. The next element in the cycle that contains *i* is *map*[*i*]. The next element after that is *map*^{2}[*i*]. As a shorthand we'll write *map*^{2}[*i*] to mean *map*[*map*[*i*]]. Then the third element is *map*^{3}[*i*] (*map*[*map*[*map*[*i*]]]), and so on, until you find some *k* such that *map*^{k}[*i*] = *i*.

As an example, part b of Figure 9.10 shows the three cycles that make up the permutation given in the *map* array of an indirect sort. In the first cycle, suppose we start *i* at 0. (We could start *i* at 3 or 6 as well.) Then *map*[0] = 3, *map*[3] = 6, and *map*[6] = 0, completing the cycle. The cycle can be written in shorthand as (0 3 6). Note that the third cycle, (4), is a trivial one.

One rather surprising but extremely useful property of permutation cycles is that once we've sorted the items in one cycle, they will be in their final resting place and will not be moved during the sorting of the other cycles.

The sorting of a cycle can be accomplished by realizing that each number is the index of the true location of the item indexed by the next number. (Did you get that?) For example, to sort the first cycle (0 3 6), first move *arr*[0] to a temporary holding cell. Then move *arr*[3] to *arr*[0], *arr*[6] into *arr*[3], and finally, move the contents of the holding cell into *arr*[6]. At this point, the corresponding items are (*a d g*), which you can see are in sorted order. The same process takes place for the cycle (1 5 2): *arr*[1] goes into the holding cell, *arr*[5] is placed into *arr*[1], *arr*[2] is placed into *arr*[5], and finally, the contents of the holding cell is placed into *arr*[1]. The corresponding items in the cycle become (*b c f*), which also is sorted. For the last permutation (4), no work needs to be done. The item *e* is already in the correct location.

This process of moving items in each cycle is implemented in the following *MapSort()* function.

```
template<class TYPE>
void MapSort(TYPE arr[], int map[], int n)
// This routine rearranges the array arr[] according
// to the permutation map[].
{
  TYPE tmp; // Temporarily holding cell
  int i, j, k;

  for (i = 0; i<n; i++) {
      if (map[i] != i) {
          // We haven't processed the cycle that contains i
          // so rearrange the elements in that cycle.
          tmp = arr[i];
          k = i;
          do {
              j = k; k = map[j];
              arr[j] = arr[k];
              map[j] = j; // Signal this item is in final place
          } while (k != i);
          arr[j] = tmp;
      }
  }
}
```

In this function note how we modify the map array in the inner loop by setting *map*[*j*] = *j*. This serves as a flag that indicates later that the cycle containing *j* already has been processed.

It's not hard to see that each item is moved only once into its final resting place (not counting the movement of the first item in each cycle into the temporary holding cell). This makes the *MapSort()* function $O(n)$ in complexity. By using *MapSort()* in conjunction with an indirect sort, you can physically sort arrays of large items rather efficiently. For example, Table 9.6 compares the use of an indirect sort to both regular shell sort and selection sort.

You can see from the table that the indirect sorting method wins every time, except for when simple integers are being sorted. For simple integers, the extra overhead of the map is of no use and, in fact, slows things down. The indirect sort is still faster than using selection sort by a wide margin, though. For larger items, indirect shell sorting wins because it takes advantage of the

Table 9.6 Relative Performance of Indirect Sorting

Algorithm	Sorting 5,000 integer keys	Sorting 2,000 64-byte strings	Sorting 400 1024-byte records with integer keys
Indirect shell sort	1.5	1.0	1.0
Selection sort	83.0	44.9	4.0
Shell sort	1.0	1.95	12.0

fact that shell sort uses only $O(n^{1.5})$ comparisons, instead of $O(n^2)$ as in selection sort, but indirect sort avoids doing $O(n^{1.5})$ exchanges. Instead, $O(n)$ exchanges are used, the same as in selection sort. Also, if you were to use indirect quick sorting or indirect heap sorting, you could expect even faster times, since the number of comparisons would now be $O(n\log n)$. Note from Table 9.6 that indirect shell sorting, coupled with mapsort, is four times faster than using selection sort for the 1024-byte record case and 12 times faster than regular shell sort for the same case.

Given all this, you can see where indirect sorting can be quite useful. Of course, it comes at a price: You must have $O(n)$ room for the *map* array. Once again, we have a classic speed versus space trade-off.

SPECIALTY SORTING ALGORITHMS

So far we've focused exclusively on sorting algorithms that use the comparison of keys as the basic sorting operation. We chose this as the driving feature because it gives us general-purpose sorting algorithms that depend little on the nature of the keys being used—something that's useful if you're going to tuck a sorting algorithm away into a library. However, for special types of keys, there are faster known sorting algorithms. Next we'll give you a sample of these algorithms. They should trigger some ideas in your head for the special sorting needs you may have.

Perhaps the ultimate special case is when the keys are n distinct integers that range from 0 to $n - 1$. If you were paying attention in the last section, you'll see an easy way to sort such data. The *MapSort()* algorithm can be transformed into the following *BucketSort()* function.

```
template<class TYPE>
void BucketSort(TYPE arr[], int n)
// This routine sorts the array of TYPE items, where TYPE
// is assumed to be a record containing an integer key field,
// such that each key is distinct and lies in the range 0..n-1.
{
  TYPE tmp;
  int i, j, k;
```

```
for (i = 0; i<n; i++) {
    if (arr[i].key != i) {
        // We haven't processed the cycle that contains i
        // so rearrange the elements in that cycle.
        tmp = arr[i];
        k = i;
        do {
            j = k; k = arr[j].key;
            arr[j] = arr[k];
        } while (k != i);
        arr[j] = tmp;
    }
}
}
```

As with *MapSort()*, this algorithm runs in $O(n)$ time. It's hard to imagine being able to sort any faster. (It seems you do have to visit each item at least once, to tell if it's sorted.) Of course, in most cases, the keys won't be unique (there will be duplicates), and what if the keys range from 0 to $m-1$, for some $m \neq n$? This more general case of bucket sort can be solved by having an array of m buckets, each of which is a linked list of items that have the same keys. To sort the data, each item is placed on the list of the appropriate bucket. Then the lists are concatenated together from the first bucket to the last. The result is a global list that has the items in order.

The generalized bucket sort (which we won't show) should sound familiar to you. It's basically the same method used in separate chaining hashing. (See Chapter 7.) Here, the hash value is simply the key itself. (Remember, we're assuming the keys range from 0 to $m-1$.) Another variation of the generalized bucket sort uses the same technique as open addressing hashing with linear probing. See Gonnett and Baeza-Yates [Gonnett 91].

Another special type of sorting is known as *radix sorting*, which also depends on the keys being numeric. Rather than use the keys as a whole for comparison, the keys are decomposed into digits of some radix (for example, base 10 or base 2), and the sorting takes place step by step using the digits of the keys. Radix sorting is quite intuitive, and in fact, people tend to use radix sorting when sorting numerical keys by hand.

As an example, consider sorting mail by zip code. Suppose the zip codes are five digits as they are in the U.S. For instance, the zip code for the author's home town of Conifer, Colorado, is 80433. When the post office sorts mail by zip code, they look at the first digit, which in this case is 8. This is a code standing for a region of the U.S. that happens to contain the state of Colorado. So all mail for this region is sorted into a bin. The 0 digit further specifies that, indeed, the state is Colorado. So the author's mail can be sorted further into a sub-bin for Colorado. The 4 digit specifies the western suburbs of Denver, and the last two digits (33) specify the post office for the town of Conifer.

Note that most computers are not decimal but binary, so radix sorting usually is done on computers by comparing the bits of the keys rather than the decimal digits. It turns out that the binary radix sort algorithm is very similar to quick sort. The algorithm has the same structure and uses a similar partitioning scheme.

Radix sorting can be much faster that quick sort. In fact, if the keys are random enough, you can achieve close to $O(n)$ performance. However, keys are hardly ever random in the bits, so in practice, the sorting is about on par with quick sort, but can be much worse. For more information on radix sorting, see Knuth [Knuth 73] or Sedgewick [Sedgewick 92].

T E N

More Sorting Algorithms

n Chapter 9 we focused on sorting algorithms that worked with arrays stored in memory. An underlying premise of these algorithms is efficient random access of the array elements. In this chapter we'll take a look at sorting algorithms that are designed with sequential access in mind. Specifically, we'll look at algorithms that work with linked lists and files, both data structures where random access is much slower (if even possible) than sequential access.

The most popular type of sorting algorithm for sequential access is known as *merge sorting*. We'll be working mostly with merge sorting in this chapter, although at the end you'll see two efficient algorithms for sorting files of fixed-length records using random access.

THE MERGE SORT ALGORITHM

At the high level, the basic merge sort algorithm is quite simple and elegant. The idea is to split a "list" (any sequential collection of objects will do) in half, sort both halves, and then merge the halves back together. This is done recursively, where each half is split into fourths, and so on. At the bottom level, lists with one element each are merged together. Here is a recursive formulation in pseudo-code.

```
MergeSort(list)
{
  if (list has more than one object) {
    Split list into halves A and B
    MergeSort(A)
    MergeSort(B)
```

```
        Merge A and B back into list
    }
}
```

Merging two sorted lists is the fundamental operation of merge sort. The front nodes of the lists are compared, and the smaller is chosen to be added to the output list. This is done until both lists are empty. Figure 10.1 illustrates the process. It's not hard to see that $n/2$ comparisons are made, where n is the total length of the two lists. Thus, merging is $O(n)$.

Like quick sort, merge sort is a divide-and-conquer algorithm, recursing approximately $\log n$ times. In fact, merge sort is $O(n\log n)$, like quick sort. There is a crucial difference between these two algorithms, however.

In quick sort, the partitioning is done in such a way that the two halves are always sorted with respect to each other. Thus quick sort requires no additional merging process. That's not the case with merge sort. Because of the extra merge operation, merge sort is slower than quick sort on average. But remember that choosing a good pivot is the Achilles heel of quick sort, making it sensitive to the data coming in. With merge sort, the list is always split right down the middle, no matter what the input is. Thus merge sort is guaranteed to be $O(n\log n)$, a claim quick sort cannot make. Merge sort joins heap sort as the only other popular sorting method with guaranteed $O(n\log n)$ performance. As a bonus, merge sort can be a stable sort if you code the comparisons carefully. Heap sort isn't stable.

In the merge sort pseudo-code, we were purposely vague about what type of data structure *List* is. It could be a linked list, disk file, or even an array. Let's suppose for a moment that the *List* type is an array. Here's how two sorted arrays could be merged together.

```
template<class TYPE>
void Merge(TYPE a[], int na, TYPE b[], int nb, TYPE c[])
// Merge the two sorted arrays a and b, into array c.
```

Figure 10.1 Merging two sorted lists.

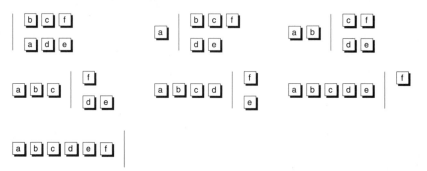

```
// It's assumed that c has been allocated to hold at least
// na + nb elements, and that the two input arrays are in
// ascending order.
{
  while(1) {
    if (na == 0) { // No more input from a, copy rest of b to c.
      while(nb--) *c++ = *b++;
      break;
    }
    else if (nb == 0) { // No more input from b, copy rest of a to c.
      while(na--) *c++ = *b++;
      break;
    }
    else { // Pick the smallest element from a and b, put it into c.
      if (*a < *b) {
        *c++ = *a++; na--;
      }
      else {
        *c++ = *b++; nb--;
      }
    }
  }
}
```

Note
All of the sorting algorithms in this chapter will be given for ascending order sorts. Making the sorts work for descending order is fairly simple, involving the reversal of the comparison of items.

The merging process for arrays is quite straightforward, but there's one catch when it's used with merge sort. During a merge sort, the array to be sorted is split in half, each half sorted, and then the two halves merged back together. Thus when *Merge()* is called, the arrays *a* and *b* are actually two halves of the same array. But where does array *c* come from? It is an auxiliary array. After merging *a* and *b* into *c*, you must either copy the elements of *c* back into the original array, or use fancy pointer manipulation and alternate back and forth between array *c* and the original array as the "current" array being sorted. Either way, $O(n)$ extra space is needed.

Because of this extra space required, merge sort is not well suited to sorting arrays. It's possible to reduce the amount of overhead required to $n/2$ elements (we won't show how), but you might wonder why you can't merge the two halves of the array in place, with no space overhead. In fact, you can, but surprisingly, all the known methods turn out to be too slow, too complicated, or both. If you are merging the array in place, nobody knows how to achieve an average time complexity of $O(n)$, which can be achieved if we allow extra space.

To give you an idea of what's involved in doing in-place array merging, Figure 10.2 shows one method that the author came up with.

Figure 10.2 In-place array merging.

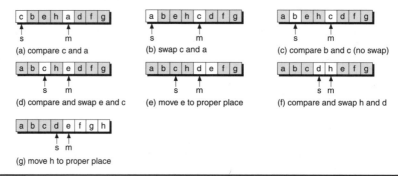

(a) compare c and a
(b) swap c and a
(c) compare b and c (no swap)
(d) compare and swap e and c
(e) move e to proper place
(f) compare and swap h and d
(g) move h to proper place

The idea is to compare the sth element (the start of the first half) with the mth element (the start of the second half), and ensure that the smallest is the first in the array, by swapping if necessary. Steps a and b of Figure 10.2 show how items a and c are swapped. Then s is incremented and compared with the mth element. For example, in Step c, b and c are compared. Since b is smaller, nothing needs to be done. When s is incremented yet another time, items e and c are compared. They are swapped, leaving the array as shown in Step d. At this point, the right half of the array is not in order, because e is not the smallest item in the half. So the items e and d are swapped.

In general, after swapping items from the two halves, the mth element will have to be inserted into its proper place in the right half, using the same method that's used in insertion sorting. Steps f and h show an example, where b is moved all the way to the end of the right half.

The following *InPlaceMerge()* function implements this merging process.

```
template<class TYPE>
void InPlaceMerge(TYPE arr[], int n)
// Merge the two sorted halves of arr in place.
{
  TYPE t;
  int m = n/2; // Index to the middle item
  int s = 0;   // Index to the first item

  while(s < m) {
    // Compare sth and mth items, swap if needed.
    if (arr[m] < arr[s]) {
      t = arr[s]; arr[s] = arr[m]; // First part of swap
      // Right half may be out of order, insert t into proper place.
      int p = m+1;
      while(p < n) {
        if (t <= arr[p]) break;
        arr[p-1] = arr[p];
        p++;
```

```
        }
        arr[p-1] = t;
    }
    s++; // Move to next item of first half.
  }
}
```

This code seems easy enough, but how good is it? Let's analyze its worst-case time complexity. In comparing the items *s* and *m*, you can see that $n/2$ comparisons are made, where *n* is the total length of the array. If we're unlucky enough, we'll have to swap items every time. After doing the swap, we must ensure the right half of the array is still sorted, moving the *m*th item to its proper place. If we're unlucky enough, we'll have to move it all the way to the end of the array every time. Figure 10.3 shows an array with this worst-case scenario. You should work through the merging to see that, indeed, a lot of operations take place. This worst-case scenario happens any time the smallest element of the left half is larger than the largest element of the right half.

Since the left and right halves are swapped $n/2$ times, and the new *m*th item must be moved to the end of the array every time, using $n/2$ comparisons, we see that the total number of comparisons is $n^2/2$. In other words, this in-place merging algorithm is $O(n^2)$ in the worst case, a far cry from the $O(n)$ complexity if an auxiliary array is used.

Of course, on average, not nearly as many comparisons will be needed. For example, on average, the items from the two halves only need to be swapped half the time, meaning $n/4$ exchanges. And out of these $n/4$ exchanges, perhaps the new *m*th item needs to be moved only a few locations, if any. Only if the *s*th item is larger than the (*m* + 1)th item do we have a problem at all. This shouldn't happen all that often, since remember, what we're swapping is the two smallest items from both halves. It's hard to estimate what the average complexity would be, but in the best case—when the array is already sorted—no movements would be needed. Thus the average case lies somewhere between $O(n)$ and $O(n^2)$.

In fact, experiments show that using the in-place merging algorithm is not much better than if we simply did a shell sort on the two halves combined. In other words, we can't really take advantage of the fact that the two halves are sorted! If we were to do the merge operation using shell sort, then, of course, the overall complexity of the merge sort algorithm would not be $O(n\log n)$, but more like $O(n^{1.5}\log n)$.

Figure 10.3 An array causing worst-case in-place merging.

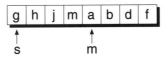

The result of this analysis is that, unless we can afford using extra space, merge sorting is not well suited to sorting arrays. The basic problem lies in the inability to rearrange the elements of the array easily. However, what if a linked list were used? Recall from the companion book that a linked list separates the logical ordering of the collection from the physical ordering. With a linked list, it's easy to rearrange elements in constant time, using only a few pointer manipulations. As it turns out, merge sorting works quite well for sorting linked lists, as you'll see next.

SIMPLE MERGE SORTING FOR LINKED LISTS

The basic merge sort algorithm is easy to formulate for linked lists, as shown in the following *MergeSort()* algorithm.

```
template<class TYPE>
void MergeSort(Snode<TYPE> *&list)
// Sort the given list in ascending order.
{
  Snode<TYPE> *other_half;

  // Must have at least two elements, or don't bother.
  if (list == 0 || list->next == 0) return;

  Split(list, other_half);
  MergeSort(list);
  MergeSort(other_half);
  Merge(list, other_half);
}
```

Note In the following sorting algorithms for linked lists, we'll assume null-terminated, singly linked lists are being used, with no headers unless specified. Later we'll show how to adapt some of the algorithms for circular lists, with or without headers.

There's one problem with our simple *MergeSort()* function. How do we split the list in half? With a linked list, we have only sequential access, so jumping to the middle seems difficult. In other words, this part of merge sort seems more suited to random access, so what gives? Actually, splitting the list is easy to code, if rather expensive. The following *Split()* function shows how.

```
template<class TYPE>
void Split(Snode<TYPE> *list, Snode<TYPE> *&half_way)
// Split the given null-terminated list in half, returning
// a pointer to the second half in half_way.
// ASSUMES at least two nodes in the list.
{
  Snode<TYPE> *a = list;
```

```
    Snode<TYPE> *b = list->next->next; // Assume at least two nodes.
    while(b) {
      a = a->next;
      b = b->next;
      if (b) b = b->next;
    }
    half_way = a->next; // Second half is after a
    a->next = 0; // Null-terminate the first half.
}
```

The trick used in *Split()* is to have two pointers chasing down the list, one that chases twice as fast as the other. When the faster pointer reaches the end of the list, the slower pointer will point right at the middle node. This is of course an $O(n)$ algorithm and thus rather expensive.

The *Merge()* function is easy to code and is a direct adaptation of the one we showed for arrays. There is a big difference, however, since we can do the merging "in place" via pointer manipulations, in $O(n)$ time.

```
template<class TYPE>
void Merge(Snode<TYPE> *&a, Snode<TYPE> *b)
// Merge the sorted null-terminated lists a and b, assumed to be
// in ascending order, returning the head of the combined sorted
// list in a.
{
  Snode<TYPE> new_list;        // Empty header
  Snode<TYPE> *p = &new_list; // Roving pointer

  while(1) {
    if (a == 0) { // Tack on rest of b list.
      p->next = b;
      break;
    }
    else if (b == 0) { // Tack on rest of a list.
      p->next = a;
      break;
    }
    else {
      if (a->info < b->info) { // First node of a is smallest.
        p->next = a; p = a; a = a->next;
      }
      else { // First node of b is smallest.
        p->next = b; p = b; b = b->next;
      }
    }
  }
  a = new_list.next;
}
```

Complete code for the simple merge sort algorithm is given on disk in the files *mlsort.h* and *mlsort.mth*. A test program can be found in *mlstst.cpp*.

Many books show this form of merge sorting for linked lists. However, even though it's $O(n\log n)$, the algorithm is not really very efficient. Having to find the middle of the list on each pass is rather expensive (even if it is done only $\log n$ times). Schemes could be devised that keep a direct pointer to the middle of the list, but this could get rather complicated, and there seems to be no way to avoid having to scan the list looking for the middle. The middle pointer has to be set somehow!

It turns out that there are several ways to do merge sorting for linked lists, without ever having to find the middle of the list, and we can double the speed of the algorithm in the process. These variations of merge sorting can be borrowed from algorithms usually given for sorting disk-based files, and these algorithms work very well on linked lists. This is a curious situation, because the same books that show the simple form of linked list–based merge sorting show the other techniques for files in later chapters. But it seems the authors of these books never put the techniques together! We're going to. Along the way, you'll get a nice introduction to the common methods used to sort disk-based files.

BALANCED 2-WAY MERGE SORTING

The problem with the simple merge sorting is that the splitting process scans through the list to split the list in two, but the fact that each node is being visited is otherwise being wasted. What if instead, while the list was being scanned for splitting, some sorting took place as well? That's the idea behind a variation of merge sorting called *balanced 2-way merge sorting*, which is illustrated in Figure 10.4.

In this method, the nodes of the original list are paired together into blocks of size 2. Each block is then sorted. The first block is placed on a temporary

Figure 10.4 Balanced 2-way merge sorting.

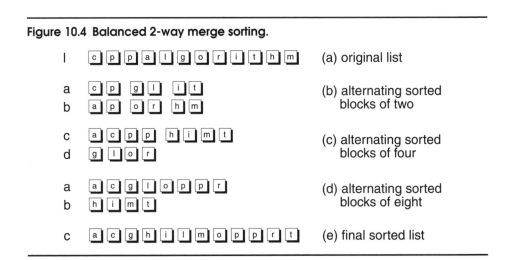

working list we'll call list *a*. The second block is placed on a temporary list *b*. All the blocks are added in the same alternating fashion. Then the first blocks from the *a* and *b* lists are grouped into blocks of 4 and sorted, and these blocks are placed on two other temporary lists *c* and *d*, in an alternating fashion. This process continues, with the sorted blocks growing by powers of 2. The *a*, *b*, *c*, and *d* lists are used alternatingly as input/output lists. Finally, the block size will encompass all the nodes being sorted, so that there will be only one output list.

You can see that four temporary lists are used with this method. However, that doesn't mean we are using four times as many nodes as are on the original list. The only overhead will be four nodes used as dummy headers for the four lists (and even these nodes could be eliminated with some awkwardness in the code). Also, the final sorted list can easily be "moved" to the original list by a few pointer manipulations. Note that during the sorting passes, you may have incomplete blocks that the code must handle properly.

Since the block size doubles each pass, it's not hard to see that log*n* passes are needed to finish the sort. In fact, like the simple merge sorting algorithm, balanced 2-way merging is $O(n\log n)$ in all cases. It's interesting to see that the list is effectively split in two each pass without wasted motion, because sorting takes place in the process. The real trick behind this idea is alternating the sorted blocks between two output lists. The following *BMLSort()* function implements the balanced 2-way merging algorithm for linked lists.

```
template<class TYPE>
void BMLSort(Snode<TYPE> *&hd, Snode<TYPE> *&tl)
// Sorts in ascending order the null-terminated list that
// has hd as the first node. Passes back the new head, as
// well as a pointer to the last node (so you that you can
// make the list circular afterward if you want to).
{
  Snode<TYPE> heads[4];
  Snode<TYPE> *tmp, *ih, *oh, *a, *b, *c, *d;
  int a_cnt, b_cnt, n;

  tl = hd;
  if (hd == 0 || hd->next == 0) return; // Fewer than two nodes

  ih = heads;   oh = heads+2;

  // Take the original list and sort it into blocks of size 2,
  // alternating the blocks between lists ih and ih+1.

  a = ih;  b = ih+1;
  while(hd) {
    Snode<TYPE> *hn = hd->next;
    if (hn == 0) {
      a->next = hd;
      break;
```

```
        }
      if (hd->info <= hn->info) {
          a->next = hd;
          a = b; b = hn;
          hd = hn->next;
      }
      else {
          a->next = hn;
          hd->next = hn->next;
          hn->next = hd;
          a = b; b = hd;
          hd = hd->next;
      }
    }
    a->next = 0; b->next = 0;

  for(n=2;;n <<= 1) { // Merge n-blocks into 2n-blocks.
    a = ih->next;  b = (ih+1)->next;
    if (b == 0) break; // Only one list. We're sorted.
    c = oh;      d = oh+1;
    a_cnt = 0;  b_cnt = 0;

    while(1) {
      // Merge n nodes from each input list into 2n nodes for an
      // output list, alternating between each output list, until
      // the input is exhausted.
      if (a_cnt >= 0) a_cnt = n; // Set to n if a list isn't empty.
      if (b_cnt >= 0) b_cnt = n; // Set to n if b list isn't empty.
      while(1) {    ▸
        // Merge n nodes from each input list into 2n nodes in
        // one of the output lists. May exhaust both input lists
        // before we get n from each.
        if (a_cnt > 0 && (b_cnt < 1 || a->info <= b->info)) {
          c->next = a;  c = a;  a = a->next;
          if (a == 0) a_cnt = -1; else a_cnt--;
        }
        else {
          if (b_cnt < 1) break; // We have enough, or EOL for both inputs.
          c->next = b;  c = b;  b = b->next;
          if (b == 0) b_cnt = -1; else b_cnt--;
        }
      }
      if (a_cnt < 0 && b_cnt < 0) break;
      tmp = c; c = d; d = tmp; // Swap output lists.
    }

    c->next = 0; d->next = 0; // Be sure to null-terminate working lists.
    tmp = ih;  ih = oh;  oh = tmp; // Swap input/output sets.
  }

  hd = a; tl = c; // Pass back head and tail of sorted list.
}
```

Complete code for the balanced 2-way merging of linked lists is given on disk in the files *bmlsort.h* and *bmlsort.mth*. A test program can be found in *bmlstst.cpp*.

In this algorithm, *n* keeps track of the current block size and *a_cnt* and *b_cnt* keep track of how many nodes are left from the current blocks being sorted from the input lists. We set these counts to –1 to signal that the blocks have been exhausted. Also, we have the input and output lists trade roles by using simple pointer swapping.

Unlike the simple merge sorting algorithm, we were able to code *BMLSort()* non-recursively, thus requiring no stack space. The code is not much larger than the combined *Split()*, *Merge()*, and *MergeSort()* functions of the simple merge sort, although *BMLSort()* is not as elegant, and is harder to follow. More significantly though, *BMLSort()* is about twice as fast as its *MergeSort()* counterpart.

Note The list-sorting functions given in this chapter can't work directly with the singly linked list classes developed in the companion book. The reason is that we've collapsed the *Snodeb* and *Snode* classes of the companion book into one *Snode* class, where the *next* pointer is public. (This is to simplify the code, for clarity's sake.) You'll have to make modifications accordingly. Also, if you wish to pass in a circular linked list from the *Slist* class, first you'll have to convert it into a null-terminated list, passing in the first real node of the list (not the dummy header). Upon return, you can use *hd* and *tl* parameters of *BMLSort()* to construct a circular linked list again. We provided no such facility for *MergeSort()*. See the comments in the code on disk for more details.

NATURAL MERGE SORTING

The blocks used in balanced 2-way merge sorting are sometimes called *runs*, since they are strings, or runs of items in ascending (or descending) order. Between runs, there may be a discontinuity in the order. The runs in balanced 2-way merge sorting are artificially limited in size to be a power of 2. This is what guarantees $O(n\log n)$ performance. However, the runs aren't always as long as they could be. Why not make the runs as long as the input ordering allows? For example, the first block shown in the first pass in Figure 10.4 could contain [*c*, *p*, *p*] instead of just [*c*, *p*].

If we allow the runs to vary in size depending on the data being sorted, the resulting variation is called *natural merge sorting*. In this variation, the runs take on their "natural" size as the input order dictates. By allowing longer runs when possible, there is a likelihood that the number of passes needed will be reduced. You may not need $\log n$ passes. In effect, you are taking advantage of the order already inherent in the lists. For example, if the list is already sorted, only one pass is needed, since the first run will be composed of the entire list!

Figure 10.5 Natural merge sorting.

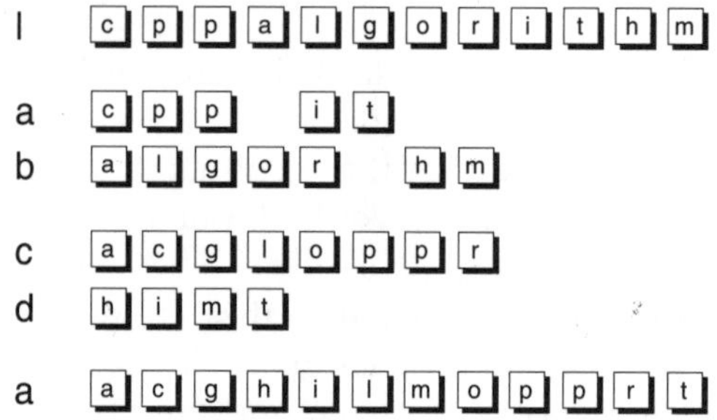

Figure 10.5 shows an example of natural merge sorting, using the same list as given in Figure 10.4 for balanced 2-way merge sorting. Note that with natural merge sorting, we need one less pass for this example.

The relaxing of the block or run size takes form in another way in natural merge sorting. In balanced 2-way merge sorting, we merge only two blocks at a time, no matter what the next blocks contain. However, in natural merge sorting, the boundaries are relaxed, and when merging two blocks together, if one block becomes exhausted, we can borrow from the next block in the same list, if the ordering allows it. For example, suppose in Figure 10.5 that the node *i* was replaced with *s* instead, so that the second block was [*s, t*]. Then the *c* list in second pass could contain two more nodes, and read [*a, c, g, l, o, p, p, r, s, t*]. The *d* list would merely contain [*h, m*].

The following *NMLSort()* function, along with a *Distribute()* helper function, implements natural merge sorting.

```
template<class TYPE>
void NMLSort(Snode<TYPE> *&hd, Snode<TYPE> *&tl)
// Sorts, in ascending order, the null-terminated list that
// has hd as the first node. Passes back the new head, as
// well as a pointer to the last node (so you that you can
/ make the list circular afterward if you want to).
{
  Snode<TYPE> heads[4];
  Snode<TYPE> *tmp, *ih, *oh, *a, *b, *c, *d;

  tl = hd;
  if (hd == 0 || hd->next == 0) return; // Fewer than two nodes

  ih = heads;   oh = heads+2;
  a  = ih;      b  = ih+1;
  Distribute(hd, a, b);

  while(1) {
    a = ih->next;  b = (ih+1)->next;
```

```
      if (b == 0) break; // Only one input list, we're done.
      c = oh;  d = oh+1;
      while(1) {
        // Make a the smallest node
        if (a->info > b->info) { tmp = a; a = b; b = tmp; }
        while(1) { // Copy run from a onto c.
          c->next = a;  c = a;  a = a->next;
          if (a == 0) { // a's contribution finished.
            Distribute(b, c, d);
            goto next_pass;
          }
          // Make a the smallest node.
          if (a->info > b->info) { tmp = a; a = b; b = tmp; }
          if (a->info < c->info) { // a's contribution is finished.
            while(1) { // Copy rest of b's run onto c.
              if (b->info < c->info) break;
              c->next = b;  c = b;  b = b->next;
              if (b == 0) {
                Distribute(a, d, c);
                goto next_pass;
              }
            }
          }
          break;
        }
      }
      tmp = c;  c = d;  d = tmp; // Swap output lists.
    }
    next_pass:
    tmp = ih;  ih = oh;  oh = tmp; // Swap input/output sets.
  }

  hd = a; tl = c; // Pass back head and tail of sorted list.
}

template<class TYPE>
void Distribute(Snode<TYPE> *s, Snode<TYPE> *&a, Snode<TYPE> *&b)
// Distributes the nodes in s by appending runs onto the ends
// of a and b in an alternating fashion. Null terminates the
// a and b lists. Returns the last nodes on the a and b lists.
{
  while(1) {
    if (s == 0) goto cap_off_lists;
    a->next = s;  a = s;  s = s->next;
    while(1) { // Copy rest of s's run onto a's.
      if (s == 0) goto cap_off_lists;
      if (s->info < a->info) break;
      a->next = s;  a = s;  s = s->next;
    }
    b->next = s;  b = s;  s = s->next;
    while(1) { // Copy rest of s's run not b.
      if (s == 0) goto cap_off_lists;
      if (s->info < b->info) break;
      b->next = s;  b = s;  s = s->next;
    }
```

```
     }
     cap_off_lists:
     a->next = 0; b->next = 0;
}
```

Complete code for natural merge sorting is given on disk in the files *nmlsort.h* and *nmlsort.mth*. A test program can be found in *nmlstst.cpp*.

The *NMLSort()* function is similar in spirit to the *BMLSort()* function. It uses the same pointer swapping tricks to alternate between input and output lists. The main difference lies in determining where the run boundaries are. Instead of maintaining simple counters, comparisons between successive nodes have to be made. This makes *NMLSort()* more awkward than *BMLSort()*, and makes it potentially slower as well. Like *BMLSort()*, *NMLSort()* is O($n\log n$) on average (and in the worst case too), but it can be much faster than that.

Our tests show that *NMLSort()* is about 10 to 20 percent slower than *BMLSort()* on randomly ordered lists. However, if the lists are already sorted, *NMLSort()* is blindingly fast, where *BMLSort()'s* speed doesn't change at all. Thus you have a trade-off between "smarter," more complicated code that runs slower at times but that also can take advantage of preexisting order.

EXTERNAL MERGE SORTING

In this section we introduce the idea of sorting data that's too large to fit into main memory. Specifically, we'll take a look at sorting records stored in a file on disk. This type of sorting is known as *external sorting*, since the records reside *external* to main memory. The type of sorting we've considered up to this point is called *internal sorting*.

The issues involved in external sorting are different from those for internal sorting. The main difference is that accessing records is more expensive for external sorting, and so data access is the dominant factor in determining sorting speed. On most operating systems, random access of records is slower than sequential access. Thus sorting algorithms that employ sequential access are likely to be significantly faster than those that use random access.

Since merge sorting employs sequential access, you might guess it makes a good candidate for external sorting, and indeed it does. The natural merge sorting algorithm given in the last section is the basis for most of the popular external sorting algorithms. Instead of working with linked lists, files are used instead, and these files are accessed sequentially. The following *NMFSort()* and *Distribute()* functions are direct adaptations of their linked-list counterparts.

```
template<class TYPE>
int NMFSort(FilePak<TYPE> *src, FilePak<TYPE> *dest)
// Sorts the src file. Passes back the name of the temp
```

```
// file that has the final sorted data. Destroys all the
// temp files except that one. Assumes src file is
// open, but closes it at the end. Returns 0 if fewer
// than one record in the file, else 1.
{
   FilePak<TYPE> heads[4];
   FilePak<TYPE> *tmp, *ih, *oh, *a, *b, *c, *d;

   if (src->Fetch() <= 0) {
      *dest = *src;
      return 0; // Fewer than one record, or some type of file error
   }

   // Let the OS create names for the temporary files.
   for (int i = 0; i<4; i++) tmpnam(heads[i].name);

   ih = heads;  oh = heads+2;
   a = ih;      b = ih+1;

   a->OpenForOutput();  b->OpenForOutput();

   Distribute(src, buff, a, b);

   src->Close();  a->Close();  b->Close();

   while(1) {
      a = ih;  b = ih+1;
      b->OpenForInput(); b->Fetch();
      if (b->Eof()) break; // Only one non-empty input file, we're done.
      a->OpenForInput(); a->Fetch();
      c = oh;      c->OpenForOutput();
      d = oh+1;   d->OpenForOutput();
      while(1) {
         // Make a the file with smallest first record.
         if (a->info > b->info) { tmp = a; a = b; b = tmp; }
         while(1) { // Copy run from a onto c.
            c->info = a->info;  c->Store();  a->Fetch();
            if (a->Eof()) { // a's contribution finished.
               Distribute(b, c, d);
               goto next_pass;
            }
         }
         // Make a the file with the smallest first record.
         if (a->info > b->info) { tmp = a; a = b; b = tmp; }
         if (a->info < c->info) { // a's contribution is finished.
            while(1) { // CopyRun(b, c);
               if (b->info < c->info) break;
               c->info = b->info;  c->Store();  b->Fetch();
               if (b->Eof()) {
                  Distribute(a, d, c);
                  goto next_pass;
               }
            }
         }
         break;
      }
```

```
      }
      tmp = c;  c = d;  d = tmp; // Swap output files.
    }
    next_pass:
    a->Close(); b->Close(); c->Close(); d->Close();
    tmp = ih;  ih = oh;  oh = tmp; // Swap input/output sets.
  }

  // Destroy the three temp files no longer needed.

  b->Close();   b->Remove();
  oh->Remove(); (oh+1)->Remove();

  // Pass back the file information of the file that
  // contains the sorted data.

  *dest = *a;
  return 1;
}

template<class TYPE>
void Distribute(FilePak<TYPE> *s, FilePak<TYPE> *a, FilePak<TYPE> *b)
// Distributes the records in s by appending runs onto the ends
// of a and b in an alternating fashion.
{
  while(1) {
    if (s->Eof()) return;
    // Write s->info to a, advance s.
    a->info = s->info; a->Store(); s->Fetch();
    while(1) { // CopyRun(s, a);
      if (s->Eof()) return;
      if (s->info < a->info) break;
      // Write s->info to a, advance s
      a->info = s->info; a->Store(); s->Fetch();
    }
    // Write s->info to b, advance s
    b->info = s->info; b->Store(); s->Fetch();
    while(1) { // CopyRun(s, b);
      if (s->Eof()) return;
      if (s->info < b->info) break;
      // Write s->info to b, advance s
      b->info = s->info; b->Store(); s->Fetch();
    }
  }
}
```

 Complete code for file-based natural merge sorting is given on disk in the files *nmfsort.h*, *nmfsort.mth*, and *filepak.h*. A test program is provided in the file *nmfstst.cpp*.

So that we could make *NMFSort()* look very similar to *NMLSort()*, we designed a *FilePak* class to help encapsulate the details of file-based access.

We're assuming here that the records being sorted are fixed length. Later on we'll talk about how variable-length records could be handled. Here is the *FilePak* class declaration. (The complete implementation is given on disk.)

```
template<class TYPE>
struct FilePak {
  char name[128];          // Name of the file
  TYPE info;               // Last record read, or next record to write
  FILE *fp;                // Associated file
  int Eof();               // Tests for end of file.
  FILE *OpenForInput();    // Opens existing file for read mode.
  FILE *OpenForOutput();   // Creates new file for write mode.
  FILE *OpenForUpdate();   // Opens existing file for read/write mode.
  void Close();            // Close the file.
  void Rewind();           // Rewind the file.
  void Remove();           // Remove the file from the system.
  int Rename(char *n);     // Rename the file.
  // Sequential access, via info field:
  int Fetch();
  int Store();
  // Sequential access, via x, bypassing the internal info field:
  int Fetch(TYPE *x, int n=1);
  int Store(TYPE *x, int n=1);
  // Random acess, via info field:
  int FetchRandom(int recno);
  int StoreRandom(int recno);
  // Random access, via x, bypassing the internal info field:
  int FetchRandom(TYPE *x, int recno, int n=1);
  int StoreRandom(TYPE *x, int recno, int n=1);
};
```

This class is designed to support reading and writing of fixed-length records from files. It isn't meant to be the "end-all" in file class design. You may want to add many other embellishments, such as better error handling, the support of variable length records, and so on. You could, for example, merge it with the *Fmgr* class given in the companion book.

The *NMFSort()* algorithm, as we've given it, runs fairly quickly, as file-based sorts go, and like the other merge sort algorithms, *NMFSort()* is $O(n\log n)$. The main drawback is that you must manage four temporary files during the sorting process. The total amount of space taken up by these files is n records. That means that you must have room on your disk for a complete copy of the file being sorted. This is analogous to the problem with using merge sorting for arrays, where an additional array must be used. Should space be a major problem for your application, in the last part of the chapter we discuss sorting algorithms that can sort a file in place.

At the end of the sorting process, the sorted data will reside in one of the temporary files. In most cases, you'll want to move this data to the original file, or at least move the data to a file with a more permanent name. In the former case, this can be accomplished by first deleting the original file and then renaming the temporary file to the name of the original file. In effect, no data actually has to be moved. Also, note that after the first distribution phase of *NMFSort()* (before the looping starts), you don't need the original file any more, so it could be deleted right then, rather than waiting until afterward.

Replacement Selection

Even though *NMFSort()* is fairly fast, it can be made faster quite easily. For file-based records, it's usually best to access records in blocks rather than individually. Most operating systems store files in clusters, such as 512 bytes or 4096 bytes. The strategy behind many external sorting algorithms is to read as many records as possible in memory, sort them with a fast internal sort routine, such as quick sort, and then write them back out, somehow merging the sorted blocks together. The larger the internal buffer, the faster the sort is going to be. When the internal buffer is sized to be a multiple of the file cluster size, the sort is likely to be even faster.

With *NMFSort()*, as written so far, we don't take into account the considerations of the last paragraph. In fact, the routine uses very little internal memory at all, only storing a few records during the merging and distribution process. How can we go about utilizing more of main memory?

Recall that the driving factor in speed for external merge sorting is the number of passes that need to be made. Also recall that the number of passes needed depends directly on how long the runs are. Suppose we can somehow use a large portion of main memory to make the initial runs longer. For example, we can read in a large block of records and then sort them, using an internal sort routine such as quick sort or shell sort. This block can be written out as one run. Using this strategy, if our internal buffer holds m records, we can have initial runs of m records. This can significantly reduce the number of sorting passes needed.

We can actually go one step better than this. It's possible to use a buffer of m records, but actually have runs, on average, of about $2m$ records! The trick is to organize the buffer as a priority queue, implemented as a min-heap. The heap is first filled up with records from the input file. Then another record is read from the file. If the record's key is equal to or larger than the minimum key on the heap, the record with the minimum key is written to the output file, and is replaced on the heap by the newly read-in record. (The heap is reorganized as needed at this point.) If the new record's key is smaller than the minimum key on the heap, it can't be a part of the current run, so the heap is sorted and then emptied onto the output list to complete the current run. Then a new run is started, with the recently read record as the first item. The heap eventually fills up, and the process continues until the input is exhausted.

This technique is called *replacement selection*, for we are selecting out of the m possibilities which record to output next and then replacing that record in the internal buffer by a new one from the input list. It's hard to predict exactly how long the runs are going to be using replacement selection, but it's not hard to see they'll be at least *m* records long. Extensive studies have been made of this, and it has been found that you can expect runs about 2*m* in length.

To modify *NMFSort()* to use replacement selection, all you need to do is pass a heap to the routine and replace the first call to *Distribute()* to a call to *HeapDistribute()* instead. The latter function is implemented as follows.

```
template<class TYPE>
void HeapDistribute(
  FilePak<TYPE> *s, Heap<TYPE> &heap, FilePak<TYPE> *a, FilePak<TYPE> *b
)
// Distributes the records in s by appending runs onto the ends
// of a and b in an alternating fashion. Uses replacement
// selection via heap.
{
  FilePak<TYPE> *t;
  TYPE *p;
  int len;

  while(1) {
    if (s->Eof()) goto finish_up;
    if (!heap.IsFull()) { // Filling up the heap
      heap.Insert(s->info); s->Fetch();
    }
    else { // Heap full, use replacement selection.
      if (s->info < *heap.Head()) { // End of run
        heap.Sort(); // Sorted in reverse order.
        len = heap.Length();
        p = heap.Head() + len - 1;
        while(len--) a->StoreDirect(p--);
        heap.Clear();
        t = a; a = b; b = t; // Swap output files.
      }
      else { // Keep run going
        a->StoreDirect(heap.Head(), 1);
        heap.Replace(1, s->info); s->Fetch();
      }
    }
  }

  finish_up:

  if (!heap.IsEmpty()) {
    heap.Sort(); // Sorted in reverse order.
    len = heap.Length();
    p = heap.Head()+len-1;
    while(len--) a->StoreDirect(p--);
    heap.Clear();
  }
}
```

Note that *HeapDistribute()* is used only at the beginning, when splitting the original file into what will become the first two input files. This means that the space taken up by the heap is not needed all the way through the sorting process. Once in progress, the merge sorting uses only a few records internally.

VARIATIONS ON A THEME

The *NMFSort()* function, modified to use replacement selection, is probably efficient enough to use for all but the most demanding sorting applications. In this section we'll discuss techniques that can be used to speed up the sorting, in case more speed is critical. All of these techniques, however, carry the higher price tag of more complexity and, in some cases, more dependency on the underlying operating system.

Multiway Merging

As we've stated before, the most effective way to speed up external merge sorting is to reduce the number of passes. We've seen where replacement selection helps do this. Another way is to split the file into more pieces, in what is known as *multiway merging*. So far we've used 2-way merging. Suppose that instead we broke the original file into four portions and then used a 4-way merge on these. By doing this, we can reduce the number of passes from $\log_2 m$ to $\log_4 m$. That's because at each step of the merge process, we're sorting four records internally rather than two. To make this work, we'll need eight temporary files rather than four. But sorting 256 records, for instance, will take roughly four passes instead of eight.

In general, to do *p*-way merge sorting, you need 2*p* temporary files, but the number of passes is reduced to $\log_p m$. If *p* is large enough, a problem occurs in merging the *p* heads of the input lists. It's more than just a single comparison. In essence, you end up doing a sort of *p* records. You can use one of the internal sorting routines for this. A better idea is to use a heap of size *p*, and use the same replacement selection technique we gave earlier.

Needing 2*p* temporary files is the main drawback to *p*-way merging. In turns out there is a way to reduce the number of files needed to just *p* + 1, using a technique known as *polyphase merging*. This technique contains a rather clever use of the Fibonnaci numbers to control the number of runs that any given file contains. By controlling the number of runs a certain way, we can reuse an input file as an output file before a complete pass is finished, thus eliminating the need for so many files. Implementing polyphase merging is a bit complicated, and we leave the details to the references. Perhaps the best place to start is Knuth [Knuth 73].

Separate Disks for Temporary Files

Another important factor in sorting speed is where the temporary files are actually located. In the code given on disk, we've stored the temporary files on the same disk as the original file being sorted. However, each time one of the files is accessed, the disk read/write head must be moved to the proper location for that particular file. In other words, we're not really utilizing sequential access very well. If you can arrange it, you should store each temporary file on a separate physical disk. That way the respective heads on the disks won't have to move as you switch between files, and you'll get truly sequential access. Also, you are exploiting more parallelism inherent in the hardware this way.

One complication of this approach is that the temporary file that ends up with the final sorted data may not be on the same physical disk that you would like the sorted data to reside on. Because of this, you can't do a simple renaming process, as we suggested earlier. Instead, you'll have to physically make a copy of the file.

Tuning the I/O Buffer Sizes

The final speedup technique we'd like to mention is in controlling the size of the I/O buffers. In the *FilePak* class, we used the *stdio* library for file access. Implicit in the routines in *stdio* is an I/O buffer associated with each open file. These buffers are given a default size, which depends on the operating system and particular compiler implementation you have. Typical values are 512 bytes and 4096 bytes. You can set the size of the buffer for any given file using the *setvbuf()* function of the *stdio* library.

You might want to investigate the default size of the I/O buffers for your implementation and make the buffers larger. If the size you give happens to be a multiple of the file cluster size, you could see a dramatic improvement in speed.

SORTING VARIABLE-LENGTH RECORDS

So far, we've considered only records that are fixed length. But suppose the records are variable length? One typical example is in sorting an ASCII file, where each line of text, terminated by a new-line character, represents one record. Within each line is one or more fields that in composite make up the key for that line. In some cases, the entire line might be treated as one large character string key.

Because merge sorting is sequential in nature, it is ideally suited to sorting variable-length record files. The unmodified *NMFSort()* function (without re-placement selection) can be adapted quite easily to sorting variable-length records. Because we encapsulated the file access in the *FilePak* class, you'll have to make modifications only to *FilePak*, and you'll need to define a class for its template *TYPE* parameter accordingly.

In the case of sorting ASCII text files, for example, you may want to create a special *Text* class that overloads the comparison operators to handle comparisons between lines of text. To be the most flexible, you should be able specify which portion of the lines are the actual keys and what type of comparison to do with these keys (for example, is it ASCII character ordering, or should the key fields be treated as integers?) This *Text* class would then become the *TYPE* parameter of the templated *FilePak* and *NMFSort()* functions.

If you wish to use replacement selection for variable-length records, things get a little trickier, because the underyling heap works only with fixed-length records. One way around this is to use an indirect heap that uses pointers or indices to the actual data. These pointers could reference variable-length records stored in some other large contiguous buffer. But a problem occurs. How do you remove variable-length records from this buffer without leaving gaps? You could allocate each record separately on the heap as it is read in, but as you might expect, this could be rather slow. Another possibility is to use a non-contiguous priority queue, where the nodes can be variably-sized. A *splay tree*, as discussed in the companion book, might work for this.

EXTERNAL RANDOM-ACCESS SORTING

As far as external sorting goes, it's hard to beat the variations of merge sorting. The one drawback of merge sorting is that you must have the equivalent of one extra copy of the file being sorted. Also, having to keep track of the temporary files is a bit of a headache. If the records being sorted are fixed length, it is possible to sort the file in place, using random access, and without needing any additional files or extra file space. Next we'll look at two techniques that are adaptations of algorithms normally intended for internal sorting. While much slower than merge sorting for moderate size and larger files, these algorithms have their uses for small files, or in situations where space or the number of open files is at a premium.

Any of the internal sorting algorithms could be used with little modification for external sorting (by replacing array access with file access), but you'll be greatly disappointed with the results. The sorting will be horribly slow. Since file access is the dominant factor, particularly when random access is employed, you want to read and write records in large blocks and do as much sorting as possible in memory. The question is: How do you fit block I/O into the internal sorting schemes? You'll see two different approaches in the next sections.

External Quick Sorting

The internal quick sort algorithm can be adapted for external sorting in an intriguing way. Recall that quick sorting uses a pivot that serves to partition the items into two halves. With internal quick sort, this pivot is a single record. However, you could generalize this pivot to be a double-ended priority queue of *m* records, presumably large enough to consume most of the memory available.

Figure 10.6 shows an example of partitioning using a generalized pivot of 3 records, in a file of 11 records. This particular method of partitioning was adapted from an algorithm given in Gonnet [Gonnet 91]. We've simplified slightly the approach used in the reference. (The approach there is optimized to handle nearly sorted files more efficiently.)

Let's walk through the steps shown in the figure:

a. There are four pointers used: two read pointers that keep track of the next records to read from both ends, and two write pointers that keep track of where the next records are to be written. The first step is to fill up the

Figure 10.6 External quick sort partitioning.

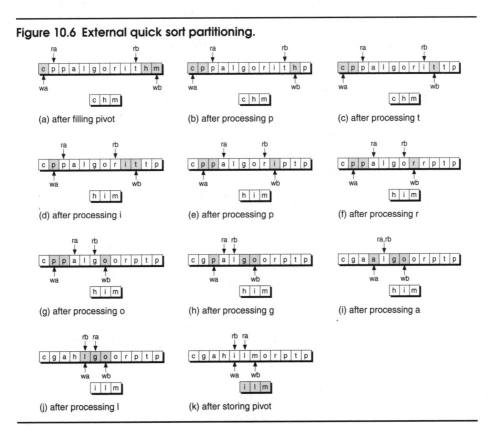

(a) after filling pivot

(b) after processing p

(c) after processing t

(d) after processing i

(e) after processing p

(f) after processing r

(g) after processing o

(h) after processing g

(i) after processing a

(j) after processing l

(k) after storing pivot

double-ended priority queue by reading records from the file. Note that we read from both ends in as balanced a way as possible. In this case, we read in the records *m, c, b* in that order. These are stored in the priority queue, which we've represented as a simple ordered list.

b. The next record is read in. In determining which end to read from, we look at the difference between the read pointers and the write pointers. We read from the end with the smallest distance. This prevents the write pointers from overtaking the read pointers. In this case we read in the record *p* from the left end. Since *p* is larger than the largest record in the priority queue, it is written out to the right end, at the location determined by the *wb* pointer. Note how we overwrite the *m* that was previously stored there.

c. The record *t* is read in from the right side using the smallest read-write distance rule. Since the record *t* is also larger than the largest record in the priority queue, it is written back out to the right side.

d. The record *i* is read in using the smallest read-write distance rule. Since *i* is smaller than the largest record in the queue but larger than the smallest record, we must add it to the queue. This is done by choosing either the minimum or the maximum record of the queue, writing it out to the appropriate side of the file, and replacing it with the record just read in. The choice is made by determining which side has the fewest records written. Once again, this helps keep the sides balanced. In this case, the left side has no records written, so we store the minimum record, *c,* on the left side, and replace it with *i*, reorganizing the queue in the process.

e. The record *p* is read, and since it is larger than the largest record in the queue, it is written out to the right side.

f. The record *r* is read, and since it is larger than the largest record in the queue, it is written out to the right side.

g. The record *o* is read, and since it is larger than the largest record in the queue, it is written out to the right side.

h. The record *g* is read, and since it is smaller than the smallest record in the queue, it is written out to the left side.

i. The record *a* is read, and since it is smaller than the smallest record in the queue, it is written out to the left side.

j. The record *l* is read. It belongs in the queue, so we write out *b* to the left side, and replace it with *l*, reorganizing the queue.

k. At this step, the read pointers have crossed, so our partitioning has finished. The final step is to write the pivot back out. Note that the pivot is completely sorted and is written out to its final location in the file.

After the partitioning has finished, the two sub-partitions, which don't contain any of the records in the pivot, are sorted in a recursive fashion.

There are several interesting aspects to this algorithm. One is that by using a pivot of larger than one record, the chance of degenerate partitioning is diminished. In fact, with our pivot of size 3, we have a similar situation to median-of-three partitioning. (It's not identical though.) Also, by keeping both sides of the pivot balanced, the chance of degenerate partitioning is reduced even further. And by using a priority queue as the pivot, with as many records as will fit in memory (you'll want to use more than three records, we can assure you), you in effect cause a fair number of records to be sorted internally in main memory, speeding up the process greatly.

While we represented the priority queue as a simple ordered list, in general, you'll want to use a data structure such as a deap. (Remember these double-ended heaps from Chapter 8?) The following *EQSort()* function implements external quick sorting, using a deap for the pivot.

```
template<class TYPE>
int ExternalQuickSort(FilePak<TYPE> &f, Deap<TYPE> &deap, int a, int b)
{
  int ra, rb, wa, wb, n;

  while(b > a) {

    n = b-a+1;
    if (n <= deap.Size()) {
      // Partition fits completely in memory, so simply sort
      // with an internal sort.
      TYPE *p = deap.Data();
      f.FetchRandom(p, a, n);
      InternalSort(p, n);
      f.StoreRandom(p, a, n);
      break;
    }

    // Use the deap as a pivot and sub-partition the partition.
    ra = wa = a;   rb = wb = b;
    while(rb >= ra) {
      // Ensure that the write ptrs don't get ahead of the read ptrs.
      if (ra-wa < wb-rb)
        f.FetchRandom(ra++);
        else f.FetchRandom(rb--);
      if (!deap.Insert(f.info)) { // Heap full
        if (f.info > *deap.Max()) {
          f.StoreRandom(wb--);
        }
        else if (f.info < *deap.Min()) {
          f.StoreRandom(wa++);
        }
        else if (wa-a < b-wb) { // Balance sides when given a choice.
          f.StoreRandom(deap.Min(), wa++);
          deap.Replace(2, f.info); // Replace minimum, rearrange.
        }
```

```
        else {
            f.StoreRandom(deap.Max(), wb--);
            deap.Replace(3, f.info); // Replace maximum, rearrange.
        }
    }
}

// End of partitioning. Write the pivot back to disk.

int len = deap.Length();
if (len != 0) {
    deap.Sort();
    f.StoreRandom(deap.Min(), wa, len);
    deap.Clear();
}

// Sort the shortest sub-file first.

wa--; wb++; // These are off by one at this point.

if (wa-a < b-wb) {
    ExternalQuickSort(f, deap, a, wa);
    a = wb; // Simulate ExternalQuickSort(f, deap, wb, b).
}
else {
    ExternalQuickSort(f, deap, wb, b);
    b = wa; // Simulate ExternalQuickSort(f, deap, a, wa).
}
    }
    return 1;
}
```

 Complete code for the external quick sorting algorithm is given on disk in the files *eqsort.h* and *eqsort.mth*. A test program can be found in *eqstst.cpp*.

This routine has the same form as the optimized quick sort algorithm from Chapter 8. While we optimized the function for tail recursion, we stopped just short of taking out the recursion altogether, although we could have.

Note how we optimized for small sub-files, by performing a simple internal sort (such as shell sort) when the partitions get below the size of the deap. Our experiments show that this has less impact than you might think. We found only about a 5 percent reduction in execution time with this optimization. This is partially due to the fact that when this portion of the code is called, the partition is rarely the full size of the deap. Instead, we found partitions of about half the size more common. This tends to the lessen the impact of the optimization.

How well does external quick sort perform? It takes roughly $\log(n/m)$ passes, where n is the number of records in the file and m is the size of the priority queue. Thus, asymptotically, external quick sorting is comparable to external merge sorting. However, the constant factor is much higher, due to the random access. Later on you'll see some actual comparisons between external quick sorting and other external algorithms. We can say right now that external quick sorting is much slower than merge sorting, even for small n. Thus, while it is an intriguing technique, you wouldn't want to use it unless you really want to sort the file in place. Even then, there is a faster algorithm for small files; we'll present it next.

External Bubble Sorting

The next in-place external sorting algorithm we'll discuss is based on an algorithm known as *bubble sort*. The bubble sort algorithm is probably the simplest sorting algorithm you can code. Here is one implementation for sorting internal arrays.

```
template<class TYPE>
void BubbleSort(TYPE arr[], int n)
// Ascending bubble sort. Smaller keys are "lighter"
// and thus float to the top.
{
  TYPE tmp;
  for (int t = 0; t < n-1; t++) {
      for (int b = n-1; b > t; b--) { // Bubble up from bottom.
          if (arr[b] < arr[t]) { // Swap if bottom lighter.
              tmp = arr[b]; arr[b] = arr[t]; arr[t] = tmp;
          }
      }
  }
}
```

The idea behind bubble sorting is to compare the first item with the last. If the last item is smaller, the two are swapped. Then the first item is compared with the next-to-last item, and the two are swapped if appropriate. This process continues through the entire array. If you think of the first item as the topmost item, then at the end of one pass, the lightest, or smallest, item will have "bubbled up" to the top. For the next pass, the second item is compared with the last, then with the second to last, and so on. After enough passes, the array will be sorted. Figure 10.7 illustrates one pass of bubble sort.

If bubble sorting is so simple, why didn't we cover it in Chapter 9? It's because bubble sorting is $O(n^2)$, and of the $O(n^2)$ algorithms, it is the slowest of the ones we've seen. Bubble sort is so slow that we weren't going to even mention it in this book. But there's a lesson to be learned in this. Even though bubble sort makes a horrible internal sorting algorithm, it turns out that it can be generalized quite nicely for in-place external sorting! In fact, for small to moderate size files, it will run circles around the more sophisticated external quick sort!

Figure 10.7 One bubble sorting pass.

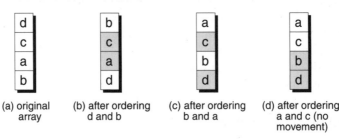

| (a) original array | (b) after ordering d and b | (c) after ordering b and a | (d) after ordering a and c (no movement) |

The external bubble sort algorithm that we present is given in Dufrene and Lin [Dufrene 92], where it is called the Dufrene/Lin Sort, based on the author's names. The basic idea is to generalize the notion of comparing two records, bubbling the smallest to the top. Instead of using two records, what if we compare m records, where m is large enough to fill up the internal memory?

To fill up these m records, we can read a block of $m/2$ records from the first part of the file and $m/2$ records from the last part. These are combined in a single array and sorted, using some internal sort algorithm. The last $m/2$ records of this array are the largest of the ones read, so we write them back out to the last block of the file. The first $m/2$ records represent the smallest records so far, so we keep them in the first half of the array. Then we read in the next to last block of $m/2$ records from the file and store them in the second half of the array. The array is then again sorted, and the largest $m/2$ records are stored back out in the file.

This process continues until all blocks have been processed. At the end of the first pass, the first half of the array contains the smallest records of the entire file, so they are written out to the first block in the file. Note that these records are now in their final location. To start the second pass, we read in the second block from the file, and combine and sort it with the last block in the file. The process continues as before. When enough passes are performed, the entire file will be sorted. Figure 10.8 illustrates one pass of the external bubble sort algorithm, where $m = 4$.

You'll notice several times in the example shown that a block is read in, only to be stored again, intact. You might be able to optimize the algorithm by detecting when no sort actually takes place, although it's debatable whether it's worthwhile.

After the first pass, the blocks are sorted individually, so an in-place merging could be used when combining them rather than a full sort. In our experiments, we tried using the in-place merging algorithm for arrays given earlier, but the results were inconsistent. Sometimes in-place merging was faster than doing a full sort, sometimes it wasn't. You are probably better off doing a full sort, using something like shell sort.

Figure 10.8 One pass of external bubble sorting.

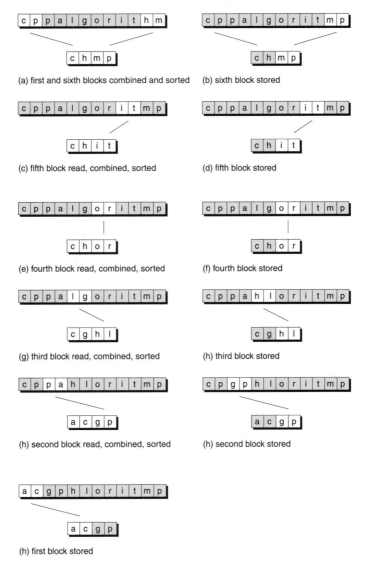

(a) first and sixth blocks combined and sorted

(b) sixth block stored

(c) fifth block read, combined, sorted

(d) fifth block stored

(e) fourth block read, combined, sorted

(f) fourth block stored

(g) third block read, combined, sorted

(h) third block stored

(h) second block read, combined, sorted

(h) second block stored

(h) first block stored

Of the external sorting routines we've given, external bubble sort is the simplest of them all, as illustrated in the following *EBSort()* function.

```
template<class TYPE>
void EBSort(FilePak<TYPE> &f, int t, int b, TYPE *buff, int buff_sz)
// External bubble sort (Dufrene/Lin Sort) for a file of records,
// from record t to record b. Uses a buffer of size buff_sz, which
```

```
// should be a power of 2 (otherwise the last record in the buffer
// will be wasted).
// NOTE: This code is written for ascending sort.
{
  TYPE *pt, *pb;
  int n, m, rlast, rt, rb, bm, bmlast;

  n = b-t+1;
  if (n <= 1) return;

  if (buff_sz >= n) { // Can do sort entirely in memory.
    f.FetchRandom(buff, t, n);
    InternalSort(buff, buff_sz);
    f.StoreRandom(buff, t, n);
    return;
  }

  m = buff_sz / 2;

  // Figure out the size and location of the blocks. Note that
  // the last block might not be full size.

  rt = t;
  bm = n % m;
  if (bm == 0) bm = m;
  bmlast = bm;
  rlast = b - bm + 1;

  pt = buff; pb = buff + m;

  f.FetchRandom(pt, rt, m);

  while(1) {
    rb = rlast; bm = bmlast;
    while(rt+m <= rb) {
      f.FetchRandom(pb, rb, bm);
      InternalSort(pt, m+bm);
      f.StoreRandom(pb, rb, bm);
      rb -= m;
      bm = m; // No more partial blocks
    }
    f.StoreRandom(pt, rt, m);
    rt += m;
    if (rt == rlast) break;
    memmove(pt, pb, m*sizeof(TYPE)); // Avoid write fby read.
  }
}
```

 Complete code for the external bubble sort algorithm is given on disk in the files *ebsort.h* and *ebsort.mth*. A test program can be found in *ebstst.cpp*.

Like the internal bubble sort algorithm, the external bubble sort is basically $O(n^2)$. The size of the internal buffer has a significant impact on the runtime. External bubble sort's constant time factor is much smaller than the one for external quick sort. This is due mainly to the fact that the external bubble sort can utilize the internal buffer more effectively, and you don't have the overhead of a double-ended priority queue. The external bubble sort algorithm reads and writes blocks in their entirety, whereas in external quick sort, the blocks are accessed record by record. Because of these reasons, the external bubble sort is much faster than external quick sort for small to medium-size files. If the files are small enough, bubble sort can even outperform merge sorting. However, because the complexity is quadratic rather than logarithmic, eventually both merge sorting and external quick sorting will win.

COMPARISON OF THE EXTERNAL SORTING ALGORITHMS

Now we'll summarize the results of this chapter. We've shown three basic algorithms for sorting records in a file. External merge sorting uses sequential access, whereas external quick sorting and bubble sorting use random access. The sequential access gives merge sorting a huge advantage in terms of sorting speed, and it can easily be adapted to sorting variable-length records. However, the external quick sort and bubble sort routines can sort the records in place, needing no extra space or temporary files. They won't work for variable-length records, however.

Of the two random access routines, external bubble sort is the method of choice, unless the files are quite large. The external quick sort will be faster for these larger files. However, it will most likely be orders of magnitude slower than external merge sort, so you wouldn't use external quick sort unless disk space is really at a premium.

To give you a comparison of the algorithms in a real situation, Table 10.1 shows the execution times (in seconds) for each of the algorithms under different file and record buffer sizes, as determined on the author's machine. The records being sorted were small—just an integer key and an integer data

Table 10.1 Execution Times of the External Sorting Algorithms (in seconds)

	Buffer size = 100				Buffer size = 1,000			
$n =$	1,000	5,000	10,000	20,000	1,000	5,000	10,000	20,000
Bubble sort	0.55	54	204	800	0.06	3.6	17	71
Quick sort	6.4	93	218	537	0.1	37	104	270
Merge sort	1.92	8.3	19	40	0.33	4.5	11	26

field. If *n* is the file size and *m* is the internal record buffer size, then we'll call *m/n* the buffer ratio. You can see from the tables that the external bubble sort routine can utilize higher buffer ratios more effectively than the other two routines.

It's important to note that disk caching was turned off during the tests and that the default I/O buffer sizes were used as given by Borland C++ 3.1. If you turn on caching, fiddle with the I/O buffer sizes, and use different size records, you are likely to get different results from what is shown here. Also, in a real application, you would probably use larger record buffers than we did in our tests. You want the buffers to be as big as you can get away with.

ELEVEN

Text Searching

I n this chapter we'll cover algorithms for searching strings of text. Usually this conjures up the image of searching a manuscript written in English or some other natural language, looking for keywords and phrases. However, text searching is much broader than this. For example, you might want to search a database of genetic codes. These codes might consist of the letters "a", "c", "t" in strings of various lengths, some quite long. Another example is searching for binary codes in an executable file, perhaps looking for certain program fragments.

The general problem is the following: Given a string of characters we'll call *text*, of length *n*, search for the string of characters *pat*, of length *m*. In most of the algorithms to follow, both *text* and *pat* can be arbitrary binary strings (any byte value is okay), or they may be null-terminated strings (the more typical case).

This leads to another consideration, that of the size of the *alphabet*, or set of legal characters that the strings can be made of. In the genetic code database, the alphabet size is three characters (with perhaps a few other characters for code delimiters, and so on). In English text, the alphabet might be as small as 27 characters (the letters plus the space character) or, more commonly, made up of the 128 ASCII characters. In general, the alphabet size is usually treated as 256 characters—all the legal values of an 8-bit byte. Keep in mind though, that increasingly more applications support languages with much larger alphabets. In the future it's quite likely that the alphabet size will play a larger role in the design of text searching algorithms, although it's been largely ignored up to now. The algorithms we'll present pay no particular attention to the alphabet used, other than to accommodate the different sizes when necessary.

273

For most of the chapter, we'll assume that the text to be searched resides completely in memory, in one long string. At the end of the chapter, we'll show how you can adapt the algorithms to search for text stored in files. Also, we assume that we're looking for an exact match of the pattern. Chapter 12 shows a completely different approach to text searching that can perform approximate searching on patterns.

STRAIGHTFORWARD SEARCHING (SFS)

The most straightforward method for searching text is to examine each position k of the text and determine whether the successive m characters match the pattern. If a mismatch occurs before the complete pattern is matched, the next text position, $k + 1$, is tried as the start of the pattern. A typical function is shown in *SFSearch()*.

```
int SFSearch1(char *text, int n, char *pat, int m)
// A straightforward text search algorithm, using subscripting.
// Returns position of first match, or -1 if no match.
{
  int i, j, k;

  for (k = 0; k<n; k++) {
      for (j = 0, i = k; j<m; j++, i++) if (pat[j] != text[i]) break;
      if (j == m) return k; // Match found at position k.
  }
  return -1;
}
```

The straightforward search (SFS) algorithm is simple, short, and intuitive. But if you examine Figure 11.1, which illustrates this type of searching, it doesn't take long to realize that SFS isn't very smart. The algorithm blindly tries each position of the text as the start of the pattern, even if it has already examined some of the characters in prior match attempts. For example, after matching the first three characters of "tomcat" and then failing on the "c", there is no reason to then try the positions starting with "o", "m", or "c", since none of these could possibly lead to a match. Instead, the search could skip down to the position containing the "a", the first position whose character is at that point unknown and thus should be examined.

Later on you'll see algorithms that perform "smart" skipping during searching. But before we discuss those algorithms, let's analyze the performance of SFS. In the worst case, SFS is quadratic—$O(nm)$—in terms of the number of character comparisons. This worst case occurs in highly repetitive text and patterns. One example is a text of n "a"'s, where the pattern is a series of $m-1$ "a"'s followed by a terminating "b". Fortunately, such pathological cases don't occur very often in practice. One counterexample, however, is in searching a database of genetic codes, where the alphabet is small and the codes highly repetitive.

Figure 11.1 Straightforward searching.

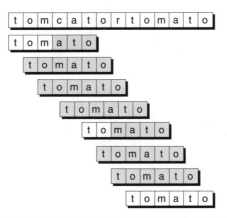

In the most common use of text searching—searching documents of English or some other language—SFS is essentially linear—$O(n)$. This can be justified by the observation that in typical text, a match of the first character of the pattern occurs infrequently, and the chances of the second character matching are even smaller, and so on. Thus the inner loop isn't likely to execute very often or for very long.

The algorithmic step of text searching is a single-character comparison. This is such a tiny step that other bookkeeping parts of the algorithm can easily mask it. As such, over the years it has been found that the time complexity evaluation for text-searching algorithms, using character comparisons as the step, doesn't mean a whole lot. It's not that it's irrelevant, but rather that other factors, such as how well the code is optimized, can have just as great an impact.

Next we'll take a look at ways we can optimize the SFS algorithm so that you can see the dramatic effect these optimizations have on the algorithm. So far we haven't talked much about optimizing code in this book. We've focused instead on choosing good algorithms. Optimization is heavily machine- and compiler-dependent and should be considered only after choosing an appropriate algorithm, and then only if the first attempt at coding the algorithm isn't good enough.

We're going to optimize SFS, even though it is known to be inferior to other searching algorithms. There is some method to this madness. First, it provides an opportunity to illustrate a few of the optimization tricks that you can apply to algorithms in general. Some of the optimizations specific to SFS turn out to be useful in the more sophisticated searching algorithms. Also, the SFS algorithm isn't completely inferior. Its simplicity can mean fewer maintenance headaches, a fact that shouldn't be overlooked.

Optimizing SFS

Probably the most basic optimization performed in C++ is to use pointer incrementing rather than subscripting. Pointer optimization is useful when you are scanning an array sequentially rather than with random access. That's because incrementing a pointer is generally faster than computing a subscript. For example, here is a modified version of the straightforward searching algorithm that uses pointer arithmetic.

```
int SFSearch2(char *text, int n, char *pat, int m)
// A straight-forward text search algorithm, using pointer
// arithmetic. Returns position of match, or -1 if no match.
{
  char *t, *et, *p, *ep, *q;

  t = text;  et = text + n;  ep = pat + m;
  while(t < et) {
    p = pat;  q = t;
    while(p < ep && *p++ == *q++) { ; }
    if (p == ep) return t - text;
    t++;
  }
  return -1;
}
```

This form of the algorithm is slightly harder to follow than one using subscripting. However, our tests show speed improvements of 20 to 50 percent. (The actual improvement depends heavily on the machine you are using and how well your compiler optimizes.)

The second optimization we show is based on the observation that the algorithm can be split into two parts: scanning for the first character in the pattern, which we'll call the *scan loop*, and matching the rest of the characters in the pattern, which we'll call the *match loop*. In the first versions of the SFS algorithm, these operations were intertwined. In scanning for the first character, part of the code for the match loop is executed. For example, the test *p == ep* is executed (testing for a full match) in *SFSearch2()* even when the first character doesn't match. This causes substantial overhead, which can be eliminated by completely separating the two loops, as shown in the next variation of *SFS*.

```
int SFSearch3(char *text, int n, char *pat, int m)
// A straight-forward text search algorithm, using pointer
// arithmetic and an optimized scan loop.
// Returns position of match, or -1 if no match.
{
  char *t, *et, *p, *ep, *q;
  char first = *pat;

  t = text;  et = text+n-m+1;  ep = pat + m;
  while(1) {
```

```
    // Do fast loop scanning for first character match.
    while (t < et && *t != first) t++;
    if (t >= et) break;
    // First character matches, so execute match loop.
    p = pat;  q = t;
    while(++p < ep && *p == *++q) { ; }
    if (p == ep) return t - text;
    t++;
  }
  return -1;
}
```

Now the match loop is much tighter, yielding a 20 percent or more reduction in runtime by our tests. Another improvement we made was to set *et*, the end of text pointer, to *text + n − m + 1*, instead of *text + n*. If there are fewer than *m* characters left in the text, no match is possible, so you might as well terminate the search early.

Separating the scan loop and match loop may seem like a relatively insignificant optimization. But it's an important technique to remember, especially in more sophisticated algorithms, where the setup time for the match loop can be non-trivial. In general, it makes sense for any searching algorithm to first quickly search for a potential match, and only then to do a more complicated and complete test.

Using Sentinels

We can improve the scan loop still further by eliminating the end-of-text test. To do this, we can use the first character of the pattern as a sentinel by placing it at the end of the text. Thus the scan loop is guaranteed to always terminate by using the match test alone. Figure 11.2 illustrates a sentinel of "t" stopping the scan loop, while searching the text "nothere" for the pattern "tomato."

Once the scan loop terminates, we can then test whether we found a match or whether we scanned too far. For example, the scan loop can be changed to the following.

Figure 11.2 Using a scan loop sentinel.

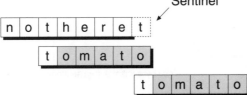

```
// Assumes pat[0] (first) is stored at the end of the text.
while (*t != first) t++;
if (t >= et) break; // Really at end, so no match.
```

This yields a very tight scan loop, that according to our tests, should give you at least a 10 percent speedup. Higher speedups are possible on some machines, since they can perform this simple scan loop in one instruction.

As it happens, you can also make use of the same sentinel in the match loop. (This trick was discovered by the author.) For example:

```
// Assumes pat[0] stored at the end of the text
// and that the first character position has already been tested.
p = pat;   q = t;
while(*++p == *++q) { ; }
if (p == ep) return t - text; // Return match position.
```

For the sentinel to work in this case, two conditions must hold. First, the pointer *t*, which points to the character in the text that matches the first character in the pattern, must not be closer than *m* bytes from the end of the text. (It won't be, due to the test for *t* >= *et* before the match loop.) Second, the pattern must be null-terminated. Figure 11.3 illustrates how these conditions work.

When *t* is greater than *m* bytes from the end of the text, then the match loop terminates either on finding a true mismatch in the pattern or when the null byte of the pattern is reached. In the latter case, we have a complete match. Suppose that *t* is exactly *m* bytes from the end of the text. If there is a complete match, the null byte of the pattern corresponds to the position in the

---

Figure 11.3 Using a match loop sentinel.

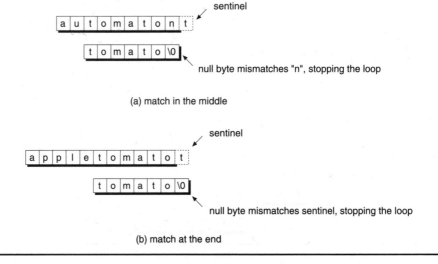

(a) match in the middle

(b) match at the end

text where the sentinel resides. Since the sentinel is equal to the first character in the pattern, it can't be null, and thus a mismatch occurs, stopping the loop. The following *SFSearch4()* function shows the sentinel being used for both the scan loop and match loop.

```
int SFSearch4(char *text, int n, char *pat, int m)
// Straight-foward searching with sentinel optimization.
// ASSUMES text has room for at least n+1 bytes.
// ASSUMES pattern is null-terminated, but the text doesn't
// have to be. Returns position of match, or -1 if no match.
{
  char *t, *et, *p, *ep, *q;
  char first, save_ch;
  int rv = -1;

  // Set up sentinel at the end of the text, saving what was there.
  first = *pat;  save_ch = text[n];  text[n] = first;

  t = text;  et = text+n-m+1;  ep = pat + m;
  while(1) {
    while (*t != first) t++;  // Scan loop
    if (t >= et) break;
    p = pat;  q = t;
    while(*++p == *++q) { ; } // Match loop
    if (p == ep) { rv = t - text; break; }
    t++;
  }

  text[n] = save_ch; // In case you want to restore last byte of text
  return rv;
}
```

Using the sentinel in the scan loop has a much greater impact than using it in the match loop. There are two reasons for this. First, if you are lucky, the match loop won't be executed very often. Second, even if the match loop is executed, then performing an extra test in the match loop has little effect when searching for short patterns. However, if the text you are scanning is highly repetitive, with a small alphabet, then the likelihood of matching the first character is high, so the match loop may be executed quite often. And if the pattern is long, then many tests will be performed in the match loop. For these cases, the match loop sentinel will pay off. In our tests, involving typical English text with patterns an average of seven characters in length, the improvement was only around 3 percent.

Throughout most of this book we have avoided using sentinels. That's because they are very error prone. It's easy to forget to put in a sentinel, causing infinite loops that can be hard to track down. When you use a sentinel, several assumptions are made that must be kept in mind. It's easy to forget these assumptions. In the case of the sentinel used in *SFSearch4()*, it's assumed that *text* points to at least $n + 1$ bytes of room. If *text* points to a null-terminated string, then it's the null byte that gets replaced by the sentinel. If *text* is some arbitrary binary string, then you must make sure to allocate one extra byte at the end.

If you choose to use the sentinel in the match loop, then another assumption is that the pattern is null-terminated. In typical applications, it will be, but if you are searching for a general binary pattern, then be sure *not* to use the sentinel in the match loop. Also, the *text* string cannot contain nulls in the middle, but you can use the sentinel in the scan loop for both general binary patterns and text.

Sooner or later you will get burned by the assumptions made for the sentinel used here. We almost guarantee it. So why bother using sentinel optimization? When used by itself it offers speedup of only 10 to 15 percent. However, all the optimizations add up. For example, *SFSearch4()* runs almost twice as fast as *SFSearch1()*. The biggest improvements are due to using pointer incrementing and separating the scan loop from the match loop. Note that the latter is necessary if you use a sentinel.

CREATING A SEARCH CLASS

In the code for SFS we've given so far, we haven't taken advantage of C++. The routines shown could easily be in C. Next we show how encapsulating the search mechanism into a class is helpful. Two observations lead to a search class design: We need some way of hiding the sentinel manipulation, and we need a way to search for any other occurrences of the pattern in the text, not just the first occurrence. The following SFS class shows how we can handle both problems.

```
class SFS {
private:
  unsigned char *text, *pat, *start_pos;
  int n, m;
  unsigned char save_ch, first;
  void SetSentinel();
public:
  SFS();
  void SetPattern(char p[], int m_);
  void SetText(char t[], int n_);
  void RestoreText();
  int Search();
};


SFS::SFS()
// Reset the searcher to no pattern or text.
{
  text = 0;  n = 0;  start_pos = 0;
```

```
    pat  = 0;  m = 0;
}

void SFS::SetSentinel()
// Sets the last byte of text (usually where the null-terminator
// goes) to the first byte of the pattern.
{
  if (n == 0 || m == 0) return; // Not ready to set sentinel.
  text[n] = pat[0];
}

void SFS::SetPattern(char p[], int m_)
// Record the pattern of length m_ > 0 to be used.
{
  pat = (unsigned char *)p;  m = m_;  first = *pat;
  SetSentinel();
}

void SFS::SetText(char t[], int n_)
// Record the text to search, of length n_ > 0. The text
// should have room for n_ + 1 bytes. (You'll have this
// many bytes if the text is null-terminated.)
{
  text = (unsigned char *)t;  n = n_;  start_pos = text;
  save_ch = text[n]; // In case sentinel is used.
  SetSentinel();
}

int SFS::Search()
// Searches the previously recorded text for the previously
// processed pattern. Returns pos >= 0 if there's a match,
// else -1.
{
  unsigned char *t, *et, *p, *ep, *q;

  t = start_pos;  et = text+n-m+1;  ep = pat + m;
  while(1) {
    // Scan loop, using sentinel.
    while (*t != first) t++;
    if (t >= et) break;
    // Match loop, using sentinel.
    p = pat;  q = t;
    while(*++p == *++q) { ; }
    if (p == ep) {
       start_pos = t + 1; // Where next search is to begin
       return t - text;   // Where match is
    }
    t++;
  }

  return -1;
}
```

```
void SFS::RestoreText()
// Restores the last byte of the text that may have
// had a sentinel stored there.
{
  text[n] = save_ch;
}
```

 The *SFS* class is given on disk in the files *sfs.h* and *sfs.cpp*. Test programs can be found in the files *sfstst.cpp* and *sfstst2.cpp*.

In using the *SFS* class, the basic idea is to set up pointers to the pattern and text using *SetPattern()* and *SetText()*. Setting the sentinel in the text is performed behind the scenes by calling *SetSentinel()*. Then a loop can be coded that searches for each occurrence of the pattern. Here is an example.

```
char *text = "a tomato automaton";
char *pat = "tomato";

SFS searcher;

searcher.SetText(text, strlen(text));
searcher.SetPattern(pat, strlen(pat));

while(1) {
  int posn = searcher.Search();
  if (posn == -1) break;
  cout << "Match found at " << posn << '\n';
}

cout << "No more matches found\n";

searcher.RestoreText(); // In case null byte of text should be restored.
```

You'll notice that using the *SFS* class is reminiscent of the generators of Chapter 3. On disk you'll see generator classes like *SFS* for each of the search algorithms in this chapter. However, here we'll show only the functional forms.

BOYER-MOORE SEARCHING

While the SFS algorithm is simple and straightforward, it turns out there are algorithms, not too much more complicated, that can easily run two times faster or better. These algorithms are all based on an algorithm known as the *Boyer-Moore (BM) algorithm,* first published in 1977 [Boyer Moore77]. We'll discuss the general BM algorithm first and then later discuss variations that are simpler and just as fast, if not faster, than BM.

The BM algorithm attempts to make the searching smarter when a mismatch occurs. Rather than blindly try a match with the next position in the text, as illustrated in Figure 11.1 given earlier, the BM algorithm takes advantage of the known properties of the pattern to compute skips larger than 1. The algorithm looks both at what character in the text caused the mismatch and also what sub-pattern matched before the entire match failed. From these two pieces of knowledge, the BM algorithm computes two skip values, choosing the largest one to skip by. We'll take a look at these two skip values in the next sections, starting with the one extracted from the mismatched character.

Like the SFS algorithm, the BM algorithm can be broken down into two loops: the scan loop and the match loop. In the scan loop, BM searches not for the first character of the pattern but rather the last. If a match of the last character occurs, then the pattern and text are examined in reverse order in the match loop, looking for a full match. Matching the pattern in reverse order is rather unintuitive, but Boyer and Moore found that this enabled them to calculate big jumps in the searching when a mismatch occurs.

Scan Loop Skipping

Consider what happens when a mismatch on the last character of the pattern occurs in the scan loop. Two cases emerge, as illustrated in Figure 11.4. If the mismatched character in the text does not occur anywhere in the pattern, then there is no way that any of the characters leading up to the mismatch can be involved in a match, so you might as well skip down m characters in the text rather than just one character. This is illustrated in Figure 11.4b, where "g" does not occur in the pattern "tomato".

If the mismatched character *does* occur in the pattern, then it's not safe to jump m positions, but you can certainly jump enough positions so that the rightmost occurrence of the mismatched character, as it appears in the pattern (if at all), is aligned at the current text position. For example, in Figure 11.4a, the mismatched text character is "m", which also occurs in the middle of the pattern, so the pattern is shifted down to align the two "m"'s. Likewise in Figure 11.4c, the two "t"s are lined up.

Figure 11.4 Delta1 skipping during the scan loop.

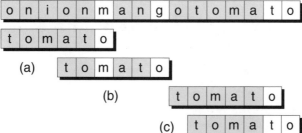

The skip computed due to the mismatched character is called the *delta1 skip*. Each character in the alphabet is assigned a *delta1* value, and these values can be precomputed before the searching starts by using just the pattern, as shown in the following code.

```
// Compute delta skip values
for(k = 0; k < MAX_ALPHABET_SIZE; k++) delta1[k] = m;
for(k = 0; k < m; k++) delta1[pat[k]] = m-k-1;
```

For those characters that occur in the pattern, the skip value is based on the rightmost position of the character in the pattern (in case the character occurs more than once in the pattern). The positions are counted from the right, starting with 0. For example, in the pattern "tomato", the *delta1* value for "o" is 0, for "t" it's 1, "a" has a value of 2, and "m" a value of 3. For those characters that don't occur in the pattern, the *delta1* value is equal to *m*, the length of the pattern, which in this case is 6.

Note that the last character in the pattern has a *delta1* value of 0. This can be used as a test to determine when the last character has been found, as shown in the following modified scan loop.

```
t = start_pos+m-1; // t points to possible match of last pattern character.
et = text+n;

while(t < et) {
  // Scan loop using delta1 skip.
  k = 0;
  while((k = delta1[*(t += k)]) && (t < et)) { ; }
  if (t >= et) break; // Really past the end
  // Do match loop.
  ...
}
```

As we did in the SFS algorithm, we can use sentinel optimization to get rid of the explicit bounds test, *t < et,* in the scan loop. However, we must use the last character of the pattern as the sentinel instead of the first, and we must place *m* sentinels at the end of the text to handle the fact that the skip may be as large as *m* characters. For example:

```
t = start_pos+m-1;   et = text+n;
memset(et, pat+m-1, m); // Place m sentinels after the text.

while(t < et) {
  // Scan loop using delta1 skip.
  k = 0;
  while((k = delta1[*(t += k)]) { ; }
  if (t >= et) break;
  // Do match loop.
  ...
}
```

Using *delta1* skips rather than always shifting by 1 can yield a significant reduction in the search time. It's likely that *delta1* will be much larger than 1. The actual amount depends on the length of the pattern and where the character occurs in the pattern. For example, in the case when the mismatched text character does not occur in the pattern at all, the skip is *m* positions. Obviously, the longer the pattern, the bigger the likely skip. It's not uncommon to see a factor of two or better speedup using *delta1* skips.

We can gain further speed by using an optimization known as *loop unrolling*. The idea is to explicitly list several of the cycles of the loop as separate statements, eliminating the branching and testing and any other loop overhead for these statements. For example, we can unroll the scan loop by a factor of three (i.e., eliminate three branches per loop), as shown in the following code.

```
t = start_pos + m - 1;  et = text + n;

while(t < et) {
  // Unrolled scan loop using sentinels at the end of the text.
  k = delta1[*t];
  while(k) {
    k = delta1[*(t += k)];
    k = delta1[*(t += k)];
    k = delta1[*(t += k)];
  }
  if (t >= et) break; // After end of text
  // Do match loop.
  ...
}
```

This loop unrolling relies on having the *m* sentinels at the end of the text and on the fact that when *k* is 0 in either the first or second statement of the scan loop, executing the rest of the statements in the scan loop has no effect (other than redundant processing). As before, a test is made after the scan loop to determine if there was really a match, or if we went beyond the end of the text.

Whether loop unrolling has any effect and how many times to unroll the loop is highly machine dependent. Extensive studies by D. Sunday [Sunday 91] show that unrolling by a factor of three is a good place to start. According to his statistics, you could easily see a 50 percent speed improvement due to the use of sentinels and loop unrolling.

Skipping after the Match Loop

When a potential match is found after the scan loop, then the match loop is entered. In the BM algorithm, this match takes place in reverse, as shown in the following code.

```
// Match loop: Assumes t points to the text character that matches the
// last character of the pattern, which itself is pointed to by ep.
```

```
p = ep;  q = t;
while(--p >= pat && *p == *--q) { ; }
if (p < pat) return q - text; // We have a match!
// Else, compute an amount to shift down by.
t += some_skip_amount;
```

In the case where the matching fails, we must compute a new position in the text to start scanning with. We can use the *delta1* skip values to help us. For example, suppose we are matching the pattern "banana" that's lined up with the text "havana", as shown in Figure 11.5. During the reverse match loop, the match fails on the letter "v". Since "v" is not in the pattern, we can shift the entire pattern down to bypass the "v", resulting in a skip of three positions.

Note that the *delta1* value for "v" is actually 6, since the pattern length is 6. Then why can we skip only three positions? The reason is that the *delta1* values are computed assuming that the mismatch occurs at the last character in the pattern, not somewhere in the middle. Thus we must make a correction based on the relative position of the mismatch. An easy way to do this is simply to compute a new value for t based on q. (The variable t points to the text position that corresponds with the tail of the pattern, and q points to the mismatch position.) For example:

```
t = q + delta1[*q];
```

You should verify to yourself in Figure 11.5 that t is indeed six positions past "v" after the *delta1* skip but that the absolute shift is only three positions.

There's a big problem with using *delta1* after the match loop: Suppose the mismatched text character occurs in the pattern, but only to the right of the current mismatch point? Then the new value of t will actually be to the left of where it was. Figure 11.6 illustrates this shift to the left, where the character causing the mismatch, "n", has a *delta1* value of 1 in the pattern "banana". Thus, computing the new tail of the pattern, $t = q + 1$, actually causes a shift backward by one position.

Shifting the pattern backward is a worthless move, for we've already determined that the pattern can't possibly match before the current position. That's presumably how we came to the current position in the first place! We

---

Figure 11. 5 A *delta1* skipping after the match loop.

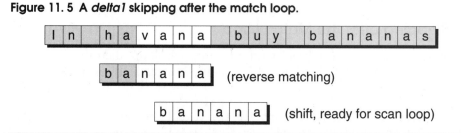

Figure 11.6 A backward *delta1* skip.

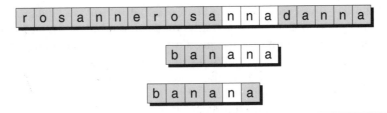

could always use a skip value of 1 if we find the *delta1* value is worthless, but is the shift as large as it could be? As it turns out, the BM algorithm uses another type of skip, called *delta2*, which can cause a bigger shift than 1.

The *delta2* skip is obtained by taking note of the string of characters that *do* match during a failed match loop. For example, in the "havana/banana" match, the subpattern "ana" at the right end of the pattern matches. We can use this knowledge in the following way: We would like to align this sub-pattern in the text with the next leftward recurrence of the sub-pattern in the pattern itself. That gives us a good possibility for a match. Figure 11.7 shows an example of this.

Note from Figure 11.7 how the second reocurrence of "ana" in the pattern is aligned with "ana" in the text after the shift. Also notice that a different character precedes the new occurrence of the sub-pattern, namely "b". This is fortunate, for had the character been a "v" (as in "vanana"), then we would know that the match would simply fail again. Having a different character shifted into the point of mismatch doesn't guarantee that we'll have a full match, but if the character *isn't* different, we know for certain that no full match is possible. In the case where the pattern contains no recurrence of the sub-pattern with these properties, then we know we can shift the entire pattern down by *m* positions.

The *delta2* skip values can be computed based solely on positions in the pattern, and they depend only on the pattern itself, not the text being searched. Thus, like the *delta1* skips, the *delta2* skips can be computed ahead of time in

Figure 11.7 *Delta2* skipping after the match loop.

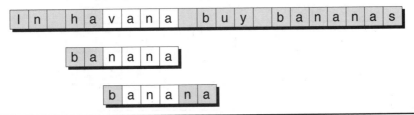

a preprocessing step. Figure 11.8 illustrates the amount to be shifted for each position of mismatch for the pattern "banana".

Summarizing Figure 11.8, you can see that the *delta2* shifts for the pattern "banana" are [6, 6, 2, 6, 4, 1]. Actually, these aren't quite correct, for if you wish to compute the amount of skip based on the point of mismatch, q, you need to adjust these values. Figure 11.9 illustrates the adjustment, and the adjusted values shown in the figure are the "true" *delta2* values.

Sometimes the computed *delta2* skip is greater than the *delta1* skip, as is the case when matching "rosanna" to the pattern "banana". Here the absolute shift due to *delta2* is 6 and the absolute shift due to *delta1* is –1. Sometimes, however, *delta1* is greater, as is the case for matching "havana" to "banana", where *delta2* gives an absolute shift of 2, but *delta1* a shift of 6. Thus, after the match loop, the BM algorithm can choose the maximum of these two values to obtain the greatest shift.

Figure 11.8 *Delta2* shifts for a sample pattern.

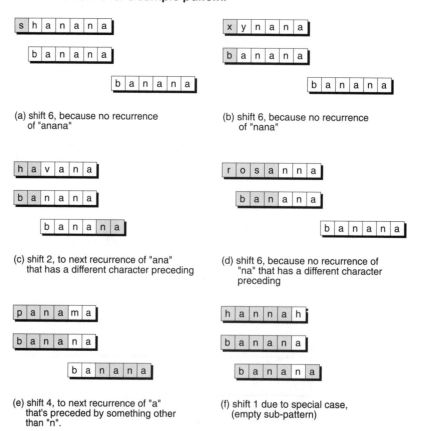

(a) shift 6, because no recurrence of "anana"

(b) shift 6, because no recurrence of "nana"

(c) shift 2, to next recurrence of "ana" that has a different character preceding

(d) shift 6, because no recurrence of "na" that has a different character preceding

(e) shift 4, to next recurrence of "a" that's preceded by something other than "n".

(f) shift 1 due to special case, (empty sub-pattern)

Figure 11.9 Adjusted *delta2* skips.

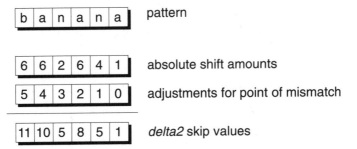

A Boyer-Moore Search Function

We are now in a position to give the complete Boyer-Moore algorithm. We show two functions. The *Preprocess()* function computes the *delta1* and *delta2* skip values, and the *BMSearch()* function does the actual searching, using the sentinel and loop-unrolling optimizations discussed earlier.

```
const int MAX_ALPHABET_SIZE = 256;
const int MAX_PATTERN_SIZE = 80;

void Preprocess(unsigned char *pat, int m, int *delta1, int *delta2)
// Compute the two Boyer-Moore deltas, based on the pattern.
{
  static int f[MAX_PATTERN_SIZE];
  int j, k, q, qp, t, tp;

  // Compute delta1, which determines how much to skip
  // based on what text character caused the mismatch
  // we mismatched on.

  for(k = 0; k < MAX_ALPHABET_SIZE; k++) delta1[k] = m;
  for(k = 0; k < m; k++) delta1[pat[k]] = m-k-1;

  // Compute delta2, which determines how much to skip
  // so we align with the next plausible occurrence of
  // the sub-pattern that did match.
  // Note: This calculation includes Rytter's correction.

  for(k = 0; k < m; k++) delta2[k] = 2*m-k-1;

  for (j = m-1, t = m; j >= 0; j--, t--) {
      f[j] = t;
      while ((t < m) && (pat[t] != pat[j])) {
        if ((m-j-1) < delta2[t]) delta2[t] = m-j-1;
        t = f[t];
      }
  }
}
```

```
    q = t;   t = m-t;

    for (j = 0, tp = 0; tp < t; j++, tp++) {
        f[j] = tp;
        while((tp >= 1) && (pat[j] != pat[tp-1])) tp = f[tp-1];
    }

    qp = 0;
    while (q < m-1) {
      for (k = qp; k <= q; k++) {
          if (m+q-k < delta2[k]) delta2[k] = m+q-k;
      }
      qp = q + 1;
      q = q + t - f[t-1];
      t = f[t-1];
    }
}

int BMSearch(unsigned char *text, int n, unsigned char *pat, int m)
// Finds first occurrence of the pattern of length m
// in the text of length n. Returns position of match,
// or -1 if can't find.
// WARNING: text must have room for n + m bytes!
{
    static int delta1[MAX_ALPHABET_SIZE];
    static int delta2[MAX_PATTERN_SIZE];
    static unsigned char save_ch[MAX_PATTERN_SIZE];
    unsigned char *t, *et, *p, *ep, *q;
    int k, k1, k2, rv = -1;

    Preprocess(pat, m, delta1, delta2);

    t  = text+m-1;   et = text+n;   ep = pat+m-1;

    // Add in m sentinels at the end of the text. Save what's
    // already there in case it needs to be restored.

    memcpy(save_ch, et, m);
    memset(et, pat[m-1], m);

    while(t < et) {
      // Unrolled scan loop looking for last character match.
      k = delta1[*t];
      while(k){
        k = delta1[*(t += k)];
        k = delta1[*(t += k)];
        k = delta1[*(t += k)];
      }
      if (t >= et) break;
      // Reverse order match loop.
      p = ep;   q = t;
      while(--p >= pat && *p == *--q) { ; }
      if (p < pat) {
         rv = q - text; // Return position of match.
```

```
        break;
    }
    // Compute amount to skip, based on what character mismatched
    // and where the mismatch occurred. Take the larger of the skips.
    k1 = delta1[*q];   k2 = delta2[p-pat];
    if (k2 < k1) k2 = k1;
    t = q + k2;
  }

  memcpy(et, save_ch, m); // Restore text where sentinels resided.
  return rv;
}
```

Code for the Boyer-Moore algorithm is given on disk in the file *bmpgm1.cpp*. Also, the algorithm is given in generator form in the files *bm.h* and *bm.cpp*, with test programs supplied in *bmtst.cpp* and *bmtst2.cpp*.

From the *Preprocess()* function, you can see that computing the *delta1* skips is easy. However, computing the *delta2* skips is non-trivial. It's interesting from a historical perspective that it took several tries before a correct *and* efficient method for computing *delta2* was published in the computer science journals. The method we give uses a correction given by W. Rytter [Rytter 80]. However, we had to convert the computation from 1-based subscripting to the 0-based subscripting used by C++. We hope we got it right!

Like the SFS algorithm, the BM algorithm is $O(n)$ in terms of the number of comparisons made for typical text and patterns. However, due to the large skips that can take place, the BM algorithm can be much faster than SFS. When searching English text for patterns of average length 7, our tests shows over a twofold improvement in speed over the optimized SFS algorithm. With longer patterns, the results are likely to be even better.

The speed of the BM algorithm can be attributed to the *delta1* skipping. The *delta2* skipping is important only in cases where the text and patterns are highly repetitive, or when a small alphabet is used. The *delta2* skipping makes the BM algorithm essentially linear, roughly $O(n + m)$, in the worst case, although if the text contains a large number of matches of the pattern, the complexity is not much better than that for the SFS algorithm.

VARIATIONS ON BOYER-MOORE

While Boyer-Moore searching can be lightning fast, it does have one drawback—that of complexity. In particular, computing *delta2* is tricky and error prone. Several variations of the BM algorithm have emerged over the years. These variations are based on the premise that it's the *delta1* skipping that gives BM most of its speed. Thus many people have proposed simplifying the *delta2* calculations, or simply dropping them altogether. It should be noted that all of these variations are $O(nm)$ in the worst case rather than $O(n+m)$, since the *delta2* calculation is dropped or simplified.

The Tuned-Boyer-Moore Algorithm

In [Sunday 91], D. Sunday gives an excellent survey of different variations of the Boyer-Moore algorithm. One variation that performs better than the original BM algorithm uses a simplified calculation for the *delta2* skips, in what Sunday calls the *mini-delta2 skip* (or *md2* for short). The *md2* skip is based on the first leftward reocurence of the last character of the pattern and is easily computed:

```
// Compute md2 skip for pattern pat of length m.

k = m-1;
unsigned char skipc = pat[k];
while(k > 0) if (pat[--k] == skipc) break;
md2 = m-k-1;
// In case there is no recurrence, shift by 1.
if (md2 == 0) md2 = 1;
```

Note that there is only one *md2* value, not an array of them.

Sunday used the *md2* skip in a modified BM algorithm he called the *Tuned-Boyer-Moore algorithm* (TBM). This algorithm is almost the same as the optimized BM algorithm we gave in the last section, except the *delta2* calculations are replaced with the *md2* calculation and, after the match loop, the next text position is computed simply as $t\mathrel{+}= md2$, rather than by using $t = q + max(delta1, delta2)$. Note that the *delta1* value is still used in the scan loop.

In our tests, the TBM algorithm was about 10 percent faster than the optimized BM algorithm. That's due to the shorter preprocessing time on the pattern, and the shorter calculation when computing a new skip after a failed match. More important though, the TBM algorithm is much simpler and less error prone, since the *delta2* skip values don't need to be computed. Note, however, that the *md2* skip doesn't fully compensate for highly repetitive text as the original *delta2* can, thus it won't be as efficient in these situations.

The TBM algorithm is given on disk in generator form in the files *tbm.h* and *tbm.cpp*. Test programs can be found in *tbmtst.cpp* and *tbmtst2.cpp*.

The Boyer-Moore-Horspool Algorithm

In 1980 R. Horspool published a paper [Horspool 80] giving a variation of the BM algorithm that uses only the *delta1* skipping. This algorithm, known as the *Boyer-Moore-Horspool* (*BMH*) algorithm, performs about as well as the Boyer-Moore algorithm but is much simpler.

Horspool also discovered that when computing the skip after the match loop, a text character corresponding to any position of the pattern could be used to compute the *delta1* skip value. That is, you don't need to use the text character at the point of mismatch. You could, for example, use the text character corresponding to the last character of the pattern. This can sometimes result in larger shifts than that used in the BM algorithm. We won't discuss the BMH algorithm in detail, since the next algorithm, based on BMH, is slightly faster and perhaps easier to understand.

The BMH algorithm is given on disk in generator form in the files *bmh.h* and *bmh.cpp*, with test programs in *bmhtst.cpp* and *bmhtst2.cpp*.

The Quick-Search Algorithm

In 1990 D. Sunday [Sunday 90] extended Horspool's idea and discovered that you can actually use the text character that occurs one position after the pattern. He observed that since every shift is 1 or greater, the text character that follows the end of the pattern must appear in the pattern in order for there to be a match after the shift. Thus Sunday proposed a different skip value to take advantage of this.

Sunday's *delta1* skip, which we'll call the *sd1* skip, is the same as the original BM *delta1* skip, except every value is incremented by 1. This means that *sd1* is never 0 as *delta1* is, so we can use it both in the scan loop and after the match loop without any help from other skip values. Figure 11.10 shows an example of *sd1* skipping during the scan loop. In the first shift, the skip character is "a", whose *sd1* value is 3. For the second shift, the skip character is "o", whose *sd1* value is 1. For the third shift, the skip character is "t", whose *sd1* value is 2. In the final shift, the skip character is "m", whose *sd1* value is 4.

Figure 11.10 *Sd1* skipping during the scan loop.

o	n	i	o	n	m	a	n	g	o	t	o	m	a	t	o

t	o	m	a	t	o

t	o	m	a	t	o

t	o	m	a	t	o

t	o	m	a	t	o

t	o	m	a	t	o

If you compare Figure 11.10 with Figure 11.4 (the same example, but using *delta1* skipping), you'll notice that for the example given, *sd1* skipping requires an extra pass during the scan loop. In this case, the original *delta1* skipping is better. However, in cases where the text character after the pattern does not occur in the pattern, the skip will be $m + 1$ positions using *sd1* compared to m positions using *delta1*. In effect, it's as if we're using a pattern length of $m + 1$ rather than m. For short patterns, the likelihood that the skip character is not in the pattern is higher than in longer patterns. Also, the effect of using $m + 1$ rather than m decreases as m gets larger. Thus the *sd1* skip is likely to perform better than *delta1* for short patterns but to have less of an advantage on longer patterns.

Using *sd1* skipping, Sunday gives an algorithm called *Quick Search* (QS) that we give in the following *QuickSearch()* function.

```
const int MAX_ALPHABET_SIZE = 256;

int QuickSearch(unsigned char *text, int n, unsigned char *pat, int m)
// Sunday's quick-search algorithm. This version returns the position
// of the first match of the pattern, or -1 if not found.
{
  int skip[MAX_ALPHABET_SIZE];
  unsigned char *t, *et, *p, *ep, *q;
  unsigned char first = *pat;
  int k;

  // Compute Sunday's delta (sd1) used in skipping. This
  // delta is Boyer-Moore's "delta1" + 1.

  for (k = 0; k<MAX_ALPHABET_SIZE; k++) skip[k] = m+1;
  for (k = 0; k<m; k++) skip[pat[k]] = m - k;

  t  = text;  et = text+n-m+1;  ep = pat+m;

  while(1) {
    // Scan loop that tries to match the first character of the pattern.
    while (t < et && *t != first) t += skip[t[m]];
    if (t >= et) break;
    // Match loop that works in the forward direction.
    p = pat;  q = t;
    while(++p < ep && *p == *++q) { ; }
    if (p == ep) return t - text;
    t += skip[t[m]];
  }

  return -1;
}
```

 The QS algorithm is given on disk in the file *qspgm1.cpp*. The algorithm is also provided in class form in the files *qs.h* and *qs.cpp*, with test programs in *qstst.cpp* and *qstst2.cpp*.

In this algorithm you'll note that we search for the first character in the pattern during the scan loop, and the match loop works in the forward direction, in contrast to the more unintuitive reverse matching of the original BM algorithm. A big advantage of using *sd1* skip values is that the order the pattern is matched is irrelevant, a fact we'll discuss further in the next section.

If you compare *QuickSearch()* with the *SFSearch3()* algorithm given toward the beginning of the chapter, you'll find that the QS algorithm is just a simple extension of the SFS algorithm, adding two statements to preprocess the pattern and changing the skip value from 1 to *sd1*. The QS algorithm is a jewel. It's not much more complicated than the SFS algorithm, yet our tests reveal it runs at least twice as fast. The QS algorithm runs about the same speed as the BMH algorithm, perhaps faster on shorter patterns, but it's slightly slower than the TBM and BM algorithms.

Part of the reason QS runs slower than BM or TBM is that we didn't optimize it with loop unrolling or sentinel optimization. Actually, loop unrolling doesn't work with the QS algorithm, but we can use sentinel optimization. We must store the first character of the pattern *m* times at the end of the text. However, it's probably not worthwhile to complicate QS further. It seems a shame to corrupt the beauty and elegance of the algorithm this way. Even if optimized, QS is still slower (if only slightly) than either BM or TBM. If you want more speed, use the optimized versions of these other algorithms.

Least-Frequent Character Optimization

There is one final optimization that you should be aware of. Rather than search for the first or last character of the pattern in the scan loop, why not search for the character in the pattern that occurs the fewest times in the text? That way the chance of a match occuring, and the slower match loop executing, is greatly reduced. Horspool proposed using the least-frequent character (LFC) this way in [Horspool 80] with respect to the straightforward search algorithm.

The LFC idea can be adapted to the Boyer-Moore algorithms by inserting a "guard test" between the scan loop and match loop. (These two loops operate as before.) During preprocessing, the LFC of the pattern is found by consulting a table of frequencies representative of the text being searched. Both the value and position of the LFC in the pattern are recorded. After each scan loop, if the text character at the position where the LFC occurs in the pattern doesn't match the LFC, you know there's no reason to execute the match loop. This idea is given in Sunday [Sunday 91].

The idea of using the LFC can be extended in another way. If you use the QS algorithm with *sd1* skipping, it turns out that it doesn't matter which way the match loop executes, forward or backward. In fact, you could use any permutation of the character positions during the matching! One advantageous permutation would be to match the characters in order of increasing character frequencies. That is, the least-frequent character is tested during the guard test, then in the match loop, the second least-frequent character is tested, followed by the third, and so on. By doing this, the chances of the match loop executing for very long are greatly diminished. Sunday gives examples of this in [Sunday 90] and [Sunday 91], showing algorithms that are competitive with the best of the algorithms given in this book.

We won't give details on LFC optimization in this book, because the resulting algorithms are more complicated and aren't as general purpose as the ones we've shown. A table of character frequencies is required, which assumes knowledge of the text being searched. However, if you are searching text with highly skewed character distributions, you might keep LFC optimization in mind, for it could have a dramatic effect on performance.

THE KNUTH-MORRIS-PRATT ALGORITHM

The Boyer-Moore algorithm and its variants are the fastest known algorithms for typical search applications where exact pattern matching is involved. But they have one drawback: They all require the text to be buffered, with a size equal to the length of the pattern. Although the details vary, in each case you must back up in the text either before or during the match loop. Even the QS algorithm, which matches the first pattern character in the scan loop and uses a forward match loop, still looks ahead by *m* characters during the skipping. Another search algorithm that doesn't need to backtrack is the *Knuth-Morris-Pratt algorithm*, (KMP) published in 1977 [Knuth 77].

Like the BM algorithms, KMP uses a skip value, similar to *delta2*, to compute shifts down the text. It's possible to code the algorithm so that each position in the text is only visited once (or skipped over), and the algorithm never backs up in the text. Thus the KMP is a good example of an *on-line algorithm*. This has advantages in certain applications, but unfortunately KMP is much slower than the BM algorithm, mainly because it doesn't have the fast *delta1* skipping. For that reason, we won't describe KMP in detail, although sample code is given on disk.

In practice, our tests show the KMP algorithm to be almost four times slower than even the lowly SFS algorithm, but it can be faster when the text is highly repetitive. To be fair, it should be mentioned that we didn't try optimizing KMP.

 The KMP algorithm is given on disk in generator form in the files *kmp.h* and *kmp.cpp*, with test programs residing in *kmptst.cpp* and *kmptst2.cpp*.

TEXT SEARCHING IN FILES

So far we've shown the text-searching algorithms in ways that assume the text being searched resides completely in memory. A more typical scenario is for the text to reside in files on disk. It turns out that you can extend the algorithms to work with files quite easily, especially when the algorithms are given in generator form, as with the SFS class given earlier. (Remember that all the algorithms are given in generator form on the disk.)

The basic idea to file-based text searching is to read the file in chunks, using a buffer whose size is presumably optimized to the operating system you are using. These chunks are then fed to the search algorithm, to find all occurrences of the pattern being matched. When searching ASCII files, it's common to also want to know which line the matches occurred on, so extra bookkeeping can be employed to keep track of this. Also, when searching ASCII files, it's typical to assume that the patterns don't span lines; when a chunk of text is read in, any partial line at the end of the buffer must be tucked away and searched with the next chunk of text.

Many existing search programs use a strategy quite similar to what we've just given. In particular, a Unix program called *grep* does file-based text searching. There are numerous variants of *grep*, and many use some variation of the Boyer-Moore algorithm as part of their processing. So that you can get an idea of how to write a *grep*-like program, we've included on the disk full source code to a program called *yagrep* (which stands for "yet another *grep*"), which allows you to easily plug in the search algorithms given in this chapter.

 Source to the *yagrep* program can be found on disk in the file *yagrep.cpp*.

SUMMARY

In this chapter we've covered various algorithms for searching text, most of which are based on the Boyer-Moore algorithm. Table 11.1 shows how well these algorithms performed using one particular test, where each algorithm (except KMP) was optimized as much as possible. Each test involved searching for 500 different patterns, with an average length of 7, through a text database of 32,000 characters. (Thus, a total of 16MB of text was processed per test.) The text consisted of words from the Bible, extracted from a larger 1MB database that can be found on the Internet (See Sunday [Sunday 91].) Of the 500 patterns, 103 of them had one or more matches in the small database we used. The machine we used was a 33mhz 486 PC running DOS, with the BC++ 3.1 compiler.

From the table you can see that the TBM algorithm is the best performer, with the optimized BM algorithm coming in second. If you are looking for utmost speed, use the TBM algorithm. If you plan on searching text with small alphabets or highly repetitive text, use the BM algorithm. If you want a nice simple algorithm that's also fast, use the QS algorithm. Also, remember that only the BM algorithm is $O(n + m)$ in the worst case. All others are $O(nm)$.

Table 11.1 Comparison of Searching Algorithms

Algorithm	Speed (MB/sec)
Knuth-Morris-Pratt (KMP)	0.8
Straightforward search (SFS)	3.1
Boyer-Moore-Horspool (BMH)	6.8
Quick Search (QS)	7.1
Optimized Boyer-Moore (BM)	7.9
Tuned Boyer-Moore (TBM)	8.6

Keep in mind that these test results were obtained with an average pattern length of 7. You are likely to get different results with different length patterns. Remember, the longer the pattern, the faster the BM algorithm and its variants are likely to be. Also, the results you obtain will vary from compiler to compiler and machine to machine. But perhaps more important, the type of data you are searching can make a big difference. However, the general trend shown in the table is likely to be the same for all of these cases.

TWELVE

Shift-OR Text Searching

I n Chapter 11 we showed various fast algorithms for searching text, where the patterns were matched verbatim, character by character. In this chapter we introduce an entirely different approach to text searching, known as the *Shift-OR algorithm,* as invented by R. Baeza-Yates and G. Gonnet [Baeza 92]. Surprisingly, this algorithm does not involve character comparisons at all. Instead, the state of the search is maintained in bit vectors by using shifting and OR'ing operations.

By not doing character comparisons directly, the Shift-OR algorithm gains the advantage of allowing more sophisticated patterns. For example, it's easy to support a don't-care code in the pattern, which allows any character to be matched at that position. Also, you can easily specify classes of characters. For example, you might want the first position of the pattern to match any of the vowels "a", "e", "i", "o", "u". The Shift-OR algorithm can also be extended to perform *approximate pattern matching.* You might allow a certain number of mismatches to occur, or allow matches that contain a number of characters that have been deleted or inserted into the pattern. The latter extension is given in a paper by Wu and Manber [Wu 92].

In this chapter we'll take a look at the basic Shift-OR algorithm and then extend it to allow don't-care codes and character classes in the patterns. Extending the algorithm to support generalized approximate pattern matching is more complicated, beyond what we can cover here. However, we will show a simple extension in the form of *wild cards,* which are codes in the pattern that mean "any number of characters can be inserted here."

THE SEARCH STATE VECTOR

The basic idea behind the Shift-OR algorithm is to keep the state of the search in a bit vector we'll call the *search state vector,* or *state vector* for short. Each bit of the state vector corresponds to one position of the pattern. If the pattern is *m* characters long you need *m* bits in the state vector. Usually, the length of the patterns is limited to 32 characters in the Shift-OR algorithm. This allows the state vector to be conveniently stored in a single 32-bit word. Longer patterns can be supported by using multiprecision bit operations, although this will tend to slow the algorithm down.

The bits of the state vector are coded as follows: If the bit in position j is 0, it means that there is a partial match of the pattern up to the jth character. If the bit is 1, it means there is no partial match. Figure 12.1 shows an example of a state vector that represents the partial match of "tom" to the pattern "tomato". Note that to determine whether a partial match exists at position j, you look only at bit j of the state vector. The values of the other bits are of no concern.

EXACT MATCHING

Next we describe how you can use the search state vector to perform exact pattern matching. Initially, the state vector starts out with all 1's, meaning that there are no partial matches. Then, as each text character is examined, the bits in the state vector are changed to reflect any partial matches found. Figure 12.2 shows how the state vector changes when matching the pattern "tomato" to the text "tomcatortomato". If you follow the zero bits, you'll see how several partial matches are found during the search before a complete match is finally found at the end. A complete match is signaled when the rightmost bit becomes 0.

The question, of course, is how to determine algorithmically the bits of the state vector at each search step. Suppose you are given a state vector at step i, where i corresponds to the ith character in the text. Bit j of the vector can be set to zero only if the following two conditions hold: (1) the jth character of the pattern matches the ith character in the text, and (2) bit $j-1$ of the state vector at step $I-1$ is also 0. In order to have a match up to position j, the current pattern and text characters must match, and there has to be a partial match of the previous $j-1$ characters in the previous step.

Figure 12.1 A sample search state vector.

t	o	m	a	t	o

pattern

1	1	0	1	1	1

search state vector representing
the partial match "tom"

Figure 12.2 State vector searching.

	t	o	m	a	t	o		
	1	1	1	1	1	1		initial state
t	0	1	1	1	1	1	0	partial match "t"
o	1	0	1	1	1	1	1	partial match "to"
m	1	1	0	1	1	1	2	partial match "tom"
c	1	1	1	1	1	1	3	
a	1	1	1	1	1	1	4	
t	0	1	1	1	1	1	5	partial match "t"
o	1	0	1	1	1	1	6	partial match "to"
r	1	1	1	1	1	1	7	
t	0	1	1	1	1	1	8	partial match "t"
o	1	0	1	1	1	1	9	partial match "to"
m	1	1	0	1	1	1	10	partial match "tom"
a	1	1	1	0	1	1	11	partial match "toma"
t	1	1	1	1	0	1	12	partial match "tomat"
o	1	1	1	1	1	0	13	complete match "tomato"

Given these observations, it turns out to be surprisingly easy to compute the state vector given its previous value. First, the bits are shifted one position to the right, replacing the leftmost bit with a 0 (that is, a 0 is shifted in). This shifting has the effect of moving the bit used in the previous step to indicate a partial match to the next position. So if bit $j-1$ was 0, now bit j is 0. Replacing the leftmost bit with a 0 handles the case of matching the first pattern character.

Second, the shifted state vector is OR'ed with a bit-mask constructed for the ith character in the text. This mask, which we'll call a *characteristic vector*, has bits of 0 at each position that the current text character appears in the pattern and bits of 1 elsewhere. After OR'ing the shifted state vector with the characteristic vector, bit j of the state vector is 0 only if there is a match up to the jth position of the pattern; otherwise the bit is a 1. It's this shifting and OR'ing that give the Shift-OR algorithm its name. Figure 12.3 shows an example of the shift and OR operations used to construct the state vector for the partial match "tom" after having matched "to" in the previous step.

Figure 12.3 Shifting and OR'ing.

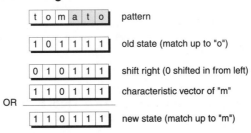

Figure 12.4 Sample characteristic vectors.

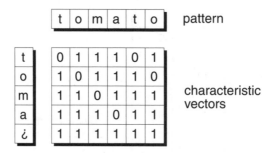

Each character in the alphabet is given a characteristic vector, and these vectors can be computed during a preprocessing step, before the actual search begins. For example, the characteristic vector for each character in the pattern "tomato" is given in Figure 12.4. For those characters not appearing in the pattern, the characteristic vectors consist of all 1s. We use the "¿" symbol to denote all characters not in the pattern.

We are now in a position to write a function that searches text using the Shift-OR algorithm. The following *SOSearch1()* function gives an example.

```
const int MAX_ALPHABET_SIZE = 256;

unsigned long zero_bit[32] = {
  // Such that bit[i] = 0, for i = 0..31;
  0x7FFFFFFFL, 0xBFFFFFFFL, 0xDFFFFFFFL, 0xEFFFFFFFL,
  0xF7FFFFFFL, 0xFBFFFFFFL, 0xFDFFFFFFL, 0xFEFFFFFFL,
  0xFF7FFFFFL, 0xFFBFFFFFL, 0xFFDFFFFFL, 0xFFEFFFFFL,
  0xFFF7FFFFL, 0xFFFBFFFFL, 0xFFFDFFFFL, 0xFFFEFFFFL,
  0xFFFF7FFFL, 0xFFFFBFFFL, 0xFFFFDFFFL, 0xFFFFEFFFL,
  0xFFFFF7FFL, 0xFFFFFBFFL, 0xFFFFFDFFL, 0xFFFFFEFFL,
  0xFFFFFF7FL, 0xFFFFFFBFL, 0xFFFFFFDFL, 0xFFFFFFEFL,
  0xFFFFFFF7L, 0xFFFFFFFBL, 0xFFFFFFFDL, 0xFFFFFFFEL
};

int SOSearch1(unsigned char *text, int n, unsigned char *pat, int m)
// Finds first occurrence of the pattern of length m
// in the text of length n. Returns position of match,
// or -1 if can't find.
{
  unsigned long cv[MAX_ALPHABET_SIZE];
  unsigned long initial_state, state, stop_mask;
  unsigned char *t, *et;
  int i;

  // Preprocess the pattern by setting up a characteristic bit
  // vector for each character in the alphabet. Basically, a
  // zero bit goes wherever the character appears in the pattern.
```

```
for (i = 0; i<MAX_ALPHABET_SIZE; i++) cv[i] = ~0; // All ones
for (i = 0; i<m; i++) cv[pat[i]] &= zero_bit[i];

// In the initial state, no matches have occurred.
initial_state = ~0; // All ones

// Set up a stop mask so that we can easily test for a 0
// in the m-1th bit of the state vector
stop_mask = ~zero_bit[m-1]; // Put a '1' in m-1th posn.

// Do the actual searching.

t  = text;  et = text+n;  state = initial_state;
while(t < et) {
  state = (state >> 1) | cv[*t++];
  if ((state & stop_mask) == 0) return (t - text) - m;
}
return -1;
}
```

The *SOSearch1()* algorithm is given on disk in the file *sospgm1.cpp*.

The *SOSearch1()* function uses an auxiliary array of bit masks, *zero_bit*, to conveniently set bits to zero when setting up the pattern. At the heart of the function is the simple statement

```
state = (state >> 1) | cv[*t++]
```

which performs the basic Shift-OR operation. Note that we are relying on the fact that in C++, a right shift causes a 0 to be shifted into the leftmost position.

In order to determine when a complete match is found, the state vector is tested against the variable *stop_mask*. This mask has a 1 in the $m - 1$th bit and 0s elsewhere. Thus, if the mask is AND'ed with the state vector, then the result will be 0 only if the $m - 1$th bit of the state vector is 0, indicating a match.

We have reversed the shifts and the order of the bits in the vectors from what is given in the original paper by Baeza-Yates and Gonnet. They used left shifts and stored the bits from right to left, the rightmost bit corresponding to the leftmost character in the pattern. Our way is more intuitive and easier to explain.

It's also possible to invert the sense of the bits, using 1s to represent a match and using an AND operation instead of an OR operation. The resulting algorithm is called the *Shift-AND* algorithm and is what is used in the paper by Wu and Manber. This is perhaps the most intuitive representation of all. Unfortunately, it requires an extra operation during the shifting, because a 1 must be placed into the leftmost position rather than a 0. This can be accomplished by OR'ing the shifted state vector with a mask having a 1 in the leftmost bit. For example:

```
// Shift-AND operation
state = ((state >> 1) | ~zero_bit[0]) & cv[*t];
```

OPTIMIZING THE SHIFT-OR ALGORITHM

As you can see from examining *SOSearch1()*, the basic Shift-OR algorithm is fairly simple. The Shift-OR operation in the inner loop is easy to compute, but it's not as fast as doing simple character comparisons. Because of this, the *SOSearch1()* function is much slower than the Boyer-Moore algorithms given in Chapter 11. In fact, our tests show *SOSearch1()* to be about four times slower than even the optimized straightforward search algorithm!

You can speed up the Shift-OR algorithm easily by splitting the basic search loop into a *scan loop* and *match loop*, a technique we discussed in Chapter 11. The idea is to search, as quickly as possible, for the first character in the pattern using character comparisons in a tight loop. Only then do you utilize the Shift-OR operations to match the rest of the pattern. Along with the scan loop, you can also place the first character of the pattern at the end of the text to serve as a sentinel. The following *SOSearch2()* function illustrates these optimizations.

```
int SOSearch2(unsigned char *text, int n, unsigned char *pat, int m)
// The Shift-OR algorithm with scan loop and sentinel optimization
{
  unsigned long cv[MAX_ALPHABET_SIZE];
  unsigned long initial_state, state, stop_mask;
  unsigned char *t, *e, *et;
  int i;

  unsigned char first = pat[0];
  unsigned char save_ch = text[n];
  text[n] = first; // Install sentinel.

  for (i = 0; i<MAX_ALPHABET_SIZE; i++) cv[i] = ~0; // All ones
  for (i = 0; i<m; i++) cv[pat[i]] &= zero_bit[i];

  initial_state = ~0; // All ones
  stop_mask = ~zero_bit[m-1]; // Put a '1' in m-1th posn.
  t  = text;  e = text + n - m;  et = text+n;

  while(1) {
    // Use fast scan loop to look for first character of pattern.
    while(*t != first) t++;
    if (t > e) break;
    // Match loop, using state vector. We get out of this loop if
    // we find a match, or we run off the end, or we've reached a
    // point where we need to search for first character again.
    state = initial_state;
    do {
      state = (state >> 1) | cv[*t++];
      if ((state & stop_mask) == 0) return (t - text) - m;
    } while(state != initial_state && t < et);
  }
```

```
    text[n] = save_ch; // Remove sentinel.

    return -1;
}
```

The *SOSearch2()* function is given on disk in the file *sospgm2.cpp*. Also, the optimized algorithm is given in generator form, similar to the search generators of Chapter 11, in the files *sos.h* and *sos.cpp*, with test programs in *sostst.cpp*, *sostst2.cpp*, and *sostst3.cpp*. The last test program lets you see the state vector as it changes during the search.

In the match loop, not only is a test made for the state vector reaching the stopping point, there is also a test for the initial state being encountered again. When the latter occurs, there are no partial matches currently taking place, so we might as well go back up to the scan loop and quickly search for the first character. Without this test, once the match loop is entered, the code would stay there for the rest of the algorithm, basically disabling the scan loop.

In Chapter 11 we noted that using a scan loop is an important optimization technique. Nowhere is that more apparent than with the Shift-OR algorithm, for our tests show nearly a *threefold improvement in speed* when a scan loop is used.

However, the Shift-OR algorithm can't compete with the fast skipping employed by the Boyer-Moore algorithms. But performing exact matching is not the reason for using the Shift-OR algorithm in the first place. As you'll see next, the algorithm can be extended easily to allow more sophisticated searching, something the Boyer-Moore algorithms can't claim.

CHARACTER CLASS MATCHING

Rather than use direct character comparisons, the Shift-OR algorithm relies on characteristic bit vectors to control the state of the search. As long as a characteristic vector has a 0 in a given position, then a match can occur at that position for the corresponding text character. In exact matching, only one characteristic vector has a 0 at a given pattern position, since there is only one character that can match at that position. However, nothing stops us from having two or more characteristic vectors with 0s at a given position. This allows a position in the pattern to be matched to a group or *class* of characters.

For example, suppose you want to match either of the two patterns "silly" and "sally". That is, in the second position of the pattern, either an "i" or "a" is allowed. To accomplish this, the characteristic vectors for both "i" and "a" should contain a 0 in the second position. Figure 12.5 shows an example. (We use the "ç" symbol to denote a character class.)

A common syntax for representing patterns with character classes is to enclose the characters in brackets. For example, the pattern "s[ia]lly" would

Figure 12.5 Character class matching.

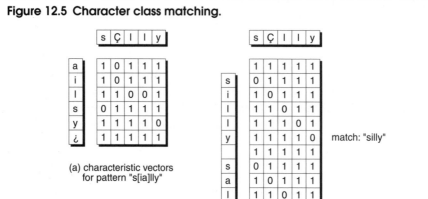

(a) characteristic vectors
for pattern "s[ia]lly"

(b) search states

match both "silly" and "sally". Sometimes you might want to indicate a range of characters in a class. That can be accomplished by using a dash. For example the class "[a-ho-z]" would match the characters from "a" to "h" and from "o" to "z".

You can extend the character class idea further and have a special *don't care* symbol, which we'll denote with "?", that represents a class containing *all* characters. Another idea is to use the complement of a character class, which we'll denote by placing a "^" symbol as the first character of a class. For example, the pattern "[^aeiou]" matches all characters that are not vowels.

Another useful idea is to allow for case-insensitive searches. We'll use a pair of "@" symbols to enclose a string of characters that can match both upper and lower case. For example, the pattern "@the@" can match the strings "The", "the", "THE", among others. A pattern to match all upper- and lower-case vowels would be denoted "@[aeiou]@". As with the other character classes, handling both upper and lower case can be accomplished easily by setting the appropriate characteristic vectors. Figure 12.6 shows an example of searching for the pattern "@g[eo]?@" in the text "GOD in Genesis".

You'll notice from Figure 12.6 that the length of the original pattern is not the same length as the characteristic and state vectors. The symbols "@", "[", and "]" are not actually searched for, and each class takes up only one bit position. The pattern shown at the top of figure, "ÇÇ?", is called the *logical pattern,* and in this case has a length of 3, so the vectors must have room for at least three bits. Note that the symbols comprising the logical pattern are of no real consequence, since they are never used in character comparisons. We chose to denote a character class with "Ç" and a *don't care* with "?".

We'll show how to write code for character class matching in the form of a generator class, called *CCS* (short for Character Class Searcher).

Figure 12.6 A complex searching example.

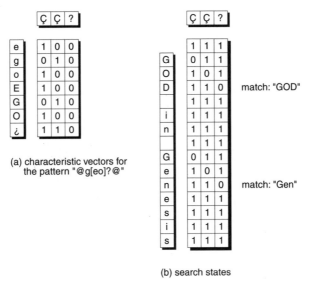

(a) characteristic vectors for the pattern "@g[eo]?@"

(b) search states

```
class CCS {
private:
  enum { MAX_ALPHABET_SIZE = 256 };
  static unsigned long zero_bit[32];
  unsigned long cv[MAX_ALPHABET_SIZE];
  unsigned char *text, *pat, *t, *e, *et;
  unsigned long state, stop_mask, initial_state, dont_care_bits;
  int run_state, n, m, orig_m;
  unsigned char first, save_ch, scan_loop_bypass, doing_at_sign;
  void SetSentinel();
  void SetCV(unsigned char c, int i);
  void SetClassFlags(unsigned char c, char *in_class);
  int ProcessClass(unsigned char *&q, unsigned char *ep, int i);
public:
  CCS();
  int SetPattern(char p[], int m_);
  void SetText(char tx[], int n_);
  void RestoreText();
  int Search();
};
```

 Complete code for the *CCS* class is given on disk in the files *ccs.h* and *ccs.cpp*, with test programs residing in *ccstst.cpp*, *ccstst2.cpp*, and *ccstst3.cpp*.

We'll show only the most relevant functions of the *CCS* class, namely, those that pertain directly to preprocessing the pattern and doing the actual search. Here is the *SetPattern()* function, which is in charge of preprocessing the pattern.

```
int CCS::SetPattern(char pa[], int m_)
// Parse the pattern and set up the state bit masks.
// Returns 1 if parse was successful, else <= 0,
// where abs(return value) indicates position of
// syntax error.
{
  unsigned char *q, *ep;
  int i, rv;

  pat = (unsigned char *)pa;  m = m_;  orig_m = m_;

  if (m == 0) return 0; // Empty pattern

  // If the first position of the pattern specifies more than
  // one character, then we must bypass the fast scan-loop
  // when searching, since we don't have a single character
  // to search for.

  scan_loop_bypass = *pat == '?' || *pat == '[' || *pat == '@';

  // Now set up the characteristic vectors.

  for (i = 0; i<MAX_ALPHABET_SIZE; i++) cv[i] = ~0;
  dont_care_bits = ~0;  // Initialize the don't care positions to "none".

  i = 0;  doing_at_sign = 0; // i keeps track of logical pattern length.

  q = pat;  ep = pat + m;
  while(q < ep) {
    switch(*q) {
      case '\\': // Escaped character
        q++;
        if (q >= ep) return pat - q; // Unexpected end of pattern
      goto ORDINARY_CHAR;
      case '?':
        dont_care_bits &= zero_bit[i++];
      break;
      case '[':
        rv = ProcessClass(q, ep, i++);
        if (rv <= 0) return rv; // Syntax error
      break;
      case ']':
      case '-':  return pat - q; // Unexpected characters
      case '@':
        doing_at_sign ^= 1; // Toggle the flag.
      break;
      ORDINARY_CHAR:
      default:
        SetCV(*q, i++);
    }
```

```
   ++q;
}

if (i == 0) return pat - q; // Empty pattern

// m will now become the length of the "logical" pattern.
// orig_m has the length of the original pattern.

m = i;

// Set up the initial state and stop mask.

initial_state = ~0;
stop_mask = ~zero_bit[m-1]; // Put a '1' in m-1th posn.

// Add in any global "don't care" bits to the characteristic vectors.

if (dont_care_bits != ~0) {
    for (i = 0; i<MAX_ALPHABET_SIZE; i++) cv[i] &= dont_care_bits;
}

first = *pat;    // To help optimize scan loop
SetSentinel();
run_state = 0; // Means we haven't done any searching at all yet.

return 1;
}
```

The *SetPattern()* function uses several helper functions to help it process the pattern. Here are the *SetCV()*, *SetClassFlags()*, and *ProcessClass()* functions.

```
void CCS::SetCV(unsigned char c, int i)
// Sets the ith bit (to zero) of the characteristic vector for
// the character c. Handles case-insensitive searching as well.
{
  if (doing_at_sign && isalpha(c)) {
    c &= ~0x20; // Force to upper case
    cv[c] &= zero_bit[i];
    c |= 0x20;  // Force to lower case
    cv[c] &= zero_bit[i];
  }
  else cv[c] &= zero_bit[i];
}

void CCS::SetClassFlags(unsigned char c, char *in_class)
// Sets the appropriate flags in the in_class array for
// the character c, handling case-insensitive searching
// if need be.
{
  if (doing_at_sign && isalpha(c)) {
    c &= ~0x20; // Force to upper case
    in_class[c] = 1;
    c |= 0x20;  // Force to lower case
```

```
      in_class[c] = 1;
    }
  else in_class[c] = 1;
}

int CCS::ProcessClass(unsigned char *&q, unsigned char *ep, int i)
// Process the pattern starting with a '[' character that's supposed
// to be pointed to by q. Returns 1 if successful doing the parse,
// else <= 0, where abs(return value) is position of syntax error.
{
  char in_class[MAX_ALPHABET_SIZE];
  unsigned int old_ch = ~0U;
  int doing_range = 0, complement = 0;
  int j;

  for (j = 0; j<MAX_ALPHABET_SIZE; j++) in_class[j] = 0;

  q++; // Skip opening bracket.

  // Look for complement character "^" as first character in class.

  if (*q == '^') { complement = 1; q++; }

  while(q < ep) {
    switch(*q) {
      case '\\': // Escaped character
        q++;
        if (q >= ep) return pat - q; // Unexpected end of pattern
      goto ORDINARY_CHAR;
      case '@':
      case '?':
      case '^':
      case '[':
      return pat - q; // Unexpected characters
      case ']':
        // At end of class, so do post-processing.
        if (old_ch == ~0U) return pat-q; // Empty class
        if (complement) {
          for (j = 0; j<MAX_ALPHABET_SIZE; j++)
            if (!in_class[j]) SetCV(j, i);
        }
        else {
          for (j = 0; j<MAX_ALPHABET_SIZE; j++)
            if (in_class[j]) SetCV(j, i);
        }
      return 1; // Parsed successfully.
      case '-': // We're doing a range.
        // We may have an unexpected '-'.
        if (doing_range || old_ch == ~0U) return pat - q;
        doing_range = 1;
      break;
      ORDINARY_CHAR:
      default:
```

```
            if (doing_range) {
                unsigned char z = (unsigned char)(old_ch) + 1;
                for (; z <= *q; z++) SetClassFlags(z, in_class);
                doing_range = 0;
            }
            else {
                SetClassFlags(*q, in_class);
            }
            old_ch = *q;
        }
        q++;
    }

    return pat - q; // Unexpected end of pattern
}
```

The complications caused by handling character classes involve only the preprocessing of the pattern. One of the unique features—and main advantages—of the Shift-OR algorithm is that the code to do the actual searching does not change whether you are doing character class matching or exact matching. That means very little slowdown when searching for more complex patterns. (The only slowdown occurs in preprocessing the pattern, which is usually negligible.)

For example, the following *Search()* function is essentially the same as shown in *SOSearch2()*, except for the fact that we give the code in generator form (so that you can handle searching for multiple occurrences, rather than just the first occurrence). The only other change has to do with the fact that you can't use scan loop optimization when the first character of the pattern involves a class of characters, since you don't have a single character to search for in that case.

```
int CCS::Search()
// Search for next position of pattern in the text.
// Returns position of the match, or -1 if no match.
{
    if (run_state == 999) return -1;        // End of text
    if (run_state == 1) goto CONTINUE_ON;   // Pick up where we left off.

    t = text; e = text+n-m;   et = text+n;
    run_state = 1;

    while(1) {
        // Fast scan loop looking for first character of pattern
        if (!scan_loop_bypass) while(*t != first) t++;
        if (t > e) break;
        // Match loop, using state vector.
        state = initial_state;
        do {
            state = (state >> 1) | cv[*t++];
            if ((state & stop_mask) == 0) return (t - text) - m;
```

```
        CONTINUE_ON:
    } while(state != initial_state && t < et);
  }

  return -1;
}
```

WILD CARD MATCHING

The *don't care* pattern "?" discussed in the last section means "exactly one character goes here, whose value we don't care about." A related type of matching is "zero or more characters can go in this position, whose values we don't care about." We'll call such a match a *wild card* and denote it with the symbol "#".

For example, the pattern "po#t" can match the strings "pot", "poet" and "polecat", among others. It's perfectly legal to have multiple wild cards in the pattern. As an example, the pattern "a#b#c#d" would match all strings that contain the first four letters of the alphabet in order, with other letters possibly intervening. In the implementation given here, we'll ignore consecutive wild cards, and we'll also throw out any leading or trailing wild cards, since they are not very meaningful. That means a pattern like "#p##t#" would be reduced to "p#t".

Handling wild cards is significantly different from handling character classes. Each character class takes up exactly one position in the logical pattern, so the match size is fixed. However, with wild cards, the match size is variable length. The wild card may take up many characters in the matching set, or none at all.

Fortunately, it's not too difficult to extend the Shift-OR algorithm to handle wild cards. The technique we will give is an adaptation (actually a simplification) of a method developed by Wu and Manbar [Wu 92]. They give a technique for extending the Shift-OR algorithm to handle generalized approximate pattern matching, where the match is allowed to have a specified number of substitutions, insertions, and deletions. This extension involves using more than one search state vector—in fact, an array of them. Handling wild cards is a special case of allowing insertions to occur in the match. As it turns out, you can modify the Shift-OR algorithm to handle wild cards by making a simple modification to the match loop and setting up one additional mask. You don't need to use an array of state vectors.

Suppose you are matching the pattern "po#t" to the text "polecat". The "#" symbol basically tells us that, once "po" is matched, then any characters that follow in the text are valid matches, with a "t" character terminating the match. For example, the characters "l", "e", "c", and "a" in the text "polecat" are valid once the "o" is matched.

Unfortunately, resetting the corresponding bits in the characteristic vectors for "l", "e", "c", "a", to indicate this wild card matching won't work, since no single bit position that can be used. However, you *can* zero the third bit in the characteristic vector for "t" to signal a complete match. This suggests that part of the solution is to use a logical pattern of "pot". That is, the wild card doesn't take up any space in the logical pattern.

To support wild card matching completely, the normal Shift-OR operation performed in the match loop can be modified as follows: First, calculate the characteristic and state vectors as before, and then AND the state vector with the result of another calculation, whose purpose is to override the search to include the effect of wild cards. This can be done by performing the calculation into two parts. The first half, which we've named *A*, is the normal Shift-OR calculation.

```
A = (state >> 1) | cv[*t]; // Normal Shift-OR
```

The bit mask *A* contains the search state as if there weren't any wild cards. A second bit mask is used, which we'll call *B*, that contains bits of 0 wherever there is a valid wild card match. This mask can be derived from the previous state vector as follows.

```
B = state | wc_mask;
```

The wild card mask, *wc_mask*, consists of all 1s, except at positions w_1, w_2, ..., w_w, where each w_j is the position j of a character that immediately precedes a wild card in the pattern. The positioning is given in terms of the logical pattern. For example, the pattern "po#t" has a logical pattern of "pot" and a wild card mask of "101". The pattern "p#t#o" has a logical pattern of "pto" and a wild card mask of "001". Note what happens if we OR the wild card mask "101" (for the pattern "po#t") with the value of the state vector from the previous step. If there was a match at the second position (that is, we had matched the "o" in the last step, or we're continuing a previous wild card match), then the state vector would also have a 0 in the second position. The result of the OR would leave this bit 0. If, however, there was no match at the second position in the previous step, the result of the OR would leave a 1 in the second bit. In other words, the OR operation serves to "enable" wild card matching at the appropriate time.

The new state vector can be computed by ANDing the two halves together.

```
state = A & B;
```

Figure 12.7 shows an example of matching the pattern "po#t" in the text "possumpolecat." Notice that the effect of the wild card is to "stretch out" the search in the second bit position. It's as if the search is put "on hold" until the terminating "t" is found.

Here is an example of a *Search()* function for a wild card search generator (which we've named *WCS*) that incorporates the use of a wild card mask. The wild card mask is easy to construct, so we won't show how to do that here.

Figure 12.7 Wild card matching.

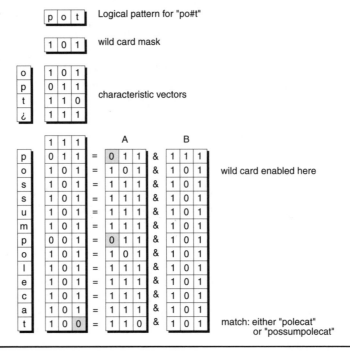

```
int WCS::Search()
// Search for next position of pattern in the text.
// Returns position of the match, or -1 if no match.
{
  long A, B;

  if (run_state == 999) return -1;
  if (run_state == 1) goto CONTINUE_ON;

  t = text; e = text+n-m;  et = text+n;
  num_probable_starts = 0;
  run_state = 1;

  while(1) {
    // Fast scan loop looking for first character of pattern.
    if (!scan_loop_bypass) while(*t != first) t++;
    if (t > e) break;
    // Match loop, using state vector
    state = initial_state;
    do {
      if (*t == '\n') { // For line oriented searching. Otherwise remove.
        t++; num_probable_starts = 0;
        break; // Restart the search at the next line.
      }
      A = (state >> 1) | cv[*t];  B = state | wc_mask;
```

```
      if ((A & one_bit[0]) == 0) {
         probable_starts[num_probable_starts++] = t - text;
      }
      state = A & B;
      if ((state & stop_mask) == 0) {
         if (doing_wild_cards) {
            int start = ClosestMatch();
            if (start == -1) return -1; // We have a problem.
            return start;
         }
         else return (t - text) - m + 1;
      }
      CONTINUE_ON:
      t++;
   } while(state != initial_state && t < et);
}

return -1;
}
```

Complete code for the *WCS* class is given on disk in the files *wcs.h* and *wcs.cpp*, with test programs in *wcstst.cpp*, *wcstst2.cpp*, and *wcstst3.cpp*.

There is one thorny problem with wild card searching. The Shift-OR algorithm makes it easy to determine that a match exists, but the problem is, you only know where the *end* of the match is. Because the actual pattern being matched is of variable length, due to the wild cards, you don't have an easy way of determining where the *start* of the match is. This is in contrast to the exact and character class matching algorithms. With these, the patterns are fixed length, so to compute the start of the match, you can look back $m'-1$ characters from the ending position, where m' is the length of the logical pattern.

In wild card patterns, there are clues to where the match start begins. For one thing, you know the match must contain at least m' characters. Also, if you look closely at Figure 12.7, you'll notice that the two plausible starts for the pattern "po#t" occur when bit 0 of the *A* mask is 0, indicating we've matched the first character of the pattern, which is "p". This suggests that a list should be kept of all the plausible starting positions during the search, and when a match is found, you can consult this list to find the extent of the match.

Note that you can't use the first bit position of the state vector itself to determine plausible starts. For example, consider what happens if the pattern is "p#t". After matching "p" in the text, then for all steps afterward, the state vector will have a 0 in the first bit position. Thus virtually all steps in the search would be considered a plausible start, when clearly they are not. The *A* mask, however, does not contain this effect of the wild card matching.

In the example in Figure 12.7, you'll notice that there two possible outcomes. The match can be either "polecat" or "possumpolecat". Both of these strings match the pattern "po#t". In general, whenever wild cards are allowed, you raise the possibility of more than one outcome and, worse yet, having matches that overlap. This can be handled several different ways. One is to report all possible outcomes. Another method is to report only the longest match. And yet another idea is to report only the shortest or closest match. All of these choices have merit at one time or another, depending on what your application is. In the *Search()* function of *WCS*, we've chosen to report the closest match, since that's the fastest and easiest one to compute.

There is a good reason we talk in terms of "plausible starts" instead of just "starts." Not every plausible starting position can lead to a match. For example, consider matching the pattern "po#t" to the text "ppot". Here there are two plausible starts: at the first and second positions, since both positions contain a "p". But only the second start position can lead to a match.

If you wish to find the closest matching position, an easy if inefficient way to determine whether a plausible start is a valid start is simply to repeat the search, starting at the closest plausible start. If the state vector shows a complete match by the time you reach the previously recorded end position, then you know you have a valid start. Otherwise, you must try the next plausible start (which would be a longer pattern). The following *ClosestMatch()* function, part of the *WCS* class, illustrates this technique in reporting the closest start position.

```
int WCS::ClosestMatch()
// This function is to be called after finding the end of a match.
// It goes backwards through the list of plausible starts to find
// the closest one that actually leads to a match. Returns offset
// of this starting position, or -1 if we have some trouble.
{
  unsigned char *ts;
  unsigned long S;
  int x = num_probable_starts;

  while(1) {
    if (x == 0) return -1; // What? No starts were valid?
    ts = text + probable_starts[--x]; // Try next start.
    S = initial_state;
    do {
      S = ((S >> 1) | cv[*ts]) & (S | wc_mask);
      if ((S & stop_mask) == 0 && ts == t) return probable_starts[x];
      ts++;
    } while(S != initial_state && ts <= t);
  }
}
```

Be sure to remember that this technique works only for finding the closest match. It is not guaranteed to work for the other possible matches.

There are probably more elegant and efficient methods for determining the start positions of a match involving wild cards, but your author does not know of them. Note that in some cases, you won't care what the start positions are. For example, you might want to add wild card matching to the search program *yagrep* that we gave in the last chapter. There, we are only concerned about what *line* a match occurred on, and the ending position is enough information to determine that.

By the way, you'll notice a test for the end-of-line character in the first portion of the match loop in the *Search()* function.

```
while(1) {
  if (!scan_loop_bypass) while(*t != first) t++;
  if (t > e) break;
  // Match loop, using state vector:
  state = initial_state;
  do {
   if (*t == '\n') { // For line-oriented searching
      t++; num_probable_starts = 0;
      break; // Restart the search at next line.
   }
   // Rest of match loop here ...
  } while(state != initial_state && t < et);
}
```

This test avoids having wild card matches that span multiple lines. More than likely, you would like the matching to restart at the beginning of each line. You could add this same test to the character class matching and exact matching algorithms as well.

SUMMARY

The Shift-OR algorithm is perhaps the fastest known algorithm for finding approximate matches of complex patterns. When doing exact matching, the basic algorithm, when optimized, is about twice as slow as the optimized SFS algorithm given in the last chapter. However, adding the ability to handle character classes reduces the speed of the algorithm only slightly. Extending the algorithm still further to handle wild cards causes only about a 10 percent further reduction in speed. All in all, the algorithm is quite fast at handling character classes and wild cards, compared to the other known techniques for handling these types of searches.

One drawback of the Shift-OR technique is that, as we've given it in this chapter, the patterns are limited to 32 characters (32 being the word size of many of today's machines). Handling longer patterns is fairly easy to do (you need to use multiprecision bit operations), but it can slow down the algorithm significantly. For many applications, however, a maximum pattern length of 32 characters is not much of a problem.

There are other methods for handling complex patterns, the most notable one being *regular expression searching*. A *regular expression* is like a mini-programming language that allows you to specify patterns elegantly. For example, the regular expression "(a|b)*c" matches strings consisting of any number of "a"'s or "b"'s, in any order, followed by a "c". This type of pattern matching is not possible using the simple forms of the Shift-OR algorithm given here.

Unfortunately, regular expressions are complicated to implement, and are beyond the scope of this book. It should be noted that Wu and Manber have extended the Shift-OR algorithm to handle regular expressions in a program called *agrep*, available on the Internet. The resulting algorithm is quite messy, though.

THIRTEEN

Graph Data Structures

For the remainder of the book we'll work with *graphs*, probably the most general of all non-contiguous data structures. A graph is a set of nodes, commonly called *vertices*, linked together by a set of arcs, commonly referred to as *edges*. An example of a graph is given in Figure 13.1.

Graphs can be used anytime you have applications involving objects that have some type of connection between them. For example, you might have a set of cities connected together by roads, and you wish to determine the least expensive path between these cities, or even whether a path exists at all. The so-called information highway could also be represented as a graph, being a network of computer sites linked together in various configurations. A common question you might have for such a network is, if a given site "goes down," which of the other sites are affected?

In this chapter we'll take a look at the basics of graphs and show some of the numerous ways of representing them. The representations we give are perhaps the most common. With a little imagination, you can probably think of

Figure 13.1 A sample graph.

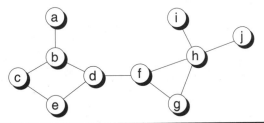

other ways to represent graphs. Understanding the data structures behind graphs will pave the way to the graph algorithms given in the chapters to come.

GRAPH TERMINOLOGY

Formally, a *graph* G is a set of *vertices* V and a set of *edges* E that connect the vertices. Each edge is a pair of vertices, possibly with a *label* associated with it. Graphs with labeled edges are called *labeled graphs*. Quite often the labels are numerical values, representing in some fashion the "cost" or "weight" associated with the edge. Graphs with these types of edges are called *weighted graphs*. For example, you might have a graph representing a network of roads between cities, where the weights are distances between cities. Figure 13.2 shows an example of a weighted graph.

The pair of vertices forming an edge can be ordered. Such an edge is called a *directed edge*, and a graph with directed edges is called a *directed graph*, or *digraph*. An example of a digraph would be a set of one-way streets in a city. The directed edges can also have weights, as illustrated in the digraph shown in Figure 13.3. In this figure, the arrows show the directions of the edges. Note that you can get from vertex A to vertex B, but not from B to A.

Note In this and the following chapters, the term *graph* means an undirected graph, unless otherwise indicated. Typically we'll use the term *digraph* when a directed graph is called for.

Figure 13.2 A weighted graph.

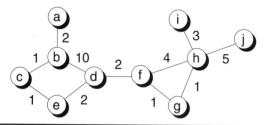

Figure 13.3 A weighted digraph.

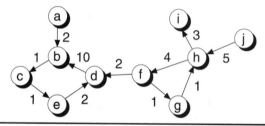

In simple digraphs, there is never more than one edge going in the same direction between two vertices. A digraph that allows more than one edge in the same direction between two vertices is known as a *multigraph*. In an undirected graph, simply having more than one edge between two vertices makes the graph a multigraph. Figure 13.4 shows an example of a directed multigraph. Note how there are two edges going the same direction from vertex 2 to vertex 3.

One application of multigraphs is in representing *finite state automata*, which we'll discuss in Chapter 16. Here each vertex is the state of a machine, and each edge represents a transition from one state to another. The labels represent inputs to the machine, and thus control which edge is traversed. For example, the graph in Figure 13.4 represents a finite state automaton that recognizes the patterns "ride" and "ryde". A transition from the state2 (vertex 2) to state 3 (vertex 3) occurs when either an "i" or a "y" is found.

Note Until Chapter 16 the data structures and algorithms we'll use assume that the graphs being processed are not multigraphs. We will use multigraphs only in Chapter 16.

One of the most fundamental queries one might make to a graph is whether one vertex is connected to another. A *path* is a set of vertices and edges connecting to vertices together. Going back to our first graph in Figure 13.1, one path between *a* and *j* is *a-b-d-f-g-b-j*. A simple path is a path where no two vertices are repeated, such as the one just given. A *cycle* is a path where the first and last vertices are repeated. For example, the path *b-c-e-d-b* forms a cycle in the graph of Figure 13.1.

If a graph has no cycles, it is known as an *acyclic graph*. If the graph is also directed, then you have a *directed acyclic graph*, or *dag* for short. Figure 13.5 shows an example of a dag.

One use of dags is in representing a set of academic courses and their prerequisites. For example, the vertices of Figure 13.5 could represent courses in a computer science curriculum. The arrows represent prerequisites. For example, the two direct prerequisites for course *e* are courses *b* and *c*.

Dags can also be used to represent the *activity network* of a project. Here the vertices are different stages of the project, the edges are tasks to be completed, and the weights on the edges are the times needed to complete the

Figure 13.4 A multigraph.

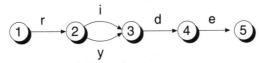

Figure 13.5 A directed acyclic graph.

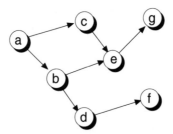

tasks. You might want to answer questions such as "What's the earliest time we can complete the project?" A whole host of algorithms and techniques can be used to work on activity networks, such as critical path analysis. We won't be covering them in this book, but the reader is referred to Horowitz and Sahni [Horowitz 90].

If no vertices in a dag have more than one incoming edge, then you have a *tree*. The dag in Figure 13.5 is not a tree, for vertex *e* has two incoming edges from vertices *b* and *c*. If we were to remove one of the edges, say from *c* to *e*, then we would have a tree, as shown in Figure 13.6.

▼ Note Trees are covered at length in the companion volume.

It turns out that, due to the basic properties of a tree, there will always be one, and only one, vertex that has no incoming edges. (It's also assumed here that the graph is *connected*, a topic we'll discuss in the next chapter.) The vertex having no incoming edges is called the *root*.

You may not recognize the graph in Figure 13.6 as a tree at first, since we've drawn the root, *a*, on the left. Normally trees are drawn with the root on top. This brings up an important point. The pictorial representation of a graph in no way defines the graph uniquely. For example, Figure 13.7 gives two different pictorial representations of the same graph.

Figure 13.6 A tree.

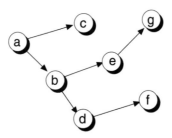

Figure 13.7 Two representations of the same graph.

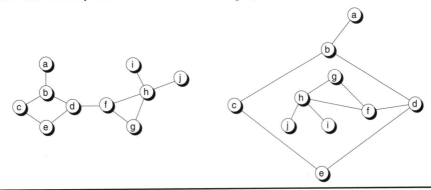

REPRESENTING GRAPHS IN C++

Next we'll investigate different ways of representing graphs in C++. There are two issues involved: how to represent the vertices and how to represent the edges. The most common way of representing vertices is to simply assign them unique numbers. These numbers can be used as handles to index other auxiliary data structures that provide more information about the vertices.

Often an auxiliary data structure, known as a *symbol table*, is used to map vertex handles into vertex names and vice versa. Along with storing the names of the vertices, the symbol table may also store other information regarding the vertex. For example, if the graph is representing cities around the globe, you might store the longitude and latitude of each city, the nationality of the city, and so forth.

A set of classes that implement symbol tables for the purposes of constructing graphs can be found on disk in the files *strbuff.h*, *strbuff.cpp*, *strarr.h*, *strarr.cpp*, *symtab.h*, and *symtab.cpp*.

In our graph representations, we won't be concerned directly with symbol tables; instead we'll be working exclusively with vertex handles. We've typed the handles as unsigned integers. It's convenient to give the handle value 0 a special meaning, representing a "null" vertex. Thus, in the code given in this book, valid vertex handles are ≥ 1.

ADJACENCY MATRICES

There are two popular ways to represent and store edges in a graph; a contiguous representation and a non-contiguous representation. We'll study the contiguous representation first.

Figure 13.8 Adjacency matrix for digraph in Figure 13.3.

	a	b	c	d	e	f	g	h	i	j
a	2									
b		1								
c				1						
d	10									
e			2							
f			2		1					
g						1				
h					4			3		
i										
j							5			

The most straightforward way to implement a graph is to use a two-dimensional array known as an *adjacency matrix*. An adjacency matrix contains $n \times n$ elements, where n is the number of vertices in the graph. Figure 13.8 shows the adjacency matrix for the digraph of Figure 13.3.

The elements of the matrix represent the edges between vertices. For example, in a graph G, $G(i,j)$ returns the edge going from vertex i to vertex j. If the graph is unlabeled, the edge labels can simply be booleans, where a 1 means the vertices are connected and a 0 means they are not. If the edges have labels, then the element types of the matrix are the label types. In a typical scenario, numerical types are used, where 0 means no connection and non-zero values give the weight or cost of an edge. Sometimes the value 0 might be a valid weight or cost. In cases like this, you could use a negative value to indicate no connection. Alternatively, a very high value could be used.

In Figure 13.8 we've represented non-existent edges with blanks, with the understanding that some appropriate numerical value should be used instead.

In an adjacency matrix it's easy to find all outgoing edges of a vertex. Merely scan the row for the corresponding vertex. Likewise, you can find all incoming edges by scanning the appropriate column. The number of outgoing edges from a vertex is known as the *out-degree* of the vertex. The number of incoming edges is known as the *in-degree* of the vertex.

An example of a class implementing adjacency matrices is given in the following *AGraph* class declaration.

```
template<class TYPE>
class AGraph {
protected:
  TYPE *wts;        // The edge data itself
  TYPE no_edge;     // Return edge value for no edge
  unsigned nv;      // Number of vertices in use
  unsigned nv_alloc; // Vertex capacity
  int is_digraph;   // Directed/undirected graph flag
```

```
  unsigned Alloc(unsigned n);
  void Copy(const AGraph<TYPE> &g);
public:
  AGraph(unsigned n, int is_digraph_ = 1, const TYPE &no_edge_ = 0);
  AGraph(const AGraph<TYPE> &g);
  ~AGraph();
  void operator=(const AGraph<TYPE> &g);
  void Clear(const TYPE &x);
  void Clear();
  unsigned NewVtx();
  void SetSize(unsigned n);
  void SetSizeToMax();
  int SetEdge(unsigned a, const TYPE &x, unsigned b);
  int RmvEdge(unsigned a, unsigned b);
  const TYPE *AdjEdges(unsigned i) const;
  const TYPE &operator()(unsigned i, unsigned j) const;
  const TYPE &NoEdge() const;
  int ValidVtx(unsigned v) const;
  unsigned NumVertices() const;
  unsigned NumEdges() const;
  void InDegrees(unsigned *in_deg) const;
  void OutDegrees(unsigned *out_deg) const;
  unsigned Capacity() const;
  int IsDigraph() const;
};
```

 Complete code for the *AGraph* class is given on disk in the files *agraph.h* and *agraph.mth*. Also, a simple test program can be found in *agtst.cpp*.

The implementation of the *AGraph* class is fairly straightforward (if you are comfortable with intermediate C++ programming, that is), so we won't show many of the details here. The basic idea is to allow the size of the graph (the number of vertices) to be specified at runtime. We do this by dynamically allocating the matrix as a one-dimensional array and then providing member functions that make the array look two-dimensional.

Since it's not convenient to resize two-dimensional matrices once they've been built, in the *AGraph* constructor you specify the maximum size of the graph (that is, the maximum number of vertices allowed). The graph starts out with no "logical" vertices in use. To add a vertex, you must use the *NewVtx()* function. This may seem odd, but it's done to make the *AGraph* class behave like its linked-list *LGraph* counterpart that you'll be seeing later on. For convenience, the functions *SetSizeTo()* and *SetSizeToMax()* are provided to specify the number of vertices in use quickly.

Note One common mistake is to forget to call either *NewVtx()*, *SetSizeTo()*, or *SetSizeToMax()* after constructing an *AGraph* object. You'll end up with a graph with no vertices. The symptoms of this are algorithms that don't seem to do anything.

The template *TYPE* parameter specifies the type you would like to use for the edge labels. It's intended that this type be numeric, although with minor modifications, this restriction can be removed. One of the places that assumes a numeric type is in the member variable *no_edge*. This variable gives the value that is used to indicate no connection between two vertices. The default value of *no_edge* is 0, but you can change it to something else more appropriate if need be.

The workhorse functions of *AGraph* are *SetEdge()*, *RmvEdge()*, *operator()*, and *AdjEdges()*. All of these functions work with unsigned integers for vertex handles. The first two functions are used to add or remove edges from the graph. The *AGraph* class can be used to represent both directed and undirected graphs. The *is_digraph* flag indicates which. In the case of undirected graphs, an edge weight is actually stored twice. For example, if there is an edge between vertices *a* and *b*, then the matrix elements *g(a, b)* and *g(b, a)* both will be set to the same edge weight. This can easily be seen by examining the *SetEdge()* function.

```
template<class TYPE>
int AGraph<TYPE>::SetEdge(unsigned a, const TYPE &x, unsigned b)
// Adds or updates an edge from a to b, with edge weight x.
// Returns 1 if successful, else 0 if a and b are invalid.
{
#ifndef NO_RANGE_CHECK
   if (!ValidVtx(a) || !ValidVtx(b)) return 0;
#endif
   wts[a*nv_alloc + b] = x;
   if (!is_digraph) wts[b*nv_alloc + a] = x;
   return 1;
}
```

Notice the subscript calculations used to compute the actual one-dimensional subscript from the two-dimensional subscripts. The matrix is stored in row-major order, similiar to that used in the matrices given in Chapter 7 of the companion volume. Also notice the calls to *ValidVtx()*. For a vertex to be valid, it must be greater than zero and less than or equal to the number of vertices in use in the graph. The calls to *ValidVtx()* are optional, being controlled by the same NO_RANGE_CHECK macro we used in the array classes in the companion volume.

The *operator()()* function makes the graph look like a two-dimensional matrix. For example, if you have an *AGraph* object *g*, then *g(i, j)* returns the edge weight between the *i*th and *j*th vertices. Here is the definition of *operator()()*.

```
template<class TYPE>
const TYPE &AGraph<TYPE>::operator()(unsigned a, unsigned b) const
// Returns the edge wt between a and b. If there is no connection
// between these two points, then no_edge is returned.
{
```

```
#ifndef NO_RANGE_CHECK
  if (!ValidVtx(a) || !ValidVtx(b)) return no_edge;
#endif
  return wts[a*nv_alloc + b];
}
```

An important feature of this function is that it returns a constant type reference. There is a good reason for this. If we allowed the edge weight that is returned to be modified by the caller, then we would have a problem with undirected graphs. Only one of the two copies of the weight would be updated. If you need to update an edge weight, the *SetEdge()* and *RmvEdge()* functions should be used instead.

In the algorithms to come, the *operator()()* function is not used that much. Using sequential access instead is more efficient. That's what the *AdjEdges()* function is for. Given a vertex handle, this function returns a pointer to the corresponding row of the adjacency matrix. By walking this row using simple pointer incrementing, you can examine all adjacent vertices to the vertex specified. All you need to do is check for edge weights that aren't equal to *no_edge*.

Here is an example of constructing the graph of Figure 13.3. In this example, we'll record the vertex handles in variables that have the same names as the corresponding vertices of the graph. In a real application, you might use a symbol table in conjunction with the graph object, as we do in the sample programs on disk.

```
AGraph<int> g(10, 1); // A digraph with a capacity of 10 vertices

unsigned a = g.NewVtx();   unsigned b = g.NewVtx();
unsigned c = g.NewVtx();   unsigned d = g.NewVtx();
unsigned e = g.NewVtx();   unsigned f = g.NewVtx();
unsigned g = g.NewVtx();   unsigned h = g.NewVtx();
unsigned i = g.NewVtx();   unsigned j = g.NewVtx();

g.SetEdge(a, 2, b);   g.SetEdge(b, 1, c);  g.SetEdge(c, 1, e);
g.SetEdge(e, 2, d);   g.SetEdge(d, 10, b); g.SetEdge(f, 2, d);
g.SetEdge(f, 1, g);   g.SetEdge(g, 1, h);  g.SetEdge(h, 4, f);
g.SetEdge(h, 3, i);   g.SetEdge(j, 5, h);
```

Here's how the *AdjEdges()* function can be used to "walk" all the edges in the graph.

```
// Print out all edges of the graph.

unsigned nv = g.NumVertices();

for (unsigned v = 1; v <= nv; v++) {
    const TYPE *p = g.AdjEdges(v) + 1; // The 1 is needed!
    for (unsigned w = 1; w <= nv; ++w, ++p) {
```

```
        if (*p != g.NoEdge()) {
            cout << v << " -[" << *p >> "]- " << w << '\n';
        }
    }
}
```

We've chosen to represent vertices as unsigned integers, with 0 being the "null" vertex. Because of this, all of the arrays involving graphs are used as if they are 1-based, instead of 0-based as is normal in C++. For example, we use *for loops* like

```
for (v = 1; v <= nv; v++) // For 1-based subscripting
```

instead of the more normal

```
for (v = 0; v < nv; v++) // For 0-based subscripting
```

At the time the graph classes were designed, using 0 to mean "null vertex" seemed intuitively appealing. For one thing, it allowed us to keep the vertex handles "in sync" with the symbol table handles, which were also designed so that 0 meant "null symbol." In retrospect, coding the handles this way may not have been a good idea, for it forces the use of 1-based subscripting, which, if you are not paying attention, can be the source of numerous errors.

We handle 1-based subscripting two different ways. Most of the time, we allocate one more element in each array to accommodate the 1-based subscripting and then simply don't use the 0th element, (or use it for other purposes). However, in the case of the two-dimensional adjacency matrix, this would amount to a lot of wasted space. (Both the first row and first column would be wasted.) Instead, we adjust the pointer to the array elements in an appropriate manner (actually, it is decremented by $nv + 1$ elements), so that $g(1,1)$ points to the 0th array element.

Triangular Adjacency Matrices

When an adjacency matrix is used to represent an undirected graph, there is a lot of wasted space. Due to the bidirectionality of the edges, the adjacency matrix is symmetric about the diagonal. All that's really necessary is to store either the upper or the lower triangle of the matrix. Figure 13.9 illustrates this for the undirected graph of Figure 13.2.

Due to the magic of encapsulation and overloaded operator functions, it's possible to store only the lower half or upper half triangle and still have the matrix look rectangular. The *UAGraph* class (which stands for "undirected graph") implements the technique.

Figure 13.9 Adjacency matrix representations for the graph in Figure 13.2.

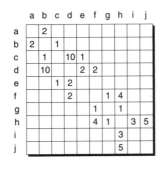

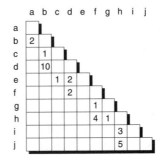

(a) symmetrical form   (b) lower triangular form

 Complete code for the *UAGraph* class is given on disk in the files *uagraph.h* and *uagraph.mth*, with a test program given in *uagtst.cpp*.

The *UAGraph* class has the same public interface as the *AGraph* class. The secret is in how the subscript calculations are performed. Like the adjacency matrix in the *AGraph* class, the elements of the matrix in the *UAGraph* class are stored in row-major order in a single one-dimensional array. However, each row varies in length. We've chosen to use the lower triangle of the symmetrical matrix, so that the first row contains one element, the second row two elements, and so on. Figure 13.10 illustrates this layout.

Given the row-major layout of the triangular matrix, it's not hard to concoct a formula for computing the actual one-dimensional array element for any pair of two-dimensional indices, as is shown in the following *operator()()* function.

```
template<class TYPE>
const TYPE &UAGraph<TYPE>::operator()(unsigned a, unsigned b) const
// Returns the edge weight between a and b. If there is no connection
```

Figure 13.10 Row-major triangular matrix ordering.

(a) triangular matrix   (b) row-major array representation

```
// between these two points, then no_edge is returned.
{
#ifndef NO_RANGE_CHECK
  if (!ValidVtx(a) || !ValidVtx(b)) return no_edge;
#endif
  if (a < b)
     return wts[((b*(b-1)) >> 1) + a];
     return wts[((a*(a-1)) >> 1) + b];
}
```

As before, we are using 1-based subscripting, and the *wts* pointer has been adjusted in other allocation functions to compensate for this. The size of the one-dimensional array is computed by the following equation:

```
nelems = (nv * (nv + 1)) >> 1;
```

Keep in mind that the *operator()()* function actually won't be used that much in the algorithms to come. The real workhorse is the *AdjEdges()* function. Given a vertex, this function computes a pointer to the appropriate row of the matrix. The idea is then to increment this pointer along the row, looking for non-null edges, as you saw in previous examples for *AGraph* objects. With a triangular matrix, though, stepping along a row is more complicated. Figure 13.11 shows what a row looks like in triangular matrix form. As soon as we "hit" the diagonal, the row turns 90 degrees and walks down a column.

It's possible to define a "smart pointer" that incorporates the logic needed to scan a row of a triangular matrix. We sketch the technique in the following *UagPtr* class.

```
template<class TYPE>
class UagPtr {
// Points to a row of a triangular adjacency matrix.
private:
  unsigned i, r, inc;
  TYPE *p;
public:
```

Figure 13.11 A row in a triangular matrix.

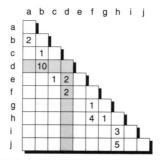

```
    UagPtr(TYPE *row_ptr, unsigned row);
    UagPtr(const UagPtr<TYPE> &s);
    void operator++(int); //Postfix
    void operator++();    //Prefix
    const TYPE &operator*() const;
    operator int() const;
};
```

Complete code for the *UagPtr* class is given on disk in the files *uagraph.h* and *uagraph.mth*.

The *UagPtr* constructor takes a pointer to the beginning of a row and also the row number.

```
template<class TYPE>
UagPtr<TYPE>::UagPtr(TYPE *row_ptr, unsigned row)
{
    i = 1; r = row; inc = 1;
    p = row_ptr;
}
```

With this information, the constructor sets up variables to support the increment operators. Here is the prefix increment operator function, for example.

```
template<class TYPE>
void UagPtr<TYPE>::operator++()
// Prefix operator. We don't return any value for efficiency's sake.
{
    p += inc;                    // inc == 1 until the diagonal is hit.
    if (++i > r) ++inc;          // Past the diagonal
    else if (i == r) inc = r; // At the diagonal
}
```

The variable *i* keeps track of when we hit the diagonal and causes the pointer incrementing to track the appropriate array element.

The *AdjEdges()* function of *UAGraph* returns a *UagPtr<TYPE>* object. Here's how *UagPtrs* can be used to find adjacent edges of a vertex *v*.

```
UAGraph<int> g(10); // Create an undirected graph.
...
UagPtr<TYPE> p = g.AdjEdges(v);
for (unsigned w = 1; w <= nv; ++w, ++p) {
    if (*p != g.NoEdge()) {
        cout << v << " -[" << *p >> "]- " << w << '\n';
    }
}
```

Using the *UAGraph* class for undirected graphs cuts the size requirements for the graph in half, as compared to using the *AGraph* class. However, you do pay a performance penalty in accessing the edges. This is yet another manifestation of the speed vs. space trade-off.

Regardless of whether you optimize adjacency matrices for symmetry, there is still a lot of wasted space for many graphs. Adjacency matrices work best when the graphs are *dense*, that is, when the vertices all tend to be interconnected. That is usually not the case. The more common situation is to have a *sparse* graph, as is the case with all the graphs you've seen in this chapter. Sparse graphs lead directly to sparse adjacency matrices. To handle sparse graphs, perhaps another data structure is appropriate, as we'll discuss next.

ADJACENCY LISTS

A more space-efficient way to represent sparse graphs is to create a linked list for each vertex. Each list has one link for each edge adjacent to the source vertex. A link stores not only the edge *weight*, but a handle to the adjacent, destination vertex as well. The graph itself is made up of an array of these linked lists. Figure 13.12 illustrates the adjacency list for the digraph of Figure 13.3.

Implementing an adjacency list class is fairly straightforward and in fact is quite similar to the hash table with separate chaining class *SHash* given in Chapter 7 (without the hashing, of course). On disk you'll find an adjacency list-based graph class in the class *LGraph*.

```
template<class TYPE>
struct EdgeParm {
  TYPE wt;
  unsigned adjvtx;
  EdgeParm();
```

Figure 13.12 Adjacency list for the digraph in Figure 13.3.

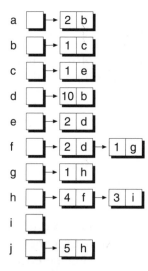

```
    EdgeParm(unsigned av, const TYPE &x);
};


template<class TYPE>
struct EdgeLink : EdgeParm<TYPE> {
    EdgeLink<TYPE> *next;
    EdgeLink(unsigned av, const TYPE &x, EdgeLink<TYPE> *n=0);
};

template<class TYPE>
class LGraph {
protected:
    EdgeLink<TYPE> **vertices;
    TYPE no_edge;
    unsigned def_grow_factor, nv_alloc, nv, ne;
    int is_digraph;
    int SetDirEdge(unsigned a, const TYPE &x, unsigned b);
    ...
public:
    LGraph(unsigned nv_, int is_digraph_ = 1, const TYPE &no_edge_ = 0);
    int SetEdge(unsigned a, const TYPE &x, unsigned b);
    const TYPE *EdgeWt(unsigned a, unsigned b) const;
    const TYPE &operator()(unsigned a, unsigned b) const;
    EdgeLink<TYPE> *AdjEdges(unsigned v) const;
    ...
};
```

 Complete code for the *LGraph* class is given on disk in the files *lgraph.h* and *lgraph.mth*, with a simple test program in *lgtst.cpp*.

The public interface to *LGraph* is virtually identical to that of *AGraph*. The main difference is in the *AdjEdges()* function, which returns the linked list of adjacent vertices rather than a pointer to array elements. Here is some sample code to print out the edges of an *LGraph*.

```
for (v = 1; v <= nv; v++) {
    EdgeLink<int> *el = graph.AdjEdges(v);
    while(el) {
        cout << v << " -[" << el->wt << "]- " << el->adjvtx << '\n';
        el = el->next;
    }
}
```

Like the *AGraph* class, if you are using an undirected graph, then two edges are actually stored for each undirected edge. These edges appear on two separate lists. Here are the *SetDirEdge()* and *SetEdge()* functions to illustrate this fact:

```
template<class TYPE>
int LGraph<TYPE>::SetDirEdge(unsigned a, const TYPE &x, unsigned b)
```

```
// Creates a new directed edge going from vertex a to b, with
// edge weight x. If edge already exists, the weight
// is updated. Returns 1 if successful, or 0 if there
// an allocation error.
{
  EdgeLink<TYPE> *p = vertices[a];
  EdgeLink<TYPE> *q = 0;

  while(p) {
    if (p->adjvtx == b) {
      p->wt = x;
      return 1;
    }
    q = p;
    p = p->next;
  }

  // Edge not currently in the list, so add it.
  p = new EdgeLink<TYPE>(b, x);
  if (p) {
    if (q == 0) vertices[a] = p; else q->next = p;
    ne++;
    return 1;
  }
  return 0;
}

template<class TYPE>
int LGraph<TYPE>::SetEdge(unsigned a, const TYPE &x, unsigned b)
// Creates a new edge going from vertex a to b, with
// edge weight x. If this is an undirected graph, then an edge
// from b to a is also added. If edge(s) already exist(s), the
// weight is updated. Returns 1 if successful, or 0 if a or b
// is invalid or there is a allocation error.
{
#ifndef NO_RANGE_CHECK
  if (!ValidVtx(a) || !ValidVtx(b)) return 0;
#endif
  if (SetDirEdge(a, x, b) == 0) return 0; // Allocation error
  if (is_digraph) return 1;
  // We're undirected, so add edge going the other way.
  return SetDirEdge(b, x, a);
}
```

Like the *AGraph* class, the *LGraph* class stores twice as many edges as are actually needed for undirected graphs. Unlike array-based graphs, though, there doesn't appear to be any simple, efficient way of suppressing the duplicate edges for list-based graphs, as we did in the *UAGraph* class.

You'll notice that we've also included an *operator()()* function in the *LGraph* class, to make the list-based graphs look like a matrix. The *operator()()* function utilizes the workhorse *EdgeWt()* function.

```
template<class TYPE>
const TYPE &LGraph<TYPE>::operator()(unsigned a, unsigned b) const
// This returns the edge weight between a and b, if they
// are connected. If they aren't connected, or either
// vertex is invalid, then no_edge is returned.
{
  const TYPE *p = EdgeWt(a, b);
  return p ? *p : no_edge;
}

template<class TYPE>
const TYPE *LGraph<TYPE>::EdgeWt(unsigned a, unsigned b) const
// Returns a pointer to the weight field of the edge between a and b.
// Returns 0 if no edge exists, or if either a or b is invalid.
{
  EdgeLink<TYPE> *p;

#ifndef NO_RANGE_CHECK
  if (!ValidVtx(a) || !ValidVtx(b)) return 0;
#endif

  p = vertices[a];
  while(p) {
    if (p->adjvtx == b) return &p->wt;
    p = p->next;
  }

  if (is_digraph) return 0; // Couldn't find.

  // We're undirected, so look the other direction.

  p = vertices[b];
  while(p) {
    if (p->adjvtx == a) return &p->wt;
    p = p->next;
  }

  return 0; // No undirected edge either
}
```

The *operator()()* function is not very efficient for list-based graphs. For most algorithms, you'll want to use the *AdjEdges()* function and walk the list of adjacent vertices instead.

We've made the array that's used to hold the linked lists resizable. This, coupled with the natural ability of linked lists to be resized, allows *LGraph* objects to be very flexible in terms of size. In fact, overall, list-based graphs are mostly superior to array-based graphs. Since the majority of graphs are sparse, the list-based form will be more compact than the array-based form, and unless you are trying to use the graph like a matrix, the list-based graphs will be just as fast as the array-based graphs.

However, if the graph is dense, then the array-based representation will be more compact than the list-based representation. Also, you'll see in later chapters that some elegant graph algorithms work best if the graphs are in array-based form.

PACKED GRAPHS

Before we leave this chapter, we'll show yet another way to represent graphs. This way is useful if your graphs are static and you want them to take up the least amount of space possible. Once a packed graph is built, you can't add or remove any vertices. This representation is a hybrid of array-based and list-based graphs. Like the list-based graphs, we store only the adjacent edges to each vertex. However, these edges are stored in a contiguous fashion rather than in a linked-list form. Thus, when walking the adjacent edges of a vertex, simple pointer incrementing is used instead of linked-list traversal. Figure 13.13 shows the packed representation of the digraph of Figure 13.3.

Three arrays are used in the packed graph representation. One array holds the edges, each of which consists of an edge weight and a destination vertex. Another array holds the *out-degrees*, which are the number of outgoing edges for each vertex. Another array holds offsets into the edge array, one offset for each vertex, which gives the starting point in the array for the set of outgoing edges of that vertex.

Figure 13.13 Packed representation for the digraph in Figure 13.3.

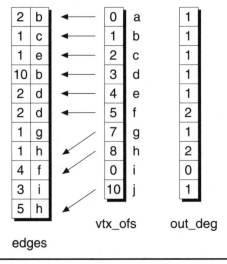

edges   vtx_ofs   out_deg

The packed graph form is implemented with the following *PGraph* class.

```
template<class TYPE>
class PGraph {
protected:
  TYPE no_edge;
  EdgeParm<TYPE> *edges;
  unsigned *vtx_ofs;
  unsigned *out_deg;
  unsigned nv, ne;
  int is_digraph;
  int Copy(const LGraph<TYPE> &g);
  ...
public:
  PGraph(const LGraph<TYPE> &g);
  const EdgeParm<TYPE> *AdjEdges(unsigned v) const;
  ...
};
```

Complete code for the *PGraph* class is given on disk in the files *pgraph.h* and *pgraph.mth*, with a simple test program in *pgtst.cpp*.

The *PGraph* class has a public interface that's very similar to the public interfaces of the other graph classes. The main difference is that there are no functions provided to add or remove vertices or edges from the graph. Another other notable difference is that *AdjEdges()* returns an *EdgeParm* pointer, which points to the first element of an array of structures, each storing an edge weight and a destination vertex.

Perhaps the most notable difference is that *PGraph*s aren't built vertex by vertex. Instead, you should build the original graph with one of the other graph classes, and then convert the graph into packed form via one of the *PGraph* constructors. For example, here is the constructor that creates a packed graph from a list-based graph. It uses the associated *Copy()* function to do most of the work.

```
template<class TYPE>
PGraph<TYPE>::PGraph(const LGraph<TYPE> &g)
// Create a packed graph from a list-based graph.
: no_edge(g.NoEdge())
{
  edges = 0; vtx_ofs = 0; out_deg = 0;
  Copy(g);
}

template<class TYPE>
int PGraph<TYPE>::Copy(const LGraph<TYPE> &g)
{
  unsigned v, ofs, cnt;
```

```
no_edge = g.NoEdge();
is_digraph = g.IsDigraph();
unsigned new_ne = g.NumEdges();
if (!is_digraph) new_ne *= 2;

if (Realloc(g.NumVertices(), new_ne)) {
    ofs = 0;
    for (v = 1; v <= nv; v++) {
        EdgeLink<TYPE> *el = g.AdjEdges(v);
        cnt = 0;
        EdgeParm<TYPE> *ep = edges + ofs;
        while(el) {
          *ep = *el;
          ep++;
          cnt++;
          el = el->next;
        }
        out_deg[v] = cnt;
        if (cnt) {
           vtx_ofs[v] = ofs;
           ofs += cnt;
        }
        else vtx_ofs[v] = 0;
    }
    return 1;
}
return 0;
}
```

Because we use offsets rather than pointers when referencing the set of adjacent edges for each vertex (using the *vtx_ofs* array), *PGraph* objects are easily relocatable, making them useful for storing static graphs on disk. The graphs can be loaded into memory in their entirety and used directly with no further adjustments needed.

TOWARD AN ADJACENT EDGE ITERATOR CLASS HIERARCHY

Since the four graph classes we've presented in this chapter have basically the same public interface, you might wonder why we do not create a hierarchy of graph classes, with an abstract base class at the top specifying the public interface we've used over and over.

In principle, such a hierarchy could be built, except for one nagging problem. The *AdjEdges()* function in each of the graph classes returns something different. In the *AGraph* class, a *TYPE* pointer is returned. The *UAGraph* version of *AdjEdges()* returns a *UagPtr<TYPE>* object. The *LGraph* class version returns an *EdgeLink<TYPE>* pointer. And finally, the *PGraph* class version returns an *EdgeParms<TYPE>* pointer.

The way we iterate through the adjacent edges of the graph depends on the type of graph we're using. This suggests that, instead of having a graph class hierarchy, what we really want is an *adjacent edge iterator* hierarchy. Functions could be provided that initialize such an iterator to use a specified graph and to start with the adjacent edges of a specified vertex. Then, a *Next()* function could be specified that would return the next adjacent edge of the vertex.

Building such an iterator class hierarchy is not terribly difficult. However, our experience shows that using such iterators in the algorithms given in the next few chapters turns out to be less than ideal. This is especially true in graph algorithms that we have designed to be iterators themselves. For example, in the next chapter you'll see an iterator (or generator, depending on how you look at it) that can walk through the edges of a graph in depth first order. We found that using adjacent edge iterator objects in the depth first iterator algorithm was inefficient (too much data needed to be placed on the recursion stacks), and the code can be hard to follow at times (too much indirection is going on).

Instead of building generalized iterators, we've "hard-coded" most of the algorithms in the next few chapters in two forms, one for *AGraph* objects and one for *LGraph* objects. While we don't provide code directly for *UAGraph* objects or *PGraph* objects, the algorithms given could be modified quite easily to handle these other types of graphs.

If we accessed the graphs as matrices, using the *operator()()* functions, and took out the *AdjEdges()* functions, it would be easy to create a graph class hierarchy. However, accessing the edges of the graphs using matrix subscripting is usually too inefficient, particularly with the *LGraph* and *PGraph* classes.

FOURTEEN

Basic Graph Traversals

I n Chapter 13 we set the groundwork for studying graph algorithms by showing various ways to implement data structures that
represent graphs. In this chapter we take a look at the basic ways
that are used to move about the vertices and edges of a graph.
While you could process graphs in an ad-hoc fashion, it's more useful to use a
systematic approach. The two most basic approaches are known as *depth-first
traversal* and *breadth-first traversal*. From these traversals, many interesting
algorithms can be spawned, as you'll see in this chapter.

By the end of this chapter, you should start seeing an underlying pattern to
graph traversals, a pattern that leads to a very general way of looking at graph
traversals and graph algorithms. This intriguing method will be the subject of
Chapter 15.

DEPTH-FIRST TRAVERSAL

Depth-first traversal is a systematic way of walking a graph, where you explore
the first path you see, as far as the path can take you. When you reach a "dead
end," you back up and try the last alternative vertex that you didn't choose
before, and take another path as far as you can. Anywhere along the way, if
you encounter a vertex you've seen before, you ignore it. This process continues recursively until all the vertices in the graph have been visited.

For example, given the graph in Figure 14.1, Figure 14.2 shows the depth-
first traversal of that graph, if we start at vertex *a*. Note that all the edges of the
graph are traversed, except for the edge from *b* to *a*, since by the time we
reach *b*, vertex *a* has already been visited.

Figure 14.1 A sample graph.

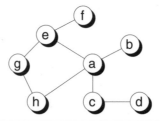

Figure 14.2 Depth-first traversal of the graph in Figure 14.1.

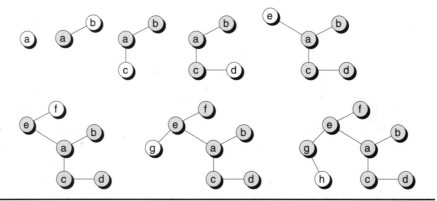

The final graph shown in Figure 14.2 can be redrawn to look like a tree. In fact, the edges visited during a depth-first traversal can be used to construct a tree known as the *depth-first spanning tree*. Figure 14.3 shows this tree for the traversal just illustrated.

Figure 14.3 A depth-first spanning tree.

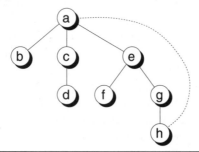

Figure 14.4 Another depth-first spanning tree.

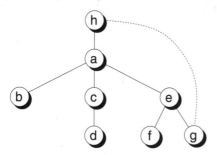

In Figure 14.3 we've drawn all the edges of the original graph, and thus the tree is just another pictorial representation of the graph. However, one of the edges, from *h* to *a*, doesn't occur in the depth-first traversal and isn't really a part of the tree. We've shown that edge using a dashed line, and it's known as a *back edge*. All the other edges are known as *tree edges*. Note that a back edge causes a cycle in the tree.

It's easy to tell the two types of edges apart when walking the graph. If you are about to traverse an edge, and the destination vertex hasn't been visited before, then you have a tree edge; otherwise you have a back edge.

The depth-first spanning tree we've shown isn't the only one possible for the given graph. If you start the traversal from a different vertex, you'll get a different spanning tree. For example, Figure 14.4 shows the depth-first spanning tree when the starting vertex is *h*.

Even if you start at the same vertex, you can still get different spanning trees, depending on which order you choose to visit the adjacent edges of a vertex. In the case of list-based graphs, this is tied directly to what order the edges were added when the graph was built.

A Depth-First Visitor Class

Now it's time to see an algorithm for depth-first traversal. Recall that one of the fundamental operations of depth-first traversal is that when a "dead end" is reached, we must back up and try the alternatives that had been left pending. This suggests that we need a data structure to hold pending choices, such as a stack, and this further suggests that a recursive algorithm is in order.

Also, we need some way of "marking" a vertex as being "visited," so that we don't try to visit a vertex twice. There is also other data that is useful to record. One type of data is known as the *depth-first number*. Each vertex can be given a number which reflects the order the vertex was visited during the traversal. You might also want to record the *depth-first level*, which is defined as the number of vertices away from the starting vertex a given vertex happens to be from the root, as given in the spanning tree. Still another piece of useful information is the parent of a vertex in the spanning tree. We'll collect this data into the following structure, which serves as an element in an array of vertex information.

```
struct LDFVinfo {
  char mark;      // Visit marker flag.
  unsigned num;   // Depth-first number
  unsigned lvl;   // Depth-first level
  unsigned mom;   // Immediate parent
};
```

In Chapter 2, we mentioned that it's sometimes useful to encapsulate a recursive routine into a class, where the recursive routine becomes a member function. This eliminates the need to pass all the parameters that aren't really needed on the stack. Instead, the object's *this* pointer is used, which points to all the non-stack-based data. The following *LDFV* class, written to traverse list-based graphs in depth-first order, shows an example of this.

```
// List-based depth-first visitor class

template<class TYPE>
class LDFV {

protected:
  LGraph<TYPE> *g;  // The graph we are traversing
  LDFVinfo *info;   // Vertex info while running
  unsigned tkt;     // Visit "ticket" dispenser.
  unsigned nv;      // Number of vertices we'll be handling
  void Dealloc();
  unsigned Realloc(unsigned n);
public:
  LDFV(LGraph<TYPE> *g_);
  virtual ~LDFV();
  virtual void Reset();
  virtual void Reset(LGraph<TYPE> *g_);
  int Ok() const;
  void Traverse(unsigned sv);
  void TraverseAll();
  virtual void Preprocess(unsigned v);
  virtual void PostProcess(unsigned v);
  int Mark(unsigned v) const;
  unsigned Num(unsigned v) const;
  unsigned Parent(unsigned v) const;
  unsigned Level(unsigned v) const;
};
```

Complete code for the *LDFV* class is given on disk in the files *ldfv.h* and *ldfv.mth*, with a test program in the file *ldfvtst.cpp*. A companion class, *ADFV*, designed for array-based graphs, is given in the files *adfv.h*, and *adfv.mth*, with a test program in *adfvtst.cpp*.

To use the *LDFV* class, you create an instance and pass in the graph you would like to traverse, using the following constructor.

```
template<class TYPE>
LDFV<TYPE>::LDFV(LGraph<TYPE> *g_)
{
  g = g_;
  info = 0;
  nv = Realloc(g->NumVertices());
  Reset();
}
```

The *Realloc()* function (not given here) dynamically allocates the *info* array, which holds *LDFVinfo<TYPE>* data. The *Reset()* function then initializes this array.

```
template<class TYPE>
void LDFV<TYPE>::Reset()
{
  for (unsigned i = 1; i <= nv; i++) {
      info[i].mom = 0; info[i].mark = 0;
      info[i].num = 0; info[i].lvl = 0;
  }
  tkt = 0; // Reset depth-first numbering ticket.
}
```

The *Traverse()* function is the main function of the class, and it's responsible for performing the recursive depth-first traversal.

```
template<class TYPE>
void LDFV<TYPE>::Traverse(unsigned u)
// Visits the vertices in the graph in depth-first order, starting
// with vertex u. Examines only those vertices that can be reached
// by u.
{
  unsigned w;

  Preprocess(u);
  info[u].mark = 1;    // "Open" the vertex.
  info[u].num = ++tkt; // Record depth-first ordering.

  EdgeLink<TYPE> *elink = g->AdjEdges(u);
  while(elink) {
    w = elink->adjvtx;
    if (!info[w].mark) {                 // Node has not been seen before.
       info[w].mom = u;                  // u is our immediate parent.
       info[w].lvl = info[u].lvl + 1;    // Record depth-first level.
       Traverse(w);                      // Here is the recursion.
    }
    elink = elink->next;
  }
```

```
    info[u].mark = 2; // "Close" the vertex.
    PostProcess(u);
}
```

Critical to the operation of *Traverse()* are the marker flags, one for each vertex. As far as the traversal is concerned, a vertex can be in one of three states. When *mark = 0*, it signals the vertex has not been seen before. When *mark = 1*, it signals that the vertex has been seen, but its descendants are still being processed. A vertex in this state is called an *open vertex*. Finally, when we're completely finished with a vertex, it is *closed*, and *mark* is set to 2.

During the traversal, the depth-first number and level of each vertex are computed. Figure 14.5 shows these two numberings for the traversal of our sample graph, starting with vertex *a*. In the figure we've replaced the vertex names with the appropriate numbers instead. Notice that the depth-first levels are based on the spanning tree and not on the graph itself. For example, vertex *b* is adjacent to the starting vertex *a*, yet we show it as having a depth-first level of 3. It *is* in the third level, if you inspect the spanning tree given in Figure 14.3.

The *Traverse()* function calls two virtual functions, *Preprocess()* and *PostProcess()*, which you can override to do other processing on the vertices, before the vertex is opened and after it is closed, respectively. By default, we have these functions print out the vertex being processed. Here is sample output from these functions, from which you can see the stacklike nature of the depth-first traversal.

```
Pre-processing a
Pre-processing b
Post-processing b
Pre-processing c
Pre-processing d
Post-processing d
Post-processing c
Pre-processing e
```

Figure 14.5 Depth-first numbers and levels.

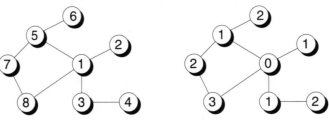

      (a) depth-first numbers   (b) depth-first levels

```
Pre-processing f
Post-processing f
Pre-processing g
Pre-processing h
Post-processing h
Post-processing g
Post-processing e
Post-processing a
```

The *LDFV* class and its sister *ADFV* class were written with generality in mind. The classes keep track of information that isn't necessary for every depth-first traversal. Sometimes all that's required is to keep track of whether a vertex has been seen before, so that the traversal works properly. You could drop the *num, lvl,* and *mom* data and keep only the *mark* data. Another alternative is to use the *num* data to serve as the *mark* data. If a vertex has a depth-first number of 0, then you know it hasn't been visited yet.

A Depth-First Iterator

Using the techniques of Chapter 3, you can turn the depth-first recursion into iteration and develop a depth-first iterator, as given in the following *LDFI* class and associated structures.

```
struct LDFIinfo {
  char mark;      // Vertex marker
  unsigned num;   // Depth-first number
  unsigned lvl;   // Depth-first level
  unsigned mom;   // Immediate parent
};

template<class TYPE>
struct LDFIParms {
  EdgeLink<TYPE> *elink; // Working edge link
  unsigned iv;           // Working inner vertex
};

template<class TYPE>
class LDFI {
private:
  LStaq< LDFIParms<TYPE> > stk; // Recursion stack
  LDFIinfo *info;               // Current vertex info
  LDFIParms<TYPE> parms;        // Recursion parms
  LGraph<TYPE> *graph;          // The graph being traversed
  EdgeData<TYPE> curr_move;     // Current edge
  unsigned tkt; // Depth-first numbering "ticket"
  unsigned ov;  // Working outer vertex
  unsigned nv;  // Number of vertices we'll be handling
  int outer_state, inner_state;
  void Dealloc();
  unsigned Realloc(unsigned n);
```

```
public:
  LDFI(LGraph<TYPE> *g, unsigned v=1);
  ~LDFI();
  void Reset(LGraph<TYPE> *g, unsigned v=1);
  void Reset(unsigned v=1, int clear=1);
  int Ok() const;
  int Eos() const;
  int EosAll() const;
  EdgeData<TYPE> *Curr();
  EdgeData<TYPE> *Step();
  EdgeData<TYPE> *StepAll();
  int Mark(unsigned v) const;
  unsigned Num(unsigned v) const;
  unsigned Parent(unsigned v) const;
  unsigned Level(unsigned v) const;
};
```

Complete code for the *LDFI* class, written for list-based graphs, is given on disk in the files *ldfi.h* and *ldfi.mth*, with a sample program in *ldfitst.cpp*. A corresponding class for array-based graphs, *ADFI*, is given on disk with a sample program in the files *adfi.h*, *adfi.mth*, and *adfitst.cpp*.

Here is the *LDFI* constructor, which requires a pointer to a graph and a starting vertex, indicating the traversal you'd like to perform.

```
template<class TYPE>
LDFI<TYPE>::LDFI(LGraph<TYPE> *g, unsigned v)
{
  graph = 0;
  Reset(g, v);
}
```

The constructor calls the following *Reset()* functions.

```
template<class TYPE>
void LDFI<TYPE>::Reset(LGraph<TYPE> *g, unsigned v)
// Resets the iterator to use graph g, and use the starting
// vertex v.
{
  if ( !graph || (nv != g->NumVertices()) ) {
    if (!graph) info = 0; // Be sure to do this!
    graph = g;
    nv = Realloc(graph->NumVertices());
  }
  else graph = g;
  Reset(v);
}

template<class TYPE>
void LDFI<TYPE>::Reset(unsigned v, int clear)
// Using the same graph as before, reset the iterator to
```

```
// start at vertex v. If clear == 1 (the default), it means
// to clear the vertex information array as well.
{
  unsigned i;

  if (clear) {
    for (i = 1; i <= nv; i++) {
        info[i].mark = 0; info[i].num = 0;
        info[i].lvl = 0;  info[i].mom = 0;
    }
  }
  stk.Clear();
  curr_move.src = 0;  curr_move.dest = 0;  curr_move.wt = 0;
  tkt = 0;  ov = v;
  outer_state = 0;  inner_state = 999;
  StepAll(); // Prime the pump.
}
```

The *StepAll()* and *Step()* functions correspond to the *TraverseAll()* and *Traverse()* functions of the *LDFV* class. (You'll learn about the *StepAll()* and *TraverseAll()* functions in the next section.) Here is the *Step()* function.

```
template<class TYPE>
EdgeData<TYPE> *LDFI<TYPE>::Step()
// Step along the current spanning tree, visiting only
// edges that can be reached.
{
  unsigned w;

  if (inner_state == 1) goto State1;
  if (inner_state == 999) return 0;
  inner_state = 1;

  curr_move.src = 0; curr_move.wt = 0;

  RecurStart:

  info[parms.iv].mark = 1;     // "Open" the vertex.
  info[parms.iv].num = ++tkt; // Record when we encountered it.
  curr_move.dest = parms.iv;
  return &curr_move;

  State1:

  curr_move.src = curr_move.dest;

  parms.elink = graph->AdjEdges(parms.iv);

  while(parms.elink) {
    w = parms.elink->adjvtx;
    if (!info[w].mark) { // Hasn't been seen before.
      info[w].mom = parms.iv; // w's parent
      info[w].lvl = info[parms.iv].lvl + 1;
```

```
        stk.Push(parms);
        parms.iv = w;
        curr_move.wt = parms.elink->wt;
        goto RecurStart;

        Place1:
        curr_move.src = parms.iv;
    }
    parms.elink = parms.elink->next;
  }

  info[parms.iv].mark = 2; // Close the vertex.
  if (stk.Pop(parms)) goto Place1;

  curr_move.dest = 0; curr_move.wt = 0;

  inner_state = 999;
  return 0;
}
```

As was the case in Chapter 3, the *Step()* function uses *gotos* and an explicit stack to "flatten" the recursion into iteration. The *Step()* function returns a pointer (which is null at the end of the traversal) to an *EdgeData* structure, giving the edge just traversed.

```
template<class TYPE>
struct EdgeData {
  TYPE wt;
  unsigned src, dest;
};
```

The *Curr()* function (not shown) also returns the edge last traversed. Here is an example of using the *LDFI* class and its *Step()* and *Curr()* functions.

```
LDFI<int> dfi(&graph); // Assume graph with int weights is already built.

cout << "Here are the edges in visitation order:\n";

EdgeData<unsigned> *ep = dfi.Curr();
while(ep) {
  cout << symbols[ep->src] << '<' << ep->src << '>' << " -[";
  cout << ep->wt;
  cout << "]- " << symbols[ep->dest] << '<' << ep->dest << '>' << '\n';
  ep = dfi.Step();
}
```

Next we give the output of this code, if the graph in Figure 14.1 is traversed. In the output, the edge weights are bracketed with "-[]-" symbols. Since the graph is not actually labeled, we've given the edge weights the arbitrary value of 1. To start the process, the first edge is a fictitious one, with a null (0) source vertex, a 0 weight, and the root of the traversal as the destination.

```
<0>  -[0]- a<1>
a<1> -[1]- b<2>
a<1> -[1]- c<3>
c<3> -[1]- d<4>
a<1> -[1]- e<5>
e<5> -[1]- f<6>
e<5> -[1]- g<7>
g<7> -[1]- h<8>
```

If you treat the edges as moves to be made, then the depth-first iterator can be viewed as a generator, generating the moves of a depth-first traversal. With these generated edges, you could, for example, take the output and build a depth-first spanning tree (without the back edges though, since they won't be generated).

GRAPH CONNECTIVITY

The sample graph we've been using is called a *connected graph*, because every vertex is accessible from every other vertex in the graph. Thus no matter what vertex we start with in a depth-first traversal, all of the vertices will be reached and will appear in the spanning tree. In general, though, it's possible for some vertices to be unreachable from others. Consider the graph shown in Figure 14.6. This graph consists of two groups of vertices, known as *connected components*. The vertices in each connected component are all reachable from one another, but no vertex from one connected component can be reached in another. So, for example, if we were to do a depth-first traversal from vertex *a*, the only vertices visited would be *a, b, c, d,* and *e*.

In order to reach all vertices of a graph with two more connected components, another function is supplied in the *LDFV* class, called *TraverseAll()*.

```
template<class TYPE>
void LDFV<TYPE>::TraverseAll()
// Does a full depth-first visit of graph g. Visits all vertices
// of all DFS spanning trees present in the graph. Starts with
// vertex 1.
{
```

Figure 14.6 A graph with two connected components.

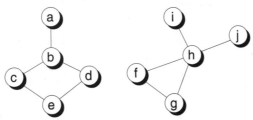

```
    for (unsigned ov = 1; ov <= nv; ov++) {
        if (!info[ov].mark) Traverse(ov);
    }
}
```

This function loops through all vertices in the graph, and if a vertex hasn't been visited yet, it is treated as the root of a spanning tree, and a depth-first traversal is performed. Note, then, that several spanning trees might result from a complete depth-first traversal. The trees as a collection are known as a *depth-first spanning forest*. Figure 14.7 shows the spanning forest that results from the graph in Figure 14.6, if we use vertices *a* and *f* as the roots of the two spanning trees.

As is the case with spanning trees, a spanning forest is yet another pictorial representation of a graph, and many different forests can be created for any given graph, depending on how the graph was built and which vertices you pick as the roots of the spanning trees.

Similar to the *TraverseAll()* function of the *LDFV* class, the *LDFI* class has a *StepAll()* function.

```
template<class TYPE>
EdgeData<TYPE> *LDFI<TYPE>::StepAll()
{
    if (inner_state != 999) {
        EdgeData<TYPE> *p = Step();
        if (p) return p;
    }

    if (outer_state == 1) goto State1;
    if (outer_state == 999) return 0;
    outer_state = 1;

    while(ov <= nv) {
        if (!info[ov].mark) { // Outer vertex hasn't been visited before.
```

Figure 14.7 A depth-first spanning forest of the graph in Figure 14.6.

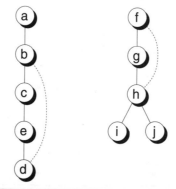

```
        parms.iv = ov;      // Set the inner vertex, which becomes
        inner_state = 0;    // the root of the next spanning tree.
        return Step();
    }
    State1:
    ov++;
}
```

In the *LDFI* class, the references to "outer" correspond to the outer looping of the *TraversallO* function of the *LDFV* class, and references to "inner" correspond to the inner looping or recursion of *TraverseO*. Here is sample output obtained by calling *StepAllO* repeatedly for the graph of Figure 14.6, assuming we've supplied suitable code to print out the edges generated.

```
<0>  -[0]- a<1>
a<1> -[1]- b<2>
b<2> -[1]- c<3>
c<3> -[1]- e<5>
e<5> -[1]- d<4>
<0>  -[0]- f<6>
f<6> -[1]- g<7>
g<7> -[1]- h<8>
h<8> -[1]- i<9>
h<8> -[1]- j<10>
```

Note how every time we start a new spanning tree, the source vertex is 0. Using the *StepAllO* function, it's easy to partition a graph into its connected components. A group ID counter (or "ticket," as we call it in the code) can be maintained, with each vertex in a component given the same group ID. This counter is incremented every time an edge is generated that has a null source vertex. Here is some code to illustrate the technique.

```
// Code to assign component or group numbers to the vertices
// of an undirected graph.

LGraph<int> graph(10);
...
unsigned *group_num = new unsigned[graph.NumVertices];
unsigned group_tkt = 0;

LDFI<int> dfi(&g); // Construct an iterator, automatically reset

EdgeData<unsigned> *ep = dfi.Curr();
while(ep) {
    if (ep->src == 0) ++group_tkt;
    group_num[ep->dest] = group_tkt;
    ep = dfi.StepAll();
}
```

With our sample graph of Figure 14.6, vertices *a, b, c, d,* and *e* would be given a group number of 1, the vertices *f, g, h, i,* and *j* would have a group number of 2. Using similar code, it would be easy to build separate graphs from each connected component. Merely start a new graph every time an edge with a null source vertex is generated.

CONNECTIVITY IN DIGRAPHS

You might wonder why the two connected components in Figure 14.6 aren't just treated as separate graphs. In practice, that would most likely be the case. However, realize that the graph data structures we gave in the last chapter don't require the vertices in the graph to be connected. Also realize that so far in this chapter, we've been using an undirected graph. If we use a digraph instead, the issue of graph connectivity becomes fuzzier.

Consider the digraph shown in Figure 14.8. On the surface, the graph appears connected. But suppose you were to start a depth-first traversal from vertex *a*. Could all the vertices be reached? The answer is no. There is no way to reach the vertices *f, g, h, i* or *j*. That's due to the directed edge from *f* to *d*, which the depth-first traversal from *a* can't cross. A digraph with this type of configuration is called a *weakly connected digraph*.

As you might guess, a weakly connected digraph can be partitioned into components where each vertex in a component is reachable by all other vertices in the component. These components are called *strongly connected components*. For the digraph of Figure 14.7, the strongly connected components are {a}, {b, c, d, e}, {f, g, h}, {i}, and {j}.

In order to reach all vertices of a weakly connected digraph, the *TraverseAll()* function of the *LDFV* class, or the *StepAll()* function of the *LDFI* class can be used. Figure 14.8 shows a depth-first spanning forest of the digraph of Figure 14.8, obtained by using either of these functions.

In the forest given in Figure 14.9, we've shown both tree edges, represented with solid lines, and back edges, represented with dashed edges. However, there is a third type of edge, called a *cross edge*, that we've drawn with dash-dot lines. A cross edge is an edge that connects two trees in a spanning forest.

Figure 14.8 A weakly connected digraph.

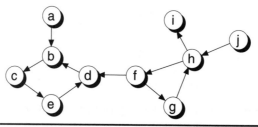

Figure 14.9 A depth-first spanning forest of the digraph in Figure 14.8.

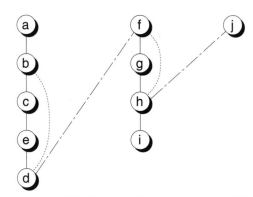

It's possible to write code to classify the edges of a spanning forest. If you encounter an edge where the destination vertex has not yet been visited, then the edge is a tree edge. If the destination vertex has been seen before and the vertex is in the same strongly connected component, then the edge is a back edge. Otherwise, the edge is a cross edge. Keep in mind that how the edges are classified depends heavily on what vertex you use as the root and how the edges were added to the graph when it was built.

Unfortunately, finding the strongly connected components of a digraph is not as easy as the similar operation of finding the connected components of an undirected graph. We might try using the depth-first iterator trick and look for generated edges with null source vertices. However, the set of "connected" components found this way will not be unique and may not be correct even if they were unique. The situation depends on where you start the depth-first traversal. For example, in the digraph of Figure 14.8, if you choose vertex *a* to start with, the proposed method will find a component having the vertices *a, b, c, d, e*. If you start with *f, h,* or *g,* then a component containing all vertices of the graph except *j* and *a* will be found. Neither one of these components is a strongly connected component. (Do you see why?)

You'll see how to find strongly connected components after the next section.

BICONNECTIVITY

Let's go back to undirected graphs for a minute. Consider the graph shown in Figure 14.10. This is the graph of Figure 14.6 with a link between *d* and *f.* Now the graph is connected. But there's another type of connectivity to consider. Suppose the vertices of this graph are computer sites on the Internet, and that the edges represent connections between the sites. In order for site *e* to talk to

Figure 14.10 A graph with articulation points.

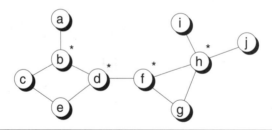

site *g*, the data must flow through the sites *d* and *f*. But what if either *d* or *f* broke down? Since there are no alternative paths from *e* to *g*, the communication would be disrupted.

Vertices like *d* and *f* are called *articulation points*; they are critical to the connection of the graph. The graph in Figure 14.10 has the four articulation points, *b*, *d*, *f*, and *h*. Vertex *d* is an articulation point because all paths from the vertices *a*, *b*, *c*, *e* to the vertices *f*, *g*, *h*, *i*, and *j*, pass through vertex *d*. Vertex *f* is an articulation point for similar reasons. Vertex *b* is an articulation point because all paths containing vertex *a* must pass through *b*. A similar situation occurs with vertex *h*, which connects vertices *i* and *j* to the rest of the graph.

In a network like the Internet, you would want to avoid having articulation points. A graph free of articulation points is called a *biconnected graph*, since there are at least two paths between any two vertices. If a graph has articulation points, then it can be broken into components called *biconnected components*, which are subsets of the graph having no articulation points. Figure 14.11 shows the biconnected components of the graph of Figure 14.10.

Surprisingly, a modified depth-first traversal of a graph can be used to find the articulation points, and the biconnected components, of a graph. Figure 14.12 shows the depth-first spanning tree of the graph in Figure 14.10, using *d* as the root vertex. (In the algorithm to follow, it doesn't matter which vertex we start with.) We've also shown the depth-first numbers of the vertices in the figure.

Figure 14.11 Biconnected components of the graph in Figure 14.10.

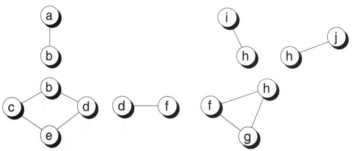

Figure 14.12 A depth-first spanning tree for the graph in Figure 14.10.

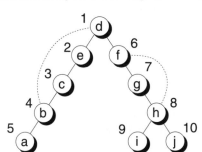

If you disconnect vertex *b* from the tree, then it isolates vertex *a* from the rest of the tree. Thus, vertex *b* is an articulation point. Likewise, disconnecting *b* isolates the vertices *i* and *j*. If *f* is disconnected, then the vertices below it in the tree are isolated. And finally, disconnecting *d* splits the tree in half, isolating both halves from each other. For these reasons, *b, d, h,* and *f* are articulation points. But you already knew that. So how can an algorithm be developed from these observations?

Any given vertex, call it *x*, is *not* an articulation point if the vertices below it in the tree have connections to vertices above *x*. In such a scenario, if *x* is removed, there is still a path to the rest of the tree for the vertices below *x*. A special case is the root, since it has no vertex above for its children to link to. If the root has at least two children, then it splits the tree in half and thus is an articulation point. If it has only one child, the root is not an articulation point.

The depth-first numberings of the vertices can help us determine whether a descendant vertex has a link above a parent vertex. When doing the depth-first traversal, you can find the highest vertex reachable by a given vertex *y* by recording the minimum depth-first number of all the reachable vertices. If the lowest number found is lower than the depth-first number of *y*'s parent, then you know there is a link higher up the tree than the parent, and thus the parent isn't an articulation point.

The following algorithm, which we've encapsulated in the class *LBicon*, uses this technique to find the articulation points of a graph.

```
struct LBiconInfo {
  unsigned num;         // Depth-first number;
  unsigned low;         // Lowest num seen by children.
  unsigned is_artic_pt; // If 1, this is an articulation point.
};

template<class TYPE>
class LBicon {
protected:
  LStaq< EdgeData<TYPE> > stk;   // Temporary stack of edges
```

```
    LStaq< EdgeData<TYPE> > bicon; // Biconnected component list
    LGraph<TYPE> *g;  // The graph we are traversing.
    LBiconInfo *info; // Vertex info while running
    unsigned tkt;      // Visit "ticket" dispenser.
    void Dealloc();
    int Realloc(unsigned n);
public:
    LBicon(LGraph<TYPE> *g_);
    virtual ~LBicon();
    virtual void Reset();
    virtual void Reset(LGraph<TYPE> *g_);
    void Run(unsigned u, unsigned mom);
    void RunAll();
    unsigned Num(unsigned v) const;
    unsigned Low(unsigned v) const;
    unsigned IsArticPt(unsigned v) const;
    const LStaq< EdgeData<TYPE> > &BiconComponents() const;
};
```

Complete code for the *LBicon* class, along with a test program, is given on disk in the files *lbicon.h, lbicon.mth,* and *lbctst.cpp.* A companion class, designed for array-based graphs, is given in the files *abicon.h, abicon.mth,* and *abctst.cpp.*

The *LBicon* class is modeled directly after the *LDFV* class. It keeps track of the depth-first numberings as well as the lowest depth-first number reachable by each vertex. This information is used to find the articulation points, which are indicated by a flag provided for each vertex. Also, a list is constructed that records the edges found during the traversal. This list can be used to find the biconnected components of the graph. The edges have a special encoding, such that an edge with a null source vertex means we've started a new component. (This is similar to what you've seen before, except in list form.)

The workhorse function is the *Run()* function, which traverses the spanning tree with the specified vertex as the root. The *RunAll()* function can also be called, to handle cases where the graph is composed of a set of separate (as opposed to biconnected) components. The *Run()* and *RunAll()* functions are the counterparts to the *Traverse()* and *TraverseAll()* functions of the *LDFV* class.

```
template<class TYPE>
void LBicon<TYPE>::Run(unsigned u, unsigned mom)
// Visits the vertices in the graph in depth-first order, starting
// with vertex u with specified parent, mom. Examines only those
// vertices that can be reached by u.
{
    EdgeData<TYPE> ed;
    unsigned w;

    info[u].num = ++tkt; // Record when we encountered the node.
    info[u].low = tkt;   // Initial value of the low number
```

```
      EdgeLink<TYPE> *elink = g->AdjEdges(u);
      while(elink) {
        w = elink->adjvtx;
        if (w != mom && info[w].num < info[u].num) {
           ed.wt = elink->wt;
           ed.src = u;  ed.dest = w;
           stk.Push(ed);
        }
        if (info[w].num == 0) { // Hasn't been seen.
           Run(w, u); // Remember, u is w's parent.
           if (info[w].low < info[u].low) info[u].low = info[w].low;
           if (info[w].low >= info[u].num) {
              if (mom) { // That is, we're not the root.
                 info[u].is_artic_pt = 1;
              }
              else {
                 // The root of the tree must have at least two
                 // children to be an articulation point.
                 EdgeLink<TYPE> *zzz = g->AdjEdges(u);
                 if (zzz && zzz->next) info[u].is_artic_pt = 1;
              }
              // Add an edge with a 0 source to indicate start
              // of next biconnected component.
              ed.src = 0; ed.wt = 0; ed.dest = u;
              bicon.AddBack(ed);
              // Now add all edges belonging to this component.
              while(stk.Pop(ed)) {
                 bicon.AddBack(ed);
                 if (ed.src == u && ed.dest == w) break;
              }
           }
        }
        else {
           if (w != mom) { // Remember u is w's parent
              if (info[w].num < info[u].low) info[u].low = info[w].num;
           }
        }
        elink = elink->next;
      }
   }

template<class TYPE>
void LBicon<TYPE>::RunAll()
// Does a full depth-first visit of graph g. Visits all vertices
// of all DFS spanning trees present in the graph. Starts with
// vertex 1.
{
   unsigned n = g->NumVertices();

   for (unsigned i = 1; i <= n; i++) {
      if (info[i].num == 0) Run(i, 0);
   }
}
```

The algorithm encoded in *Run()* is actually more complicated than a simple depth-first traversal. Besides the recursion stack, there is also another stack that is used to "tuck away" edges that belong to the same biconnected component.

The algorithm to find biconnected components runs in $O(V + E)$ time, where V is the number of vertices and E is the number of edges. Thus we have the fortuitous result that we can find biconnected components in linear time, at least for sparse graphs anyway.

Note that the algorithm in the sister class, *ABicon*, designed for array-based graphs, runs in $O(V^2)$ time. The fact that the array-based algorithm is quadratic is typical of many array-based graph algorithms. List-based graph algorithms tend to be more efficient, at least asymptotically, because you don't have to test every vertex to see if it's adjacent to a given vertex. That information is given directly in the adjacency lists.

Here is a sample output from calling *Run()* on the graph of Figure 14.10, with *d* as the starting vertex. The output lists the edges of each biconnected component. This output could be used to build separate graphs from each biconnected component. Also shown are the depth-first numberings and the numberings of the lowest reachable vertex from each vertex.

```
The biconnected components are:

Starting new component:
b - a
Starting new component:
b - d
c - b
e - c
d - e
Starting new component:
h - i
Starting new component:
h - j
Starting new component:
h - f
g - h
f - g
Starting new component:
d - f

Vertices, depth-first numbers, lows, and artic pt flags:
  a( 1) :          5,               5,          0
  b( 2) :          4,               1,          1
  c( 3) :          3,               1,          0
  d( 4) :          1,               1,          1
  e( 5) :          2,               1,          0
  f( 6) :          6,               6,          1
  g( 7) :          7,               6,          0
  h( 8) :          8,               6,          1
  i( 9) :          9,               9,          0
  j(10) :         10,              10,          0
```

Unlike the simple algorithm to find the connected components of a graph, we can't assign biconnected component or group numbers to the vertices of the graph. That's because a vertex can be in more than one biconnected component. This is the case for all vertices that are articulation points.

STRONGLY CONNECTED COMPONENTS REVISITED

Believe it or not, we're now ready to revisit the problem of finding the strongly connected components of a directed graph. It turns out that the algorithm for finding strongly connected components is very similar to the one to find biconnected components of an undirected graph. The same computation is used to find the lowest depth first numbering of any vertex reachable from a specified vertex, although it's used in a slightly different way. The following class *LStrongcon* encapsulates the algorithm.

```
struct LStrongconInfo {
  unsigned num;          // Depth-first number
  unsigned low;          // Lowest num seen by children.
  unsigned group;        // Connected group number
};

template<class TYPE>
class LStrongcon {
protected:
  LStaq< unsigned > stk;   // Temporary stack of vertices
  LGraph<TYPE> *g;         // The graph we are traversing
  LStrongconInfo *info;    // Vertex info while running
  unsigned tkt;            // Visit "ticket" dispenser
  unsigned grp_tkt;        // Group "ticket" dispenser
  void Dealloc();
  int Realloc(unsigned n);
public:
  LStrongcon(LGraph<TYPE> *g_);
  virtual ~LStrongcon();
  virtual void Reset();
  virtual void Reset(LGraph<TYPE> *g_);
  unsigned Run(unsigned u);
  void RunAll();
  unsigned Num(unsigned v) const;
  unsigned Low(unsigned v) const;
  unsigned Group(unsigned v) const;
};
```

Complete code for the *LStrongcon* class is given on disk in the files *lstrngcn.h* and *lstrngcn.mth*, with a simple test program *lsctst.cpp*. A companion class, *AStrongcon*, designed for array-based graphs, is given in the files *astrngcn.h*, *astrngcn.mth*, and *asctst.cpp*.

Unlike biconnected components, a vertex can reside on only one strongly connected component. Thus we keep track of which component a vertex belongs to by maintaining a component or group number, as given in the *Run()* function.

```
template<class TYPE>
unsigned LStrongcon<TYPE>::Run(unsigned u)
// Visits the vertices in the graph in depth-first order, starting
// with vertex u. Examines only those vertices that can be reached
// by u. Returns the lowest depth-first number of any vertex
// reachable by u.
{
  unsigned w, x, m, min;

  info[u].num = ++tkt; // Record when we encountered the node.
  min = tkt;    // Initial value of the low number
  stk.Push(u);

  EdgeLink<TYPE> *elink = g->AdjEdges(u);
  while(elink) {
    w = elink->adjvtx;
    if (info[w].num == 0) {
      m = Run(w);
    }
    else m = info[w].num;
    if (m < min) min = m;
    elink = elink->next;
  }

  if (min == info[u].num) {
    // We've got a complete group ready to process.
    info[u].group = ++grp_tkt; // Give the group a number.
    while(stk.Pop(x)) {
      info[x].group = grp_tkt;
      // Give the vertex a high number to prevent cross
      // links from being considered.
      info[x].num = g->NumVertices() + 1;
      if (x == u) break;
    }
  }

  return min;
}
```

A *RunAll()* function is also provided, in case the graph is made of up completely disconnected components. Note that it doesn't matter which vertices we use as roots of the spanning trees. The strongly connected components will still be found correctly.

The algorithm encoded in *LStrongcon* has an interesting history. Its basic form is due to *R. Tarjan* [Tarjan 72]. Like the *LBicon* algorithm, it is $O(V + E)$ and is thus linear for sparse graphs. Before *Tarjan* discovered the algorithm in

Figure 14.13 Strongly connected components of the digraph in Figure 14.8.

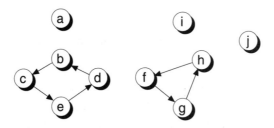

1972, researchers had been trying for years to come up with a linear algorithm to find strongly connected components, to no avail. Yet the algorithm given here is a fairly simple modification of depth-first traversal. Proving that the algorithm works, or even explaining *how* it works, is non-trivial though, so we leave that explanation to the references. Note that if used on array-based graphs, we once again have a quadratic $O(V^2)$ algorithm.

Figure 14.13 shows the strongly connected components of the digraph of Figure 14.8.

TRANSITIVE CLOSURE

In some applications, you want to know whether there is a path between any pair of vertices. One way to answer this question quickly is to build another graph from the original graph and add direct edges between vertices that have a path between them in the original. The graph that results is called the *transitive closure* of the original graph. We can create the transitive closure of a graph easily by using a slight modification of—you guessed it—depth-first traversal. The following *TransitiveCloser* class shows how.

```
template<class TYPE>
class TransitiveCloser {
private:
  LGraph<TYPE> *g;     // The graph we are traversing
  unsigned *mark;      // Visit marker.
  unsigned nv;         // Number of vertices we'll be handling
  unsigned root;       // Root of the current DFS spanning tree
  unsigned starting_tree;
  void Dealloc();
  unsigned Realloc(unsigned n);
  void Traverse(unsigned sv);
public:
  TransitiveCloser(LGraph<TYPE> *g_);
  ~TransitiveCloser();
  void Reset();
  void Reset(LGraph<TYPE> *g_);
  int Ok() const;
```

```
   void Run();
};
```

Complete code for the *TransitiveCloser* class is given on disk in the files *ltrans.h* and *ltrans.mth*, with a test program in *ltctst.cpp*.

The idea behind the *TransitiveCloser* class is to keep track of the root of the current spanning tree and to add edges to the graph between the root and all vertices encountered during a depth-first traversal. This process is repeated using all vertices as roots of depth-first spanning trees. The following *Run()* and *Traverse()* functions illustrate the process.

```
template<class TYPE>
void TransitiveCloser<TYPE>::Run()
// Use each vertex as the root of a DFS spanning tree.
{
  for (root = 1; root <= nv; root++) {
      Reset(); // So that all spanning trees are independent
      Traverse(root);
  }
}

template<class TYPE>
void TransitiveCloser<TYPE>::Traverse(unsigned u)
// Visits the vertices in the graph in depth-first order, starting
// with vertex u. Examines only those vertices that can be reached
// by u, and thus walks the DFS spanning tree with u as the root.
{
  unsigned w;

  mark[u] = 1; // Mark u as visited.

  EdgeLink<TYPE> *elink = g->AdjEdges(u);
  while(elink) {
    w = elink->adjvtx;
    g->SetEdge(root, 1, w);
    if (!mark[w]) { // Node has not been seen before.
      Traverse(w);
    }
    elink = elink->next;
  }
}
```

In the *Run()* function, it's critical that we call *Reset()*, which clears the visit marker flags. That's so we can traverse independently all the depth-first spanning trees possible from the graph.

The algorithm encoded in *TransitiveCloser* is roughly $O(V^3)$. (Think about the case when the graph is dense, when the vertices all tend to be interconnected.) Also, notice that the graph is updated in place, and at the end, it will tend to be a dense graph.

Recall that, for dense graphs, an array-based representation is better. In fact, there is a particularly elegant algorithm, also $O(V^3)$ in runtime complexity, for finding the transitive closure of an array-based graph. This algorithm is known as *Warshall's algorithm* and is given in the following *TransitiveClosure()* function.

```
template<class TYPE>
void TransitiveClosure(AGraph<TYPE> &graph)
// Does a transitive closure on the array-based graph. The
// graph is modified in place.
{
  unsigned x, y, z, n = graph.NumVertices();

  for (y = 1; y <= n; y++) {
      for (x = 1; x <= n; x++) {
          if (graph(x, y)) {
              graph.SetEdge(x, 1, y); // To normalize weights to 1
              for (z = 1; z <= n; z++) {
                  if (graph(y, z)) graph.SetEdge(x, 1, z);
              }
          }
      }
  }
}
```

 The *TransitiveClosure()* function can be found on disk in the file *atrans.h*, with a test program in *atctst.cpp*.

In the *TransitiveClosure()* function, the graph is accessed like a two-dimensional matrix, using the *operator()()* function. The graph is updated in place, with the final "matrix" representing the transitive closure. You'll note that when we add edges to achieve the transitive closure, we use a weight of 1, and in fact, we replace all connected edge weights with 1. This is to normalize the matrix, so that a weight of 1 means there is a path between the corresponding vertices and a weight of 0 means there isn't a path. (We've assumed that the graph has a *no_edge* value of 0.) Figure 14.14 shows the transitive closure matrix of the digraph of Figure 14.8.

BREADTH-FIRST TRAVERSAL

A counterpart to depth-first traversal and, in many ways, the opposite is a type of traversal known as *breadth-first traversal.* Instead of "diving in" to the first path encountered, as is done in depth-first traversal, in breadth-first traversal the movement takes place in "waves" away from the starting vertex. First, all the vertices adjacent to the starting vertex are visited, and then all the vertices adjacent to those just visited, and so on. At each stage, we process vertices

---

Figure 14.14 Transitive closure of the digraph in Figure 14.8.

	a	b	c	d	e	f	g	h	i	j
a	0	1	1	1	1	0	0	0	0	0
b	0	1	1	1	1	0	0	0	0	0
c	0	1	1	1	1	0	0	0	0	0
d	0	1	1	1	1	0	0	0	0	0
e	0	1	1	1	1	0	0	0	0	0
r	0	1	1	1	1	1	1	1	1	0
g	0	1	1	1	1	1	1	1	1	0
h	0	1	1	1	1	1	1	1	1	0
i	0	0	0	0	0	0	0	0	0	0
j	0	1	1	1	1	1	1	1	1	0

---

whose paths from the start vertex are one edge longer. If the graph being traversed is a tree, then breadth-first traversal corresponds to *level-order traversal*, as discussed in the companion volume.

Figure 14.15 shows a breadth-first traversal of the graph in Figure 14.1. The edges of the traversal form a tree known as the *breadth-first spanning tree*, as shown in Figure 14.16. This spanning tree is yet another pictorial representation of the graph. The edge shown in a dashed line is a back edge. The other edges are tree edges.

In case the graph being traversed is not strongly connected, then the traversal yields a *breadth-first spanning forest* of breadth-first spanning trees.

Figure 14.15 Breadth-first traversal of the graph in Figure 14.1.

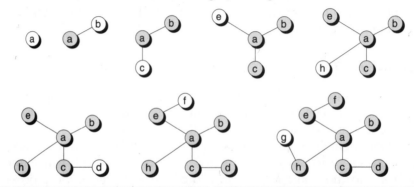

Figure 14.16 Breadth-first spanning tree of the graph in Figure 14.1.

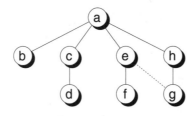

Writing an algorithm for breadth-first traversal is very similar to writing one for depth-first traversal. The only difference is the use of a queue instead of a stack. At each stage, the vertices adjacent to the given vertex are placed in the queue, to be processed in order. In this manner, the vertices are processed in level order. The following *LBFV* class, modeled closely after the *LDFV* class, implements breadth-first traversal.

```
struct LBFVinfo {
  char mark;      // Vertex marker
  unsigned num;   // Breadth-first number
  unsigned lvl;   // Breadth-first level
  unsigned mom;   // Immediate parent
};

template<class TYPE>
class LBFV {
protected:
  LGraph<TYPE> *g;  // The graph we are traversing
  LBFVinfo *info;   // Vertex info while running
  unsigned tkt;     // Visit "ticket" dispenser.
  unsigned nv;      // Number of vertices we'll be handling
  void Dealloc();
  unsigned Realloc(unsigned n);
public:
  LBFV(LGraph<TYPE> *g_);
  virtual ~LBFV();
  virtual void Reset();
  virtual void Reset(LGraph<TYPE> *g_);
  int Ok() const;
  void Traverse(unsigned sv);
  void TraverseAll();
  virtual void Preprocess(unsigned v);
  virtual void PostProcess(unsigned v);
  int Mark(unsigned v) const;
  unsigned Num(unsigned  ) const;
  unsigned Parent(unsigned v) const;
  unsigned Level(unsigned v) const;
};
```

Complete code for the *LBFV* class is given on disk in the files *lbfv.h* and *lbfv.mth*, with a test program in the file *lbfvtst.cpp*. A companion class, written for array-based graphs, is given in *abfv.h* and *abfv.mth*, with a test program in *abfvtst.cpp*.

The workhorse function *Traverse()* shows the guts of breadth-first traversal. Since we're using a queue, the algorithm admits an iterative form right from the start.

```
template<class TYPE>
void LBFV<TYPE>::Traverse(unsigned u)
// Visits the vertices in the graph in breadth-first order, starting
// with vertex u. Examines only those vertices that can be reached
// by u, and thus walks the BFS spanning tree with u as the root.
{
  LStaq<unsigned> que;
  unsigned v, w;

  Preprocess(u);
  que.Insert(u);
  info[u].mark = 1; // "Open" the vertex.
  info[u].mom = 0;  // u is the root of the spanning tree.

  while(que.Extract(v)) {
    PostProcess(v);
    info[v].mark = 2;     // "Close the vertex."
    info[v].num = ++tkt;  // Record when we closed it.
    EdgeLink<TYPE> *elink = g->AdjEdges(v);
    while(elink) {
      w = elink->adjvtx;
      if (!info[w].mark) {
        Preprocess(w);
        que.Insert(w);
        info[w].mark = 1; // "Open" the vertex.
        info[w].mom = v;  // v is our immediate parent.
        info[w].lvl = info[v].lvl + 1;
      }
      elink = elink->next;
    }
  }
}
```

Again, we made this function more general than might be needed, with code added to record the visit order of the vertices, known as the *breadth-first number*, as well as code to compute the *breadth-first level* and find the immediate parent of each vertex in the spanning tree. In case the graph isn't strongly connected, the *TraverseAll()* function (not shown) can be used to find all spanning trees of the forest.

As you might guess, you could write an iterator that generates the edges of a breadth-first traversal. We provide code for such an iterator on disk.

Complete code for a list-based breadth-first iterator class, *LBFI*, is given on disk in the files *lbfi.h* and *lbfi.mth*, with a test program in *lbfitst.cpp*. A companion class written for array-based graphs, *ABFI*, is given on disk in the files *abfi.h*, *abfi.mth*, and *abfitst.cpp*.

TOPOLOGICAL SORTING

One algorithm that uses a modified form of breadth-first traversal is known as *topological sorting*. The idea behind topological sorting is to order the vertices of a directed graph according to how many predecessors they have. Those vertices having no predecessors come first.

Topological sorting is useful in applications such as course scheduling in an academic curriculum. In this situation, each vertex represents a course, and the directed edges indicate the prerequisites of a given course. Consider the directed acyclic graph (dag) in Figure 14.17. We've given the courses (the vertices) single letters for names. Course *a* is a prerequisite for course *d*, which in turn is a prerequisite for course *f*, and so on. Since it doesn't make sense for a course to be a prerequisite of itself, the graph representing the curriculum should not have cycles in it. Thus topological sorting makes sense only for dags.

An algorithm for performing topological sorting can be developed as follows: First, compute the number of incoming edges for each vertex. Recall from Chapter 13 that this is called the *in-degree* of a vertex. Then, find all the vertices that have no incoming edges and place them on a queue. The vertices from this queue are then processed and output. At this point, it's arbitrary which order the vertices are processed. That is, any vertex with an in-degree of 0 can be the first in the ordering.

In order to continue the sorting, when a vertex is taken from the queue and output, the vertices adjacent to the output vertex are inspected, and the in-degrees of these adjacent vertices are decremented by one. This gives us the

---

Figure 14.17 A dag representing a set of course prerequisites.

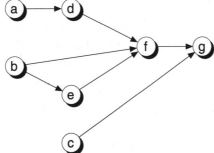

same effect as removing the output vertex from the graph. If any of the adjacent vertices now have in-degrees of 0, they are placed on the queue. This process continues until the queue is empty, at which time the sort is complete. The following *TopSort()* function, written for list-based dags, implements this technique.

```
template<class TYPE>
int TopSort(
  const LGraph<TYPE> &graph, // The graph (presumably a dag) to be sorted
  unsigned in_deg[], // The in-degrees of the vertices (gets mangled)
  unsigned num[],    // The returned topological orderings
  unsigned lvl[]     // The returned topological levels
)
// Returns 1 if successful, else 0 if the graph has at
// least one cycle accessible from sv.
{
  LStaq<unsigned> que; // To store vertices ready to be plucked
  unsigned i, u, w, n = graph.NumVertices();
  unsigned order_tkt = 0;

  for (i = 1; i <= n; i++) { // Find the initial vertices.
    if (in_deg[i] == 0) {
      que.Insert(i);  lvl[i] = 1;
    }
  }

  while(que.Extract(u)) { // Now do the bulk of the sorting.
    num[u] = ++order_tkt;
    EdgeLink<TYPE> *el = graph.AdjEdges(u);
    while(el) {
      w = el->adjvtx;
      if (--in_deg[w] == 0) {
        que.Insert(w); lvl[w] = lvl[u]+1;
      }
      el = el->next;
    }
  }

  return order_tkt == n; // Otherwise, we have cycle.
}
```

 The list-based version of *TopSort()* is given on disk in the file *ltopsort.h*, with a test program in *ltsort.cpp*. An array-based counterpart is given in the files *atopsort.h* and *atsort.cpp*.

Rather than actually "output" the vertices in sorted order, the *TopSort()* function keeps track of the topological numbering of the vertices in the *num* array. This numbering is computed in the same fashion that depth-first and breadth-first numbers are computed.

Note the check that's made upon function return. If the queue is emptied before we've visited all the vertices in the graph, then the graph has cycles. That's because if a cycle exists, then at least one vertex will have its in-degree count decremented more times than it should, and the vertex (or vertices) will be placed on the queue prematurely. The reason all this happens is that we don't prevent a vertex from being visited twice as we do with breadth-first traversal.

Here is one topological ordering of the graph in Figure 14.17.

```
a - b - c - d - e - f - g
```

The topological ordering of a graph is not unique. Many such orderings are possible. For example, here's another topological ordering of our sample graph.

```
a - c - b - e - d - f - g
```

Because the ordering isn't unique, perhaps more useful information can be gleaned from the topological sorting algorithm. Note that we keep track of the vertex levels in *TopSort()*, using the *lvl* array. These levels are analogous to the breadth-first levels of a graph. If we group the vertices according to level numbers and sort the groups based on the levels, we have another way of representing topological ordering. For example, here are the groups in our sample graph, given in order.

```
{a, b, c}  {d e}  {f} {g}
```

Using these groups, we can construct another type of graph, known as a *reduced graph*, that shows the vertices grouped into *topological components*. Figure 14.18 gives an example.

The vertices in each component are at the same level and don't depend on each other. Thus, if we wish to generate topological orderings, we can permute the vertices in each component and then combine the results. In our example, the first component has three vertices, meaning there are $3! = 6$ permutations of that component. The second component has $2! = 2$ permutations, and the third and fourth components have one permutation apiece. Thus there are $6 \times 2 \times 1 \times 1 = 12$ different possible topological orderings of the graph.

Figure 14.18 A reduced graph of topological components.

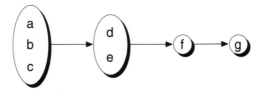

Toward More General Traversals

In the last section, we kept alluding to the fact that the topological sorting is similar to breadth-first searching. It is similar in the sense that a queue is used to hold pending vertices. However, the queue we're using is actually more like a priority queue, even though we aren't using a priority queue explicitly. We're treating the in-degree of a vertex as the "value" of the vertex, and we use a priority queue implicitly, such that those vertices with lower in-degrees have higher priorities and are thus processed first.

In this chapter we've gone from traversals that use a stack, (depth-first), to traversals that use a queue (breadth-first), and finally on to traversals using a priority queue (topological sorting). In the next chapter we'll explore the idea of using a priority queue further, and you'll see that many interesting algorithms result.

Graph Minimization Algorithms

I n this chapter we look at optimization algorithms that involve finding some kind of minimum, and where graphs are used to represent the problem space. For example, you'll see algorithms to find the *shortest paths* between pairs of vertices and to find *minimum spanning trees* of graphs. We'll also cover searching algorithms based on graphs, culminating in the A* *algorithm*, which can search for minimum solutions in an optimal way.

The most interesting aspect of this chapter is that all of the algorithms discussed can be formulated by using the same basic framework, called *priority-first traversal*. This type of traversal is a generalization of the depth-first and breadth-first traversals covered in Chapter 14.

PRIORITY-FIRST TRAVERSAL

In depth-first traversal, a stack is used as a way of ordering vertices that are to be "held up" during the traversal, pending the complete traversal of their dependent vertices. In breadth-first searching, a first-in first-out (FIFO) queue is used instead. Another type of data structure can act as both a stack and a FIFO queue, and can also act "in between" these two. This data structure is a priority queue, which is similiar to the heap data structure covered in Chapter 8.

A priority queue can act like a stack in the following way: If you arrange the priorities such that those vertices with a higher visit order number have a higher priority, then the most recently added vertices will be the first to be removed, just like a stack. If you give vertices with a lower visit order number a higher priority, then the older vertices will be favored, and you'll get the same effect as a FIFO queue. Thus by simply changing what we use for the priority, we can get depth-first traversal or breadth-first traversal.

But emulating depth-first or breadth-first traversal is not the most interesting thing you can do with a priority queue. Any arbitrary function can be used to compute the priorities. We'll call such a function an evaluation function, and denote it as f^*. Since we'll be dealing with minimization problems, we'll use the convention that those vertices with lower f^* evaluations have higher priorities. (If we implement the priority queue with a heap, this means a min-heap will be used.) Using f^*, a priority-first traversal algorithm can be stated as follows.

Priority-First Traversal Algorithm

1. Compute f^* for the starting vertex, mark the vertex as "open", and place it on the priority queue.
2. Remove the minimum vertex, u, from the queue, and mark it as "closed." Then, for each vertex w that's adjacent to u and not already closed, compute the f^* value for w, and place w on the priority queue, marking it as being "open". If w is already on the queue, (that is, it's already marked as "open"), and if and only if the new f^* value is lower, update the priority queue to reflect the change.
3. Repeat Step 2 until the queue is empty.

There are two critical aspects to Step 2. The first is that once a vertex is closed, it is never reopened. This makes algorithms originating from priority-first traversal *greedy.* You encountered a greedy algorithm in the *Knapsack problem* of Chapters 2 and 3. Recall that in a greedy algorithm, once you find a partial solution, you commit to it and assume it will be part of the final solution. Most of the algorithms in this chapter are greedy, but at the end you'll see a prime example of a non-greedy algorithm in the A* algorithm. For the A* algorithm, we'll be making a modification to priority-first traversal such that a closed vertex may be reopened if necessary.

The other critical aspect of Step 2 is what to do if a vertex is already open and is being visited again. Note that its priority is updated on the queue if and only if the new priority is higher (corresponding to a lower f^* value.) This not only prevents unnecessary computation from taking place (there's no sense in trying a solution that's known to be worse than one we've already found), but it also prevents the algorithm from going into indefinite loops. For example, if we are emulating depth-first or breadth-first behavior, the f^*s of the vertices never change once computed, so there's no way to obtain lower f^*s, and cycles are prevented.

Contrast this behavior to that of the depth-first and breadth-first traversals of Chapter 14. In these traversals, a vertex is simply never visited twice, avoiding the whole problem of indefinite loops.

A PRIORITY-FIRST VISITOR CLASS FOR LIST-BASED GRAPHS

Using the priority-first traversal algorithm, we can develop a class that visits all the nodes of a list-based graph using a priority-first traversal. The class, *LPFV*, is very similar to the *LDFV* and *LBFV* classes of Chapter 14.

```
enum GraphProbe {
  depth_first, breadth_first,
  min_span_tree, shortest_path, user_defined
};

struct LPFVinfo {
  char mark;      // Vertex marker
  unsigned num;   // Priority-first number
  unsigned mom;   // Immediate parent
};

template<class TYPE>
class LPFV {
protected:
  LGraph<TYPE> *g;    // The graph we are traversing
  LPFVinfo *info;     // Vertex info while running
  TYPE *fstar;        // Fstar evaluation for each node
  unsigned tkt;       // Priority-first "ticket dispenser"
  unsigned nv;        // Number of vertices we'll be handling
  char greedy;        // Are we using a greedy algorithm?
  virtual void Dealloc();
  virtual unsigned Realloc(unsigned n);
  TYPE FstarDFS(unsigned sv, EdgeParm<TYPE> *p);
  TYPE FstarBFS(unsigned sv, EdgeParm<TYPE> *p);
  TYPE FstarMST(unsigned sv, EdgeParm<TYPE> *p);
  TYPE FstarSSP(unsigned sv, EdgeParm<TYPE> *p);
  virtual TYPE FstarUSR(unsigned sv, EdgeParm<TYPE> *p);
  TYPE (LPFV<TYPE>::*Fstar)(unsigned sv, EdgeParm<TYPE> *p);
  virtual void SetMom(unsigned c, unsigned m);
public:
  LPFV(LGraph<TYPE> *g_, GraphProbe gp = depth_first);
  virtual ~LPFV();
  void SetGreedyBehavior(int be_greedy);
  void SetFstar(GraphProbe gp);
  virtual void Reset();
  virtual void Reset(LGraph<TYPE> *g_);
  int Ok() const;
  virtual void Preprocess(unsigned v);
  virtual void PostProcess(unsigned v);
  virtual void Traverse(unsigned sv);
  void TraverseAll();
  int Mark(unsigned v) const;
  unsigned Num(unsigned v) const;
  TYPE FstarVal(unsigned v) const;
  unsigned Parent(unsigned v) const;
};
```

Complete code for the *LPFV* class is given on disk in the files *lpfv.h* and *lpfv.mth*, with a test program in *lpfvtst.cpp*.

The workhorse function of the *LPFV* class is *Traverse()*.

```
template<class TYPE>
void LPFV<TYPE>::Traverse(unsigned u)
// Visits the vertices in the graph in priority-first order, starting
// with vertex u. Examines only those vertices that can be reached by u.
{
  fiHeap<TYPE> que(fstar, nv, 0, MIN_HEAP); // Fully indirect heap
  unsigned v, w;

  Preprocess(u);

  // We must compute the fstar value for u. We do this by using
  // a fake edge from the null vertex to u, which has 0 cost.

  EdgeParm<TYPE> fake_edge;
  fake_edge.wt = 0;  fake_edge.adjvtx = u;
  fstar[u] = (this->*Fstar)(0, &fake_edge);

  que.Insert(u);
  info[u].mark = 1; // Mark the vertex as "open".
  SetMom(u, 0);     // u is the root of the spanning tree.

  while(que.Extract(v)) {
    PostProcess(v);
    info[v].mark = 2;    // Mark the vertex as "closed".
    info[v].num = ++tkt; // Record what order we closed.
    EdgeLink<TYPE> *elink = g->AdjEdges(v);
    while(elink) {
      w = elink->adjvtx;
      if (info[w].mark == 0) { // Vertex w has not been seen before.
        Preprocess(w);
        fstar[w] = (this->*Fstar)(v, elink);
        que.Insert(w);
        info[w].mark = 1; // Mark the vertex as "open".
        SetMom(w, v);     // v is our immediate parent.
      }
      else {
        // For greedy algorithms, once a vertex is closed
        // it is never reopened.
        if (info[w].mark != 2) { // Nope, not closed
          // Thus, we must still be in the queue, so update
          // the queue if we now have a lower eval function
          // for going from v to w.
          unsigned new_f = (this->*Fstar)(v, elink);
          if (new_f < fstar[w]) {
            // We use the 0th element as a temporary holding
            // cell for the new fstar value for w. The call to
            // Replace() indirectly sets fstar[w] = fstar[0].
```

```
              fstar[0] = new_f;
              que.Replace(w, 0);
              SetMom(w, v);   // v has a new mom.
          }
        }
      }
      elink = elink->next;
    }
  }
}
```

▼ **Note** We've simplified the *Traverse()* function from what's given on disk, showing only the code for greedy traversals. Later you'll see another variation of *Traverse()* written for non-greedy traversals. The code on disk has both variations combined.

In *Traverse()*, a fully indirect heap is used to implement the priority queue. This heap indirectly uses the *fstar* array, which holds the *f** values for each vertex. The *fiHeap* is similar to the *iHeap* class given in Chapter 8. However, in addition to a *map* array, which maps an element's heap position to a physical position, another *invmap* array is used, which maps physical positions back into heap positions. The latter array is necessary because we need the capability of replacing the priority of a vertex on the queue, and to do that, we need to somehow find where the vertex is located on the queue.

To get the process started, we compute *f** for the starting vertex, using *Fstar,* which is actually a pointer-to-member function. *Fstar* is set to point at the desired evaluation function. (You may want to study your favorite C++ textbook if you are not familiar with pointer-to-member functions.) Passed to *Fstar* is the current edge, in the form of the source vertex, and then an *EdgeParm* variable, which gives the edge weight and destination vertex. Since at the start there is no edge, a fake one is created that has a null source vertex, a zero cost, and with the starting vertex as the destination.

The following *SetFstar()* function is used to set the desired evaluation function.

```
template<class TYPE>
void LPFV<TYPE>::SetFstar(GraphProbe gp)
{
  switch(gp) {
    case depth_first:
      Fstar = &LPFV<TYPE>::FstarDFS;
      greedy = 1; // Should be a greedy algorithm
    break;
    case breadth_first:
      Fstar = &LPFV<TYPE>::FstarBFS;
      greedy = 1; // Should be a greedy algorithm
    break;
```

```
  case min_span_tree:
    Fstar = &LPFV<TYPE>::FstarMST;
    greedy = 1; // Should be a greedy algorithm
  break;
  case shortest_path:
    Fstar = &LPFV<TYPE>::FstarSSP;
    greedy = 1; // Should be a greedy algorithm
  break;
  case user_defined:
  default:
    Fstar = &LPFV<TYPE>::FstarUSR; // Pointing to a virtual function
    greedy = 0; // Default is a non-greedy algorithm.
  }
}
```

Here are the two functions used to emulate depth-first and breadth-first traversals. Later we'll take a look at the other *FstarXXX()* functions.

```
template<class TYPE>
TYPE LPFV<TYPE>::FstarDFS(unsigned /* sv */, EdgeParm<TYPE> * /* p */)
// Compute fstar for the given edge to cause a depth-first search to
// take place. In this case, we don't need to know the edge explicitly.
{
  return TYPE(nv - tkt); // Thus, "younger" vertices have priority.
}

template<class TYPE>
TYPE LPFV<TYPE>::FstarBFS(unsigned /* sv */, EdgeParm<TYPE> * /* p */)
// Compute fstar for the given edge to cause a breadth-first search to
// take place. In this case, we don't need to know the edge explicitly.
{
  return TYPE(tkt); // Thus, "older" vertices have priority.
}
```

Back to the *Traverse()* function, there is a little trickery going on when we need to update the priority of a vertex on the queue. You'll notice the following code.

```
fstar[0] = new_f;    // Put the new fstar for w in 0th item.
que.Replace(w, 0);   // Indirectly sets fstar[w] = fstar[0].
```

Due to the way the indirect heap class works, we can't simply set the new f^* for vertex w by using *fstar[w] = new_f*. Instead, we must put the new value in a temporary holding cell, and this cell must be part of the same array as the other values on the queue. (That is, the new value must be in the *fstar* array.) This latter restriction is because the indirect heap class uses array subscripting to perform the heap operations. Once the proper location for w on the heap has been determined (remember, the heap is indirect), *fstar[w]* is updated with the new value by the *Replace()* function. We use the otherwise unused 0th item of *fstar* as the temporary holding cell. (This is one good reason why we use 1-based subscripting.) After the call to *Replace()*, *fstar[0]* can be used for other things.

Minimum Spanning Trees

The edges recorded during a priority-first traversal form a tree called the *priority-first spanning tree*. One difference in building this tree is that unlike the depth-first and breadth-first traversals, an open vertex may be visited more than once, and if a lower f^* is found, the parent of that node will change. You'll notice this in the *Traverse()* function, where w is given a new "mom" upon encountering a lower f^*. The function *SetMom()* is used for this purpose. Note that *SetMom()* is made virtual to accommodate the A^* algorithm, as you'll see later.

By using the following *FstarMST()* function to evaluate f^*, the traversal produces what is known as the *minimum spanning tree*. In this tree the sum of the weights of the tree edges are minimized. We'll call the algorithm used to obtain a minimum spanning tree the *Minimum Spanning Tree (MST)* algorithm.

```
template<class TYPE>
TYPE LPFV<TYPE>::FstarMST(unsigned /* sv */, EdgeParm<TYPE> *p)
// Compute fstar for the given edge to obtain a minimum spanning tree.
{
  return p->wt;
}
```

In the *FstarMST()* function, a vertex w's f^* value is based on the weight of the edge from the parent of w to w. Thus by selecting the minimum vertex from the priority queue, the edge with the lowest weight is in effect traversed. Amazingly, selecting the minimum edge at any point and committing to it will yield a spanning tree where the total of the edge weights is minimized. In other words, a greedy algorithm can find the minimum spanning tree.

There is one important stipulation about the weights used in a minimum spanning tree traversal. The weights must not be negative. Because vertices are revisited if a lower cost is found, endless loops will be created if the graph has cycles with negatively weighted edges. The algorithm will spin around and around, because after each cycle, the cost gets smaller and smaller.

What are minimum spanning trees good for? One example is in wire-wrapping a breadboard for an electronic circuit. Suppose the vertices of the graph are all the pins that need to carry ground voltage. The edge weights are the lengths of wire needed to connect any two pins (the lengths could be the "air distance" or the distance "as the crow flies" from one pin to another). By finding the minimum spanning tree, the length of wire needed is minimized.

To illustrate the construction of minimum spanning tree, consider the undirected graph in Figure 15.1. Figure 15.2 shows the minimum spanning tree being constructed edge by edge.

Figure 15.1 A sample undirected graph.

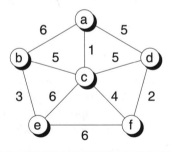

After the traversal of the graph, the *f** values for the vertices, when totaled up, will yield the total weight of the spanning tree. In this example, the total weight is 15. Note that for an undirected graph, you can use any vertex as the starting point; the same set of edges and total cost will result.

For digraphs, completely different edges may be traversed depending on the starting point and whether the digraph is strongly connected or not. In fact, if the graph, directed or otherwise, is not strongly connected, you'll obtain a *minimum spanning tree forest*. To obtain all the trees in the forest, the following *TraverseAll()* function can be used:

```
template<class TYPE>
void LPFV<TYPE>::TraverseAll()
// Does a full priority-first visit of graph g. Visits all vertices
// of all PFS spanning trees present in the graph. Starts with
// vertex 1.
```

Figure 15.2 Construction of a minimum spanning tree.

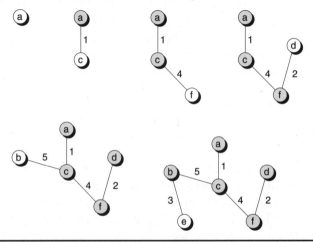

```
{
  for (unsigned i = 1; i <= nv; i++) {
      if (!info[i].mark) Traverse(i);
  }
}
```

The shape and number of spanning trees in the forest depends on how the graph was built and what vertices are picked as starting points, whether the graph is strongly connected, and so forth.

It can be shown that the MST algorithm using priority-first travesal is $O((E+V) \log V)$ for sparse graphs based on adjacency lists, where E is the number of edges in the graph and V is the number of vertices. Thus a reasonable logarithmic algorithm is obtained.

SINGLE SOURCE SHORTEST PATHS

In Chapter 14 you saw algorithms that could determine whether there was any path between two vertices. Sometimes, you are not interested in just any path, but rather, the shortest one. It turns out that with just a slight modification to the *FstarMST()* function, you can obtain an evaluation function that will find the shortest path between a source vertex and all other vertices in the graph. The algorithm obtained is known as the *single source shortest path (SSSP)* algorithm. The following *FstarSSSP()* function makes the necessary computation in the priority-first traversal framework to compute the *SSSP* algorithm.

```
template<class TYPE>
TYPE LPFV<TYPE>::FstarSSP(unsigned sv, EdgeParm<TYPE> *p)
// Compute the priority for the given edge to cause the shortest
// paths from the root of this spanning tree to be created.
{
  return fstar[sv] + p->wt;
}
```

The only change we made from the *FstarMST()* function is to use not just the weight of the edge being traversed but the sum of the weights of the edges that have been traversed to reach the given vertex. Like the MST algorithm, the SSSP algorithm is also greedy. Once we've closed a vertex, the shortest path to that vertex from the starting vertex remains frozen for the duration of the algorithm.

Like the MST algorithm, the SSSP algorithm will not work correctly for graphs that have cycles containing negative weights. Thus, be sure to use only graphs with positive weights.

It's important to realize that the SSSP algorithm finds not just the shortest path between two vertices but the shortest path to all vertices from a given starting point. It's almost as easy to find the shortest path for all vertices from a given starting point as it is to find the shortest path to a single goal. Unless you have additional information to help you find the goal vertex, other than the graph connections, it's likely that many of the other vertices will be investigated during the traversal, so you might as well record their shortest paths too. Later in the chapter you'll see the A*algorithm which can take advantage of information outside the graph, to lead the algorithm more directly to the goal vertex.

Figure 15.3 shows the edges on the shortest path from vertex *a* of our sample graph to all the other vertices. Again, a spanning tree results. (There is no special name for this tree.) Note that completely different paths (and thus trees) result for different source vertices and that the algorithm works for directed graphs as well as undirected ones. As before, if the graph is not strongly connected, a spanning forest will result.

At the end of the algorithm, the *fstar* array will contain the total cost to reach a given vertex from the start vertex. The *mom* array will give the immediate parent of each vertex on the shortest path to that vertex. For example, here are the *fstar* and *mom* arrays for the tree shown in Figure 15.3.

```
             a   b   c   d   e   f
fstar[i] = [0,  6,  1,  5,  7,  5]
mom[i]   = [0,  1,  1,  1,  3,  3]
```

With the *mom* array, you can reconstruct the shortest path from the start vertex to a given vertex. For example:

```
template<class TYPE>
void PrintPath(LPFV<TYPE> &pfv, unsigned v)
{
  if (v == 0) return 0;
  PrintPath(pfv, pfv.Parent(v));
  cout << symbols[v] << " ";
}
```

Here *symbols* is a symbol table object that has an *operator[]* function that returns the name of a vertex given its handle. The *Parent()* function of *LPFV* accesses the *mom* array.

Figure 15.3 Shortest paths from a single source.

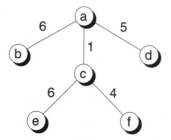

Finding the shortest paths in a graph has obvious applications. For example, if you have a road network and a home city, you can compute the shortest route between the home city and any other city. However, the SSSP algorithm can be used for more than just roads. Many optimization problems can be represented with graphs, and the SSSP algorithm can find minimum solutions for some of these problems.

Like the MST algorithm, the SSSP algorithm is $O((E + V) \log V)$ in complexity, which isn't bad for the amount of information it collects.

A Priority-First Visitor Class for Array-Based Graphs

The *LPFV* class is written for list-based graphs, where it's intended that the graph is sparse. If you have a dense graph with lots of connections, then an adjacency-matrix representation is better. We could adapt the *LPFV* class quite easily to handle array-based graphs, but there is actually a better way to implement the priority-first traversals if arrays are used.

During each step of a priority-first traversal, the minimum cost vertex is extracted, and then the costs for all adjacent, non-closed vertices are computed. Recall that in an adjacency-matrix, if you want find all the adjacent edges to a given vertex, you must examine an entire row of entries in the matrix. You pass through all the vertices in the row, looking for non-null edges. Since all the vertices of the graph will be seen, you might as well do some extra processing along the way. For example, you could compute the minimum cost vertex directly, updating it as you scan each row. In this manner, you don't need an explicit priority queue. The result is a simpler algorithm with less overhead.

This strategy is the basis of the *APFV* class, written for array-based graphs. This class is virtually the same as the *LPFV* class, except in how the *Traverse()* function is coded. As such, we won't show the class here. We'll give only the *Traverse()* function.

```
template<class TYPE>
void APFV<TYPE>::Traverse(unsigned u)
// Visits the vertices in the graph in priority-first order, starting
// with vertex sv. Examines only those vertices that can be reached
// by u.
{
  EdgeParm<TYPE> edge_parm;
  unsigned new_u, w;
  TYPE cost_uw, new_f;

  Preprocess(u);

  // We must compute the fstar value for u. We do this by using
  // a fake edge from the null vertex to u, that has 0 cost.
```

```
edge_parm.wt = 0;  edge_parm.adjvtx = u;
fstar[u] = (this->*Fstar)(0, &edge_parm);

while(u) {
  // Vertex u has the minimum cost. Mark it as closed, and then
  // examine all the vertices again to find the next minimum.
  PostProcess(u);
  info[u].mark = 2;     // Mark the vertex as closed.
  info[u].num = ++tkt;  // Record what order we closed.
  new_u = 0;
  const TYPE *ep = g->AdjEdges(u) + 1; // The 1 is needed!
  for (w = 1; w <= nv; w++, ep++) {
    // For greedy algorithms, you don't reexamine a closed vertex.
    if (greedy && info[w].mark == 2) continue;
    cost_uw = *ep; // graph(u, w);
    if (cost_uw != g->NoEdge()) { // There is an edge from u to w.
      edge_parm.wt = cost_uw;  edge_parm.adjvtx = w;
      new_f = (this->*Fstar)(u, &edge_parm);
      if (info[w].mark == 0) { // w hasn't been visited yet.
        Preprocess(w);
        info[w].mark = 1; // Mark as open and "on" the queue.
        fstar[w] = new_f;
        SetMom(w, u);
      }
      else if (new_f < fstar[w]) { // There's a smaller value for w.
        fstar[w] = new_f;
        info[w].mark = 1; // Mark the vertex as open again.
        SetMom(w, u);     // w has a new parent.
      }
    }
    // Update the implicit priority queue by keeping track of
    // the vertex with the minimum cost in new_u. w must be
    // "on" the queue to vote.
    if (info[w].mark == 1) {
      if (new_u == 0 || fstar[w] < fstar[new_u]) {
        new_u = w; // We have a new minimum vertex.
      }
    }
  }
  u = new_u;
}
}
```

Complete code for the *APFV* class is given on disk in the files *apfv.h* and *apfv.mth*, with a test program in *apfvtst.cpp*.

This variation of *Traverse()* does not use an explicit priority queue. Instead, we compute the minimum-cost vertex and record the result in the *new_u* variable, using the *fstar* array directly and examining each vertex that is still open. This is done every time through the outer loop. In a sense, an implicitly priority queue is being used. The marker flags are used to indicate whether a vertex is "on" this implicit queue or not.

Unlike the *LPFV* class, code to handle greedy vs. non-greedy traversals is trivial. Notice the simple test in the statement that reads "if (greedy && info[w].mark == 2) continue." For greedy algorithms, we don't revisit a vertex that's closed. If the algorithm isn't greedy, then the test for a closed vertex is essentially short-circuited.

PRIM'S ALGORITHM AND DIJKSTRA'S ALGORITHM

The *APFV* class can be used to perform any kind of priority-first traversal, including finding minimum spanning trees and single-source shortest paths. For the latter two algorithms, we can simplify the code and write a single function to perform the algorithms, as shown in the following *PrimDijk()* function.

```
template<class TYPE>
int PrimDijk(
  const AGraph<TYPE> &graph, unsigned src,
  TYPE *costs, unsigned *moms, int do_mst
)
// If do_mst is true, this is Prim's algorithm for finding the
// minimum spanning tree with src as the root. Else, it's Dijkstra's
// algorithm for finding the shortest paths from src to all other
// vertices. The paths are encoded in moms (which has the parent
// of each vertex), and costs has the minimum cost to each vertex.
// WARNING: The costs and moms arrays should be 1-based.
// Returns 1 if successful, else 0 if can't allocate
// a visit marker array.
{
  unsigned i, u, new_u, w, n = graph.NumVertices();
  TYPE cost_uw, new_cost;

  char *mark = new char[n+1];

  // Set up the initial costs, ancestors, and marker flags.

  for (i = 1; i <= n; i++) { costs[i] = 0; moms[i] = 0; mark[i] = 0; }

  u = src;
  while(u) {
    // Vertex u has the minimum cost. Mark it as closed, and then
    // examine all the vertices again to find the next minimum.
    mark[u] = 2;
    new_u = 0;
    for (w = 1; w <= n; w++) {
        if (mark[w] == 2) continue; // Don't reexamine a closed vertex.
        cost_uw = graph(u, w);       // Retrieve edge wt.
        if (cost_uw != graph.NoEdge()) { // We actually have an edge.
          new_cost = cost_uw;
          if (!do_mst) new_cost += costs[u]; // For SSSP
          if (mark[w] == 0) {
              // Haven't visited w yet. Set initial cost for w.
              mark[w] = 1; // Mark the vertex open and "on" the queue.
```

```
            costs[w] = new_cost;
            moms[w] = u;
          }
          else if (new_cost < costs[w]) {
            // There is a lower cost for vertex w.
            costs[w] = new_cost;
            moms[w] = u;
          }
        }
        // Update the implicit priority queue by keeping track of
        // the vertex with the minimum cost in new_u. w must be
        // "on" the queue to vote.
        if (mark[w] == 1) {
          if (new_u == 0 || costs[w] < costs[new_u]) {
            new_u = w; // We have a new minimum cost vertex.
          }
        }
      }
    }
    u = new_u;
  }

  delete[] mark;

  return 1;
}
```

If *do_mst* is true (that is, we're finding a minimum spanning tree), then the algorithm encoded in *PrimDijk()* is known as *Prim's algorithm*. This is a rather famous algorithm, and it's interesting we derived it from priority-first traversal. If *do_mst* is not true (that is, we're finding single-source shortest paths), then the algorithm encoded in *PrimDijk()* is known as *Dijstrka's algorithm*, also a famous algorithm.

Both of these algorithms are $O(V^2)$. While this isn't as good as the $O((E + V)$ log $V)$ complexities of the list-based algorithms, the array-based ones may run faster, particularly with dense graphs, as there is less overhead involved.

FLOYD'S ALGORITHM

Dijkstra's algorithm finds the shortest path between a start vertex and all other vertices. Sometimes it's useful to compute the shortest paths between *all* pairs of vertices. This is known as the a*ll pairs shortest path* (*APSP*) problem. The *APSP* problem can be solved by running the *SSSP* algorithm multiple times, once with each vertex as the starting point. For array-based graphs, however, there is another particularly simple and elegant algorithm, known as *Floyd's algorithm*. This algorithm is a generalization of Warshall's algorithm for finding transitive closures that we studied in Chapter 14.

```
template<class TYPE>
void Floyd(AGraph<TYPE> &graph, AGraph<unsigned> &moms)
```

```
// Finds the shortest paths from each vertex to all other vertices.
// Updates the graph in place to indicate the cost of each shortest
// path, and also fills up the moms graph with information on the
// actual shortest paths themselves. ASSUMES moms is empty coming in.
// Uses R. W. Floyd's algorithm.
{
  unsigned x, y, z, n = graph.NumVertices();

  for (y = 1; y <= n; y++) {
    for (x = 1; x <= n; x++) {
      if (graph(x, y)) {
        for (z = 1; z <= n; z++) {
          if (graph(y, z)) {
            TYPE old_wt = graph(x, z);
            TYPE new_wt = graph(x, y) + graph(y, z);
            if (old_wt == 0 || new_wt < old_wt) {
              graph.SetEdge(x, new_wt, z);
              moms.SetEdge(x, y, z);
            }
          }
        }
      }
    }
  }
}
```

Floyd's algorithm updates the adjacency matrix in place. After the algorithm terminates, the matrix contains the costs of the shortest paths between each pair of vertices. Figure 15.4 shows the results using our sample graph.

Also shown in Figure 15.4 is another matrix that Floyd's algorithm computes: the *shortest path ancestor matrix*. Each row of the matrix has the parent information for each vertex, along the shortest path using the vertex corresponding to the row as the starting vertex. We chose to use the *AGraph* class to implement this matrix, since it's readily handy. But remember that the resulting matrix is not meant to be interpreted as a graph.

Figure 15.4 Matrices computed by Floyd's algorithm.

	a	b	c	d	e	f
a	0	6	1	5	0	0
b	6	0	5	0	3	0
c	1	5	0	5	6	4
d	5	0	5	0	0	2
e	0	3	6	0	0	6
f	0	0	4	2	6	0

(a) original adjacency matrix

	a	b	c	d	e	f
a	2	6	1	5	7	5
b	6	6	5	10	3	9
c	1	5	2	5	6	4
d	5	10	5	4	8	2
e	7	3	6	8	6	6
f	5	9	4	2	6	4

(b) shortest path cost matrix

	a	b	c	d	e	f
a	3	0	0	0	3	3
b	0	5	0	3	0	3
c	0	0	1	0	0	0
d	0	3	0	6	6	0
e	3	0	0	6	2	0
f	3	3	0	0	0	4

(c) shortest path ancestor matrix

Using the ancestor matrix, you can print out the all the shortest paths found, using code like the following *PrintPath()* function.

```
void PrintPath(const AGraph<unsigned> &moms, unsigned i, unsigned j)
// Given the path matrix p, and a src (i) and destination (j),
// this will print out the intermediate vertices along the shortest
// path between the i and j. It doesn't print i and j themselves.
{
  unsigned k = moms(i, j);
  if (k == 0) return;
  PrintPath(moms, i, k);
  cout << symbols[k] << ' ';
  PrintPath(moms, k, j);
}
```

Here are all the shortest paths found by Floyd's algorithm for our sample graph.

```
From a to a (cost=2) : a c a
From a to b (cost=6) : a b
From a to c (cost=1) : a c
From a to d (cost=5) : a d
From a to e (cost=7) : a c e
From a to f (cost=5) : a c f

From b to a (cost=6) : b a
From b to b (cost=6) : b e b
From b to c (cost=5) : b c
From b to d (cost=10) : b c d
From b to e (cost=3) : b e
From b to f (cost=9) : b c f

From c to a (cost=1) : c a
From c to b (cost=5) : c b
From c to c (cost=2) : c a c
From c to d (cost=5) : c d
From c to e (cost=6) : c e
From c to f (cost=4) : c f

From d to a (cost=5) : d a
From d to b (cost=10) : d c b
From d to c (cost=5) : d c
From d to d (cost=4) : d f d
From d to e (cost=8) : d f e
From d to f (cost=2) : d f

From e to a (cost=7) : e c a
From e to b (cost=3) : e b
From e to c (cost=6) : e c
From e to d (cost=8) : e f d
From e to e (cost=6) : e b e
From e to f (cost=6) : e f
```

```
From f to a (cost=5) : f c a
From f to b (cost=9) : f c b
From f to c (cost=4) : f c
From f to d (cost=2) : f d
From f to e (cost=6) : f e
From f to f (cost=4) : f d f
```

Floyd's algorithm is $O(V^3)$. For sparse, list-based graphs, you may be better off making repeated calls to the SSSP algorithm. For dense, array-based graphs, however, Floyd's algorithm may win due to its extreme simplicity and low overhead.

STATE-SPACE SEARCHING

So far our discussion of graph algorithms has been biased toward algorithms whose purpose is to visit every vertex in the graph. Another class of algorithms involves starting at some initial vertex, which we'll call the *starting vertex*, and finding a particular vertex, called the *goal vertex*. Another stipulation often used is that the algorithm must minimize some function along the way, such as the cost of the path to the goal or the cost of actually *finding* the goal, or both.

Some applications involve finding a solution from one of many possible "states." This is known as *state-space searching*. These types of problems can be represented using a graph, where each vertex represents a state. One of the vertices is the state we start with, and another vertex is the goal state. The edges represent all the possible ways you can move from one state to the another when performing the search. That is, the edges represent choices you can make.

As a simple example, consider a traveler on a roadway who wants to get from one place to another. The graph vertices can represent landmarks on the roadway, and the edge costs are the distances between these landmarks. Such a roadway is illustrated in Figure 15.5. Figure 15.6 gives a graph representation of this roadway, showing the edge costs (that is, the distances between landmarks) as well. This example was adapted from one given in VanLe [VanLe 93].

Suppose a traveler wants to get from point i to point f, preferably by the shortest path possible. We could help the traveler out by applying one of our graph algorithms, but which one? The SSSP algorithm would work and would even find the shortest path. But remember that this algorithm finds *all* paths from a starting vertex (in this case vertex i), not just the the path to the goal (vertex f). Thus the SSSP algorithm may be doing more work than necessary.

If the traveler isn't picky about the traveling distance, we could also apply a depth-first or breadth-first traversal, stopping when the goal vertex is encountered. For these algorithms, the visitor classes *LDFV*, *LBFV*, and even *LPFV* could be used. However, the *Traverse()* functions of these classes would have to be modified to stop when the goal node was found.

Figure 15.5 A sample roadway.

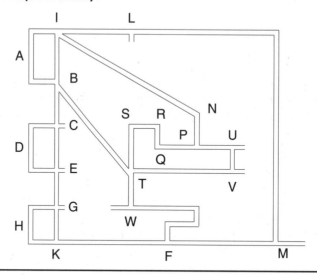

A more convenient approach is to use an iterator class, for example, *LDFI, ADFI, LBFI,* or *ABFI.* You could also develop priority-first iterators, as we've done in the *LPFI* and *APFI* classes given on disk. With the latter two iterators, you can easily choose what type of traversal you would like to perform. Both are based are their *LPFV* and *APFV* cousins.

Figure 15.6 A graph representing the roadway in Figure 15.5.

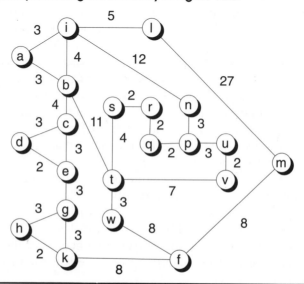

Complete code for the list-based priority-first iterator class *LPFI* is given on disk in the files *lpfi.h* and *lpfi.mth*, with a test program in *lpfitst.cpp*. A companion class for array-based graphs, *APFI*, is given on disk in the files *apfi.h*, *apfi.mth*, and *apfitst.cpp*.

When using priority-first searching, the state space is searched according to the *f** evaluation of each vertex. This type of searching is called *ordered-state space searching*. Before we talk about good ways to define *f** for this purpose, let's see how to use priority-first searching with the less interesting depth-first and breadth-first traversals. Here is an example of using an *LPFI* iterator to find the goal *f* starting at state *i*, applying a depth-first search.

```
LPFI<int> pfi(&graph, depth_first);

unsigned start = symbols["i"];
unsigned goal  = symbols["f"];

pfi.Reset(start);

EdgeData<int> *ed = pfi.Curr();
while(ed) {
  cout << ed->adjvtx << ' '; // Show the path.
  if (ed->adjvtx == goal) break;
  pfi.Step();
}
cout << '\n';

if (ed) cout << "goal found\n"; else cout << "goal not found\n";
```

To perform breadth-first searching, only the constructor call in the first statement needs to be changed. Here is the path taken for our sample roadway using both depth-first search and breadth-first search.

```
Depth-first search:   i a b c d e g h k f
Breadth-first search: i l m f
```

Remember that these solutions paths are not unique. The nodes chosen depend on what order the edges were added to the graph and how our algorithms choose the order of visiting adjacent edges of a vertex.

In the example just given, breadth-first searching found a path containing fewer vertices than depth-first searching did. In fact, breadth-first searching will always find a solution path that contains the fewest number of vertices. That's due to the level-order nature of breadth-first searching. However, the path with the fewest number of vertices isn't necessarily the shortest. The path *i-l-m-f* has a cost of 40, whereas the path *i-a-b-c-d-e-g-h-k-f*, produced by depth-first searching, has a cost of 31. Neither of these is the shortest path, which is *i-b-c-e-g-k-f*, with a cost of 25.

Heuristic Searching

It's a matter of luck whether depth-first or breadth-first searching finds you a better answer, and how fast the answer is found. Both breadth-first and depth-first searching are called *blind searching*. Neither algorithm has any clue where the goal is, both merely follow the edges presented to them locally. In this section, we discuss a technique that uses additional information that, it is hoped, can guide us more quickly to the goal node.

The information used to guide us to the goal is often stored outside the graph. Consider our roadway example. Pretend that we know not only the connections of the roadway but also the coordinates of each landmark. Using these coordinates, we could compute the air distance between any two landmarks. When choosing the next edge to traverse, we could use the rule of thumb that an edge that brings us closer in air distance to the goal node is a good candidate. Such rules of thumbs are called *heuristics*, and the searching that results is called *heuristic searching*.

Optimal Searching with the A* Algorithm

An extremely useful algorithm that performs heuristic searching is known as the A* algorithm. It was developed by Hart, Nilsson, and Raphael. (See Barr [Barr81].) This algorithm is considered one of the classics and is often categorized as a fundamental technique of artificial intelligence (AI). Not only can the A* algorithm guide you to a goal node, it can also solve some sort of minimization problem as well, and in the ideal case, it can solve the problem in an optimal way, using a minimum of computation.

As you have probably come to expect from this chapter, the A* algorithm can be understood in terms of priority-first traversal. The idea is to define f^* in such a way that heuristic information can be used. The definition of f^* for the A* algorithm is made up of two components, g^* and h^*:

$$f^*(v) = g^*(v) + h^*(v)$$

In this equation, v represents the vertex that we're evaluating. The g^* component represents the actual cost of going from the starting vertex to v, and h^* represents the estimated cost of going from v to the goal node. Thus, this definition of f^* is based not only on known information, but also on estimates. The estimates are where the heuristics come into play, and the h^* component is the carrier of this heuristic information.

By using both the g^* and h^* components, the A* algorithm determines what is the best candidate to expand in the next step. Because of this, the type of searching performed by A* is often called *best-first searching*. If you were to drop the h^* component, then the A* algorithm reduces the SSSP algorithm.

When applied to searching for a particular goal, the SSSP algortihm is some-times called *best-cost searching*.

One important restriction on the values of g^* and h^* is that they must never be negative, for reasons similar to the restriction that edge weights should not be negative when using the SSSP and MST algorthims. It's to prevent the algorithm from looping indefinitely around a cyclic path that is given a smaller (and thus more desirable) f^* evaluation each cycle.

Using either the *LPFI* or *APFI* class, it's quite easy to derive an iterator that uses the A* algorithm. The following *LAstar* class, written for list-based graphs, gives an example.

```
template<class TYPE>
class LAstar : public LPFI<TYPE> {
protected:
  TYPE *gstar;    // Collects actual costs of vertices.
  unsigned new_g; // Stores actual cost of vertex just visited.
  unsigned start, goal; // Desired path endpoints
  virtual void Dealloc();
  virtual unsigned Realloc(unsigned n);
  virtual TYPE Gstar(unsigned sv, EdgeParm<TYPE> *p);
  virtual TYPE Hstar(unsigned sv, EdgeParm<TYPE> *p);
  virtual TYPE FstarUSR(unsigned sv, EdgeParm<TYPE> *p);
  virtual void SetMom(unsigned c, unsigned m);
public:
  LAstar(LGraph<TYPE> *g_);
  virtual void Reset(LGraph<TYPE> *g, unsigned v=1);
  virtual void Reset(unsigned v=1, int clear=1);
  virtual int Run(unsigned start_, unsigned goal_);
  TYPE GstarVal(unsigned v) const;
};
```

 The *LAstar* class, written for list-based graphs, is given on disk in the files *lastar.h* and *lastar.mth*. A companion class for array-based graphs, *AAstar*, is given in the files *aastar.h* and *aastar.mth*.

In the *LPFI* class, a pointer-to-member function, *Fstar*, is used to point to the appropriate f^* function. One of the choices available is the user-defined function, *FstarUSR()*. This function is a virtual function, meant to be overridden by a derived class. The *LAstar* class overrides *FstarUSR()* to compute f^* accord-ing to the A* algorithm. The *FstarUSR()* function in turns calls the *Gstar()* and *Hstar()* functions.

```
template<class TYPE>
TYPE LAstar<TYPE>::FstarUSR(unsigned sv, EdgeParm<TYPE> *p)
// Computes the user-defined f* for the A* algorithm
{
  return Gstar(sv, p) + Hstar(sv, p);
}
```

```
template<class TYPE>
TYPE LAstar<TYPE>::Gstar(unsigned sv, EdgeParm<TYPE> *p)
// This computes g*, the actual cost from the starting
// vertex to the vertex p->adjvtx. You can override this
// function if you need to make the computation something
// else, although what's here is the most typical method.
{
  return new_g = gstar[sv] + p->wt;
}


template<class TYPE>
TYPE LAstar<TYPE>::Hstar(unsigned /* sv */, EdgeParm<TYPE> * /* p */)
// This computes h*, the estimated cost from p->adjvtx
// to the goal vertex. This is a virtual function that is
// meant to be overriden by a derived class.
{
  return 0;
}
```

The *Gstar()* function computes the known cost to a given vertex (*p->adjvtx*). This function is in fact identical to the one used to compute shortest paths, *FstarSSSP()*. The *Hstar()* function, which computes the estimated cost to the goal, is itself a virtual function, meant to be overriden further by derived classes that implement specific versions of the A* algorithm.

The value computed by *Gstar()* is recorded temporarily in the *new_g* variable. If a vertex obtains a smaller f^* value, such that the vertex has a new parent along the solution path, then *SetMom()* is called, and the *gstar* array, which stores the actual costs to each vertex from the starting vertex, is updated accordingly.

```
template<class TYPE>
void LAstar<TYPE>::SetMom(unsigned c, unsigned m)
// Along with recording the parent m of child c, we
// also update the actual cost to the node c.
{
  LPFI<TYPE>::SetMom(c, m);
  gstar[c] = new_g;
}
```

The *LAstar()* constructor sets the *Fstar* pointer to point to the *FstarUSR()* function by using the *user_defined* parameter.

```
template<class TYPE>
LAstar<TYPE>::LAstar(LGraph<TYPE> *g_)
// Constructs an A* iterator object. The user-defined Fstar
// function is used in the A* algorithm, and the traversal
// must not be greedy.
: LPFI<TYPE>(g_, user_defined)
{
  greedy = 0; // Must be non-greedy.
}
```

Another important task of the constructor is to set the *greedy* flag to 0. Unlike all the other graph algorithms we've encountered to this point, the A* algorithm is *not* a greedy algorithm. This means that it may reopen a vertex that is closed, should it find a lower-cost path leading to that vertex. We now show you how the *Step()* function of the *LPFI* class is defined for non-greedy algorithms. This function is based on the *Traverse()* funciton of *LPFV*. Keep in mind that on accompanying disk, the *Step()* function combines the greedy and non-greedy approaches, with the *greedy* flag controlling which is used.

```
template<class TYPE>
EdgeData<TYPE> *LPFI<TYPE>::Step()
// This shows the non-greedy approach
{
  EdgeParm<TYPE> fake_edge;
  unsigned w;

  if (inner_state == 1) goto State1;
  if (inner_state == 999) return 0;

  // At state 0:

  // We must compute the fstar value for ov. We do this by using
  // a fake edge from the null vertex to ov, that has 0 cost.

  fake_edge.wt = 0;  fake_edge.adjvtx = ov;
  fstar[ov] = (this->*Fstar)(0, &fake_edge);

  que.Insert(ov);
  info[ov].mark = 1; // Mark the vertex as open and on the queue.
  SetMom(ov, 0);     // ov is the root of the spanning tree.
  inner_state = 1;   // We'll soon be in state 1.

  num_expansions = 0;

  while(que.Extract(curr_move.dest)) {

    num_expansions++; // Collecting statistics

    curr_move.src = info[curr_move.dest].mom;
    curr_move.wt = *graph->EdgeWt(curr_move.src, curr_move.dest);
    info[curr_move.dest].mark = 2;    // Mark the vertex as closed.
    info[curr_move.dest].num = ++tkt; // Record order we closed.
    return &curr_move;

    State1:

    curr_move.src = curr_move.dest;
    elink = graph->AdjEdges(curr_move.src);

    while(elink) {
      w = elink->adjvtx;
      if (info[w].mark == 0) { // Never been seen before
```

```
            fstar[w] = (this->*Fstar)(curr_move.src, elink);
            que.Insert(w);
            info[w].mark = 1; // Mark the vertex as "open".
            SetMom(w, curr_move.src); // Record parent.
        }
        else {
          // Since we're not a greedy algorithm, then we must
          // revisit w even if its already closed, so that
          // we can reevaluate its fstar function.
          unsigned new_f = (this->*Fstar)(curr_move.src, elink);
          if (new_f < fstar[w]) { // Yep, lower eval
            if (info[w].mark == 1) {
                // We're on the queue, so update priority
                fstar[0] = new_f; // Using 0th element for temp
                que.Replace(w, 0);
            }
            else { // Not on the queue (w is closed), so reopen.
              fstar[w] = new_f;
              que.Insert(w);
              info[w].mark = 1; // Mark the vertex as "open" again.
            }
            SetMom(w, curr_move.src); // w has a new mom.
          }
        }
        elink = elink->next;
      }
  }

  inner_state = 999;
  curr_move.dest = 0; curr_move.wt = 0;
  return 0;
}
```

If the graph you are traversing is a tree, then it doesn't matter whether you use a greedy or non-greedy approach. Both will yield the same solution. Remember that in a tree, a vertex can have only one parent. Thus there are no alternate paths that can be tried that might yield a lower f^*.

Instead of calling the *Step()* function directly, the *Run()* function should be called, with the desired end points of the path passed in as parameters.

```
template<class TYPE>
int LAstar<TYPE>::Run(unsigned start_, unsigned goal_)
// Runs the A* algorithm, doing a best-first search from
// the start vertex to the goal vertex, computing a minimum
// cost path in as optimal a way as possible.
{
  start = start_;  goal = goal_;

  Reset(start);
  EdgeData<TYPE> *ep = Curr();
  while(ep) {
    if (ep->dest == goal) return 1; // There is a solution.
    ep = Step();
```

```
  }
  return 0; // No solution
}
```

If no path exists from the start to the goal, then no solution exists either. The *Run()* function returns a flag accordingly.

After *Run()* finishes, the *mom* array contains the immediate parents of each vertex along the solution path. The *gstar* array contains the actual costs from the starting vertex to each vertex along the solution path.

Solving the Roadway Problem

The *LAstar* class is a base class for A* iterating. You should derive classes from it to solve a particular instance of the A* algorithm. This involves overriding the *Hstar()* function to plug in the appropriate estimates. The following *LRouteFinder* class shows an example for our sample roadway problem.

```
struct LRouteCoords {
  int x, y;
};

template<class TYPE>
class LRouteFinder : public LAstar<TYPE> {
protected:
  const LRouteCoords *coords;
  virtual TYPE Hstar(unsigned sv, EdgeParm<TYPE> *p);
public:
  LRouteFinder(LGraph<TYPE> *g_, const LRouteCoords *c);
};

template<class TYPE>
LRouteFinder<TYPE>
::LRouteFinder(LGraph<TYPE> *g_, const LRouteCoords *c)
: LAstar<TYPE>(g_), coords(c)
{
}

template<class TYPE>
TYPE LRouteFinder<TYPE>::Hstar(unsigned /* sv */, EdgeParm<TYPE> *p)
// Computes the estimated cost from p->adjvtx to the goal vertex,
// using the "air distance" heuristic.
{
  TYPE dx = coords[p->adjvtx].x - coords[goal].x;
  TYPE dy = coords[p->adjvtx].y - coords[goal].y;
  return sqrt(dx * dx + dy * dy);
}
```

The *LRouteFinder* class can be found on disk in the files *lrue.h* and *lrue.cpp*, with a test program in *lroad.cpp*. A companion class for array-based graphs, *ARouteFinder*, can be found in the files *arue.h* and *arue.cpp*, with a test program in *aroad.cpp*.

The *LRouteFinder* class uses the physical coordinates of each landmark on the roadway in order to compute the air distance between a vertex and the goal. This distance is used as the h^* value. Here are sample coordinates for the roadmap of Figure 15.5.

```
a: 0 15    b: 2 13    c: 2 9    d: 0 7    e: 2 6    f: 10 0
g: 2 3     h: 0 1     i: 2 17   k: 2 0    l: 7 17   m: 18 0
n: 11 11   p: 11 8    q: 9 8    r: 9 10   s: 7 10   t: 7 7
u: 14 8    v: 14 6    w: 7 3
```

Figure 15.7 shows all the edges explored by the A* algorithm in the roadway example, where the start and goal vertices are *i* and *f*, respectively. For comparison, the figure also shows the explorations of the depth-first, breadth-first, and shortest path traversals, using the *LPFI* class (with greedy algorithms). Figure 15.8 shows the final solution paths for each traversal.

Figure 15.7 Solving the roadway problem.

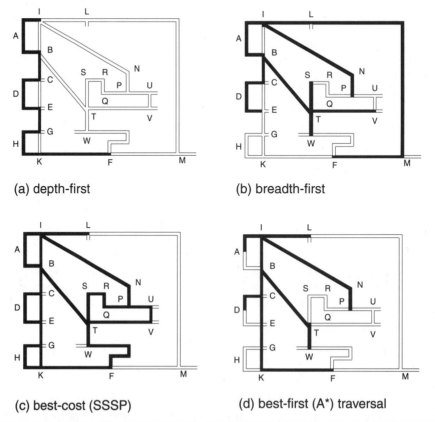

(a) depth-first

(b) breadth-first

(c) best-cost (SSSP)

(d) best-first (A*) traversal

Figure 15.8 Final solution paths for the roadway problem.

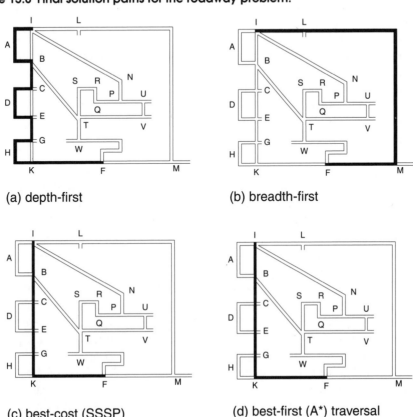

(a) depth-first

(b) breadth-first

(c) best-cost (SSSP)

(d) best-first (A*) traversal

Table 15.1 summarizes the results of each search strategy. The number-of-expansions column in the table lists how many times vertices were extracted from the priority queue and thus gives an indication of how efficient the searching was. (This can also be seen from Figure 15.7.)

Table 15.1 Summary of the Different Search Strategies

Traversal	Solution Path	Path cost	Number of expansions
Depth-first	*i a b c d e g h k f*	31	10
Breadth-first	*i l m f*	40	15
Best-cost (SSSP)	*i b c e g k f*	25	20
Best-first (A*)	*i b c e g k f*	25	14

In the depth-first search, the solution was found in the fewest expansions. This was due to pure luck, and also because no attempt was made to find the optimum path. The path chosen by depth-first search is not very good in this example. Notice in Figure 15.7 the number of unnecessary side trips the depth-first search makes. Characteristic to depth-first searching, the algorithm takes the first edge it sees, no matter how senseless, and follows the subsequent edges as far as possible, until the goal is encountered.

In the breadth-first search, the path found has the fewest number of vertices, but it is the most expensive path, pointing out the difference between the topology given by the graph and the actual physical topology. The breadth-first search took more expansions than the depth-first search. In many situations the number of expansions would be much higher. The characteristic trait of breadth-first searching is to work slowly, spreading out from the source. If the goal is far away in terms of level order, then breadth-first searching takes a long time. If, however, the goal is close to the start, breadth-first searching finds the goal quite rapidly. In this same situation, a depth-first search might spend a lot of time traveling down a long fruitless path, before trying the short path to the goal that might have been close by all along.

The best-cost search algorithm was able to find the shortest path, as you would expect, but unfortunately, the algorithm took the most expansions in finding the solution. That's because single-source, shortest-path searching knows nothing about where the goal vertex is. The algorithm is likely to spend a lot of time exploring paths totally unrelated to the goal vertex.

Contrast best-cost searching with the best-first searching of the A* algorithm. The "air distance" guides the algorithm to the goal node. Because of this, the traversal required only 14 expansions, yet the optimum path was found. In this case we were rather lucky, because only under certain conditions is it guaranteed that the A* algorithm will find the optimum solution.

The A* algorithm takes into account two factors during its searching. Not only does it try to find the best answer, but it also tries to minimize the amount of effort spent in finding that answer. The algorithm often makes a compromise and comes up with a solution that may not be the best but is at least reasonably good. In large graphs, the A* algorithm is likely to spend far less time finding the goal than the other algorithms. But unlike the SSSP algorithm, it doesn't guarantee that the best answer will be found.

How well the A* algorithm performs, and how close it comes to finding the best answer, depends on how good the heuristic function is. In the roadway problem, the "air distance" heuristic is fairly good, but it's not foolproof. For example, as the crow flies, the distance from w to f is smaller than what the road path really is. The heuristic underestimates the cost because the road from w to f winds around instead of heading directly to f.

If the heuristic function is able to estimate the cost exactly, then the A* algorithm yields an optimal solution (it finds the best path) in an optimal way (the fewest number of vertices are expanded.) The optimal scenario in our road example would be to find the path *i-b-c-e-g-k-f* and take only seven expansions in the process.

It can also be proven that if *h** never overestimates the actual cost from a vertex *v* to the goal, then A* will find an optimum solution. This is known as the *admissability condition*. As we mentioned before, it's critical that the *h** values be positive as well as the *g** values.

Let's see whether our sample problem is admissable. Table 15.2 gives the estimated costs and actual costs from each vertex to the goal *f*.

From this table you can see that the "air distance" underestimates most of the costs, but it also overestimates the cost from *h* to *f*, if only by a slight amount. Thus our sample problem is not strictly admissable, but we were lucky enough to find the optimum solution anyway.

Table 15.2 Estimated vs. Actual Costs for the Roadway Problem

Vertex	Estimated cost	Actual cost
i	18.79	25
a	18.02	24
b	15.26	21
c	12.04	17
d	12.20	16
e	10.00	14
f	0.00	0
g	8.50	11
h	10.05	10
k	8.00	8
l	17.26	30
m	8.00	8
n	11.05	24
p	8.06	21
q	8.06	19
r	10.05	17
s	10.44	15
t	7.62	11
u	8.94	20
v	7.21	18
w	4.24	8

You may wonder why overestimating is so bad. If the h^* value of a vertex is overestimated, then the f^* value will be too high, and the vertex may not be picked as the best candidate when it should. The path the vertex lies on may be ignored entirely in favor of paths that are actually worse. This happens if the goal is found before the actual costs of the paths that *are* tried become larger than the estimated cost of the path ignored.

You should understand that it's okay for h^* to underestimate the actual cost. If the actual cost is higher, the algorithm will eventually figure this out and update g^* and f^* accordingly. The goal will never be reached too soon by an underestimated path because the actual cost of that path will be computed before the goal is chosen. The worst that could happen is that the algorithm may spend more time than necessary looking for the optimum path. The algorithm may be misled in trying paths that it doesn't discover until later are not the best. In our example, 14 expansions were used, whereas in the ideal case only seven are needed.

Another factor to consider is how hard it is to compute h^*. If the computation is expensive, you may be better off with less accurate estimates, even though it means more vertices are expanded. Also, in some problems, finding the optimum answer means in effect that an exponential number of paths must be explored, which may take far too much time to complete. In these scenarios, it might be better to let h^* overestimate the actual costs. In doing so, the algorithm may terminate quicker, with the trade-off being that you may not get the optimum answer.

INTRACTABLE MINIMIZATION PROBLEMS

The algorithms given in this chapter can solve a fair number of optimization problems, and they do so in reasonable amount of time. The algorithms are logarithmic, quadratic, or cubic, the latter being for the rather expensive *All Pairs All Paths* problem. Surprisingly, slight variations of the problems we've seen prove to be very difficult to solve. It's not that there aren't algorithms to solve these problems, but rather that there are no known algorithms to find the answers in a reasonable amount of time.

An example is the *traveling salesman problem*. In this problem, a salesman has n cities to visit and would like to tour each city once and only once, using the shortest distance possible, and returning back to the starting city. This problem is similar to the shortest path problem, but the most straightforward algorithm to find a solution is $O(n!)$. This means that even for the modest size of 10 cities, over 3 million permutations of the n cities have to be examined. For a tour of 25 cities, the number of permutations is astronomical.

The traveling salesman problem turns out to be *NP-complete*. We briefly touched on this concept in Chapter 1. The term *NP* stands for "non-deterministic polynomial." Basically, a problem is *NP-complete* if there are no deterministic algorithms for it that have polynomial complexities. For example, $O(n!)$ is non-polynomial. So is $O(2^n)$. In non-polynomial algorithms, even input sizes of relatively small n yield computations that can take hundreds, thousands, millions, or billions of years, even on the fastest computers of today. *NP*-complete problems are called *intractable* because there is no practical way to solve them in the general case.

APPROXIMATE ALGORITHMS

Even if you do have a problem that is *NP*-complete, all is not lost. What makes many *NP*-complete problems hard is the stipulation that the very best answer be found. If you are willing to relax that requirement, and find an answer that's "good enough", then many of these problems can be solved relatively quickly. For example, it's possible to solve the travelling salesman problem using the A* algorithm. By using the right heuristics, you can cut the search space down dramatically. Unfortunately, a solution for the travelling salesman problem using the A* algorithm is somewhat complicated, and we don't have the room to show it here. An example can be found in VanLe [VanLe 93], and the method given is based on a technique by R.E. Gomory [Gomory 66].

The A* algorithm is just one of many approaches that can be used to trade off computational complexity for precision of answer. Other approaches that have been gaining popularity are *simulated annealing* (see for example [Press 88]), and *genetic algorithms*, (see for example [Koza 92]). Unfortunately, we don't have the space to discuss any of these alternatives in this book.

Finite State Machines

In this chapter we take a look at another application of graphs: A *finite state machine* is a device whose operation is composed of modes called *states*, of which there are only a finite number. With the appropriate input, the machine can transition from one state to another. Finite state machines are also called *finite state automata*, or simply *finite automata*. The states of a finite automaton can be depicted with an illustration known as a *state transition diagram*, and that's where graphs come in, because these diagrams are actually directed graphs. Because finite state machines are intimately tied to state transition diagrams, we'll be using the two terms synonymously in this chapter.

Finite state machines can be used to model many types of devices. In fact, we've been using simple forms of finite state machines all along in the iterator and generator classes sprinkled throughout this book. (You may have noticed that we used state variables quite frequently in these classes.) Also, the Boyer-Moore string searching algorithm is a state machine in disguise. So is the Knuth-Morris-Pratt algorithm and even the Shift-OR searching algorithm. (The latter is a very fancy form of state machine.)

As other examples, you might model the "user-interface" of a digital watch with a state machine. Another example would be a vending machine. But pattern-matching machines will be the primary focus of this chapter. We'll look at a particular kind of machine known as the *Aho-Corasick* machine, which can be used to match many strings in parallel. Thus we pick up where we left off in Chapter 12 with yet another string searching technique.

STATE TRANSITION DIAGRAMS

Figure 16.1 shows a simplified state transition diagram for a digital watch. The state machine we use to represent our watch has four states: one that displays the time, one that displays the date, one puts the watch into "stopwatch" mode, and a fourth state that represents the stopwatch when it is running and counting time.

State transition diagrams are actually labeled digraphs, with the vertices representing states and the edges representing transitions between states. The labels are the inputs to the machine. Typically the inputs are encoded as integers, which might represent offsets into other arrays that give more information regarding these inputs. For example, if the inputs are character strings, we might use a symbol table to store the strings and use the symbol table handles as the input codes. The set of unique inputs that are handled by the machine is called the *machine alphabet*.

Our digital watch has three kinds of input that represent button presses. When in the "display time" mode, the user can press the *d* button to transition to the "display date" mode. Pressing the *d* button again toggles the watch back to "display time" mode. Similarly, pressing the *t* button toggles back and forth between "display time" mode and "stopwatch" mode. In the "stopwatch" mode, pressing the *s* button starts the stopwatch, transitioning to the "display running time mode," and pressing *s* again stops the timer, transitioning back to the "stopwatch" mode, with the elapsed time being displayed.

The edges of a transition diagram show the only valid transitions for the state machine. For example, you can't go directly from "display date" state to either the "stopwatch" state or the "display running time" state since there are no edges to these states. If an input occurs for which there is no valid transition, a failure occurs. An example is if the user presses any button other than *d* while in the "display date" mode. In some state machines, failures cause the machine to go to a special fail state. In this example, though, invalid inputs are simply ignored.

Figure 16.1 A state transition diagram of a digital watch.

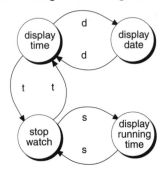

Figure 16.2 A pattern-matching machine.

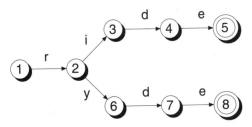

Let's take a look at another example of a state machine. Figure 16.2 shows a machine that can be used to recognize the strings "ride" and "ryde" (the latter being an Old English spelling). This machine is composed of eight states.

The labels on the pattern-matching machine are the valid characters in the input string that cause a transition to occur. We scan the input string, pulling off one character at a time. If an outgoing edge is found with a matching label, the machine can transition to the next state. For example, state 1 is known as the *start state*. In this state, only the input character "r" is allowed. For any other characters, a failure is indicated. At state 2, either an "i" or "y" is valid, causing the machine to move to either state 3 or 6. If during the scan we make it all the way to either state 5 or 8, the machine has found a match and is said to *accept* the input. States 5 or 8 are known as *final* or *accepting* states. We've indicated these states by using double circles in the diagram.

Suppose we didn't care which spelling of "ride" was matched, only that one of them was. We could modify our pattern matcher to accept either "ride" or "ryde" and have only one final state. Figure 16.3 illustrates this modified machine.

In Figure 16.3, note the two edges between states 2 and 3. Both go in the same direction, emanating from state 2. Either an input character of "i" or "y" will work. The resulting digraph is known as a *multigraph*, a term we first introduced in Chapter 13. In all of the graphs you've seen up to this point, none had more than one edge in the same direction between two vertices.

It's important to remember that the directionality of the edges matters when determining whether a digraph is a multigraph. The machine in Figure 16.1 is not a multigraph, even though it has vertex pairs with two edges between them. What this machine lacks is multiple edges going in the *same* direction.

Figure 16.3 A modified version of the machine in Figure 16.2.

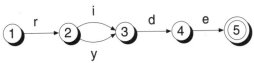

HARD-WIRED FINITE AUTOMATA

Because finite automata are represented with state transition diagrams, we can implement the automata using the graph data structures given in previous chapters. But before we show how to do this, let's look at simpler implementations.

Finite automata can be "hard-wired" directly in C++, using standard control structures. The *switch* statement is particularly convenient for this purpose. Also, it makes sense to encapsulate the machine into a class. The following *MyFSA* class, which implements the machine in Figure 16.3, gives an example.

```
class MyFSA {
private:
  unsigned state;
public:
  MyFSA()                     { Reset();          }
  void Reset()                { state = 1;        }
  unsigned State() const { return state;      }
  int Accept() const       { return state == 5; }
  unsigned Input(char x);
};
```

A test program using the *MyFSA* class is given on disk in the file *cb16_1.cpp.*

The *MyFSA* class keeps track of the state of the machine in the *state* variable, which is reset to the start state when the constructor or the *Reset()* function is called. The workhorse function is *Input()*, which takes a character and tries to move to the next state of the machine.

```
unsigned MyFSA::Input(char x)
// Given input x and the current state, this transitions
// to the next state. Returns the new state, or 0 if x is
// not a valid input.
{
  unsigned new_state = 0; // Default to "failure."

  switch(state) {
    case 1:
      if (x == 'r') new_state = 2;
      break;
    case 2:
      if (x == 'i' || x == 'y') new_state = 3;
      break;
    case 3:
      if (x == 'd') new_state = 4;
      break;
    case 4:
      if (x == 'e') new_state = 5;
      break;
  }
```

```
    return state = new_state;
}
```

The *switch* statement uses the current state to select which comparisons to execute in order to find the next state to transition to. There is one case for each state. Notice the case for state 2, which accepts either an "i" or a "y". This is how to implement multiple edges having the same label or input.

If there is no valid transition, the state is set to 0. We use the convention that state 0 is the null state, which indicates a failure. The null state is not properly a part of the machine. (It's not shown in the transition diagram.) In this example, when the null state is encountered, it means that the match has failed. We have several choices of what to do next. We may "punt" and close the machine down, or we can reset the machine back to the start state and start the matching over. The following code shows an example of using *MyFSA* in this fashion.

```
char buffer[80];

main()
{
  MyFSA fsa;

  cout << "Enter input string: ";
  cin.getline(buffer, 80);

  char *p = buffer;
  int x = 0, len = 0;

  while (*p) {
    if (fsa.Input(*p) == 0) { // Match failure
      if (len == 0) { ++x; ++p; } // Not even a partial match
      fsa.Reset(); // Start matching over.
      len = 0;
    }
    else if (fsa.Accept()) {
        cout << "Match of 'ride | ryde' found at " << (x-len) << '\n';
        fsa.Reset(); // Start matching over.
        len = 0;
    }
    else { // Continue partial matching.
      ++p; ++x; ++len;
    }
  }

  cout << "No more matches found\n";
  return 0;
}
```

In this code, a string is read in and then the *Input()* function of *MyFSA* is called repeatedly for each character. If the new state is the null state, the machine is reset and the matching process is started over. If the new state is the accept state, the matching position is output. We then reset the machine and start over. Here is an example of the output of this program, for the input string "*ryde'em cowboy, with your horse ridden hard and put up wet*":

```
Match of 'ride | ryde' found at 0
No more matches found
```

Note that a partial match is found at the word "ridden". The machine gets all the way to state 4 before failing. A partial match is also found in the word "horse" at the letter 'r'. Here the machine only gets to state 2 before failing.

STATE TRANSITION MATRICES

A more general way to implement a state machine is to use a two-dimensional array called a *state transition table*, or *state transition matrix*. Not surprisingly, this table is similar to an adjacency matrix, since the machine can be represented as a digraph. There is an important difference, though, between an adjacency matrix and a state transition matrix, as we'll discuss next.

Remember that a state transition diagram may in general be a multigraph. But multigraphs cannot be implemented with an adjacency matrix. That's because the matrix has only one entry for each pair of vertices. There's no way to store two edges between a pair of vertices going in the same direction.

In a state transition matrix, multigraphs are handled by using the entries differently. Instead of having one column per vertex, we have one column per input. The matrix entries are the corresponding states to transition to given the current state, represented by the rows, and the current input, represented by the columns. Thus a state transition matrix is an inverted form of an adjacency matrix. Figure 16.4 shows the state transition matrix for the machine in Figure 16.3.

An array-based version of a state transition matrix is fairly easy to implement, as shown by the following class *AStm*.

```
class AStm {
protected:
  unsigned **states;
```

Figure 16.4 State transition matrix of the machine in Figure 16.3.

	r	i	y	d	e
1	2				
2		3	3		
3				4	
4					5
5					

```
    unsigned def_grow_factor, nv_alloc, nv, ne;
    unsigned fail_trans;
    unsigned alphabet_size;
    int Grow(unsigned ga=0);
public:
    AStm(unsigned nv_, unsigned alphabet_size_=256);
    virtual ~AStm();
    void Clear();
    void SetGrowthFactor(unsigned dgf);
    void SetFailTrans(unsigned f);
    unsigned NewState();
    int DetachState(unsigned s);
    int SetEdge(unsigned a, unsigned x, unsigned b);
    int DelEdge(unsigned a, unsigned x, unsigned b);
    void DelAllEdges(unsigned a, unsigned b);
    int ValidEdge(unsigned a, unsigned x, unsigned b);
    unsigned operator()(unsigned src, unsigned x) const;
    int ValidState(unsigned s) const;
    unsigned *AdjStates(unsigned s);
    unsigned NumStates() const;
    unsigned NumEdges() const;
};
```

 Complete code for the *AStm* class is given on disk in the files *astm.h* and *astm.cpp*, with a test program in *astmtst.cpp*.

The *AStm* class is similar in spirit to the *AGraph* class you've worked with in the last few chapters. We decided to implement the underlying matrix differently, though. Instead of using a single one-dimensional array to store all the entries, we use an array of arrays. The *states* pointer points to an array of row pointers, and each row is allocated separately. This allows the array to grow easily to an arbitrary number of rows, although the number of columns in these rows is fixed. (In other words, you still have a maximum size for the matrix, but all unused trailing rows are not stored, reducing the potentially wasted space.) We could have implemented the *AGraph* class using an array of arrays, but we arbitrarily chose not to.

The columns in *AStm* are numbered from 1 to *alphabet_size*. In fact, the class uses 1-based subscripting, as was the case in the graph classes. Again, 0 is used to represent a null vertex, which in this case means the null state. As far as the columns are concerned, it's assumed that the inputs are given unique, consecutive integer handles, starting at 1. For example, in the matrix given in Figure 16.4, it's assumed the characters "r", "i", "y", "d", "e" are stored in a symbol table and assigned handles consecutively.

Because of the similarity to the *AGraph* class, we won't show many details of the *AStm* class. The following *SetEdge()* and *operator()()* functions should give you an idea of the differences.

```
int AStm::SetEdge(unsigned a, unsigned x, unsigned b)
// Creates a new edge going from state a to b, with
// input symbol x. Returns 1 if successful, or 0 if a or b
// is invalid.
{
#ifndef NO_RANGE_CHECK
   if (!ValidState(a) || !ValidState(b)) return 0;
#endif

  unsigned *p = states[a];
  if (p[x] == null_state) ne++;
  p[x] = b;
  return 1;
}

unsigned AStm::operator()(unsigned src, unsigned x) const
// Goes from the src state to a new state, via input symbol x.
// fail_trans is returned if no such transition exists or
// or the src state is invalid.
{
#ifndef NO_RANGE_CHECK
   if (!ValidState(src)) return fail_trans;
#endif
  unsigned *p = states[src];
  if (p[x] == null_state) return fail_trans;
  return p[x];
}
```

The *operator()()* function makes the class look like a two-dimensional matrix and is the main function of the class. It is used to determine what state to transition to, given the current state and input symbol. You'll notice that the *fail_trans* state is returned on failure. This variable is typically set to 0, but you'll see later on that it's sometimes convenient to set *fail_trans* to something else. By setting *fail_trans* equal to the start state, for instance, you can in effect make the machine automatically reset itself upon failure.

As an example of using the *AStm* class, here is another version of the *MyFSA* class. This class, which we've renamed *MatchFSA*, uses a state transition matrix and is thus more general than the *MyFSA* class. To get a different machine, all you need to do is feed *MatchFSA* a different matrix. Note that the class uses a symbol table to map characters into input symbol handles. Also note that the machine can accept only one final state. It can't be used with machines like the one given in Figure 16.2, which has two final states.

```
class MatchFSA {
private:
  AStm &stm;        // the state transition table
  SymTab &symbols;  // the input symbol table
  unsigned state;   // the current state
  unsigned istate;  // the start state
  unsigend fstate;  // the final state
```

```
public:
  MatchFSA(AStm &stm_, unsigned is, unsigned fs, SymTab &symbols_);
  void Reset()            { state = istate;           }
  unsigned State() const { return state;              }
  int Accept() const      { return state == fstate; }
  unsigned Input(char x);
};

MatchFSA
::MatchFSA(AStm &stm_, unsigned is, unsigned fs, SymTab &symbols_)
// Construct a finite state automata, to recognize strings,
// using the state transition matrix and symbol table specified,
// with the given initial state is and initial state fs.
: stm(stm_), symbols(symbols_), istate(is), fstate(fs)
{
  Reset();
}

unsigned MatchFSA::Input(char x)
// Given the input character x and the current state, this transitions
// to the next state. Returns the new state, or 0 if x is not a valid
// input.
{
  char two_buff[2];

  two_buff[0] = x; two_buff[1] = 0; // Must convert char into string.
  unsigned input_symbol = symbols[two_buff];

  return state = stm(state, input_symbol);
}
```

The *MatchFSA* class is given in template form in the files *matchfsa.h* and *matchfsa.mth*. The template parameter, not shown here, is used to plug in different types of state transition matrices, such as the list-based and hash-based versions you'll see shortly.

Here's the string recognizer example we gave earlier, rewritten to take advantage of the *AStm* class.

```
char buffer[80];

  // Construct an STM to hold the 'ride | ryde' recognizer

  AStm stm(5, 5);

  // Reserve room for the five states.
  stm.NewState(); stm.NewState(); stm.NewState();
  stm.NewState(); stm.NewState();

  // Build the input symbol table

  SymTab symbols(211, 8192);
```

```
unsigned r_ch = (unsigned) symbols.Insert("r");
unsigned i_ch = (unsigned) symbols.Insert("i");
unsigned y_ch = (unsigned) symbols.Insert("y");
unsigned d_ch = (unsigned) symbols.Insert("d");
unsigned e_ch = (unsigned) symbols.Insert("e");

// Construct the edges in the graph

stm.SetEdge(1, r_ch, 2);
stm.SetEdge(2, i_ch, 3);
stm.SetEdge(2, y_ch, 3);
stm.SetEdge(3, d_ch, 4);
stm.SetEdge(4, e_ch, 5);

MatchFSA fsa(stm, 1, 5, symbols);

// Okay, let's run this machine.

cout << "Enter input string: ";
cin.getline(buffer, 80);

char *p = buffer;
int x = 0, len = 0;

while (*p) {
  if (fsa.Input(*p) == 0) { // Match failure
    if (len == 0) { ++x; ++p; } // Not even a partial match
    fsa.Reset(); // Start matching over.
    len = 0;
  }
  else if (fsa.Accept()) {
    cout << "Match for 'ride | ryde' found at " << (x-len) << '\n';
    fsa.Reset(); // Start matching over.
    len = 0;
  }
  else { // Matching so far
   ++p; ++x; ++len;
  }
}

cout << "No more matches found\n";
```

We use a symbol table to map the input characters into contiguous handles, starting at 1. But using a symbol table implies some type of lookup, which can slow down the machine significantly. Our implementation of a symbol table uses hashing with separate chaining, so lists must be walked to find a given symbol.

In the case of a character string matcher, we can speed up the machine by using a 256-column matrix, enough to cover all possible 1-byte characters. We could then dispense with the symbol table and use the character value directly to index the columns of the matrix. The resulting matrix will be much larger and much sparser. A lot of space is wasted. Once again, we see a time/space trade-off.

LIST-BASED TRANSITION MATRICES

Another way to solve the problem of non-contiguous input handles and sparse
transition matrices is to use adjacency lists, as we did in previous chapters. In a
sense, we're utilizing the linked lists of a hash table to scan the columns of a
row, but without the hashing. Instead of hashing for the appropriate bucket
(that is, the row), we directly access the row using a subscript, just as in the
AStm class. The result is a more compact structure. Figure 16.5 shows our
sample state transition matrix in linked-list form.

Included on disk is the *LStm* class, which implements state transition
matrices in adjacency list form. This class has a virtually identical interface to
the *AStm* class, so we won't show the details. The implementation of *LStm* is
quite similar to the *LGraph* class developed in Chapter 13. The main difference
is how edges are found, as shown in the following *operator()()* function.

```
unsigned LStm::operator()(unsigned src, unsigned x) const
// Goes from the src state to a new state, via input symbol x.
// fail_trans is returned if we can't find a transition
// or the src state is invalid.
{

#ifndef NO_RANGE_CHECK
  if (!ValidState(src)) return fail_trans;
#endif

  LStmLink *p = states[src];
  while(p) {
    if (p->label == x) return p->goto_state;
    p = p->next;
  }

  return fail_trans;
}
```

Complete code for the *LStm* class is given on disk in the files *lstm.h* and
lstm.cpp, with a test program in *lstmtst.cpp*.

Figure 16.5 List-based version of the matrix in Figure 16.4.

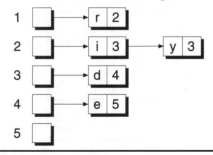

The *operator()()* function of *LStm* searches for an edge based on a specified source state (vertex) and input symbol (edge label). Contrast this with the *operator()()* function of *LGraph*, which uses as its search criteria the source vertex and destination vertex, and returns the edge label in between them. Once again we've inverted the way edges are handled, in part to accommodate the fact that the underlying digraph might be a multigraph.

Because we don't need contiguous handles for the input symbols of a linked-list based state transition matrix, a symbol table for the inputs is not required. The inputs could be used directly, assuming they are unsigned integers. Even that restriction could be removed by modifying the type assumed for the inputs in the *LStm* class.

HASH-BASED TRANSITION MATRICES

Another way to implement a state transition matrix is to use a hash table directly. Recall that the basic operation of a transition matrix is to find a new state given the current state and an input symbol. We could combine the latter two items into a key and use it during the hashing. For example, suppose we implemented a hash table with separate chaining. Figure 16.6 shows a possible configuration of the hash table for our sample state transition matrix. We've arbitrarily chosen to use seven buckets in the table.

The following *HStm* class gives one implementation of a hash-based state transition matrix.

```
union HStmKey { // Used as the hash key
  unsigned long all;
  struct { unsigned a, b; } half;
};
```

Figure 16.6 Hash-based version of the matrix in Figure 16.4.

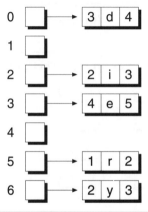

```
class HStmChain {
// Hash bucket chain link class
public:
  HStmKey key;
  unsigned trans_state;
  HStmChain *next;
  HStmChain(HStmKey k, unsigned s, HStmChain *n=0);
};

class HStm {
protected:
  unsigned *istate_goto; // To optimize the initial state
  HStmChain **buckets;
  long num_collisions, num_comparisons;
  unsigned num_edges, num_buckets;
  unsigned num_states;
  unsigned fail_trans;
  unsigned alphabet_size;
  char ok;
public:
  HStm(unsigned tsize, unsigned alphabet_size_ = 256);
  virtual ~HStm();
  void Clear();
  unsigned NewState();
  int SetEdge(unsigned a, unsigned x, unsigned b);
  unsigned operator()(unsigned src, unsigned x) const;
  void SetFailTrans(unsigned f);
  int ValidState(unsigned s) const;
  unsigned NumStates() const;
  unsigned NumEdges() const;
  unsigned NumBuckets() const;
  long NumCollisions() const;
  long NumComparisons() const;
  int OK() const;
};
```

Complete code for the *HStm* class is given on disk in the files *hstm.h* and *hstm.cpp*, with a test program in *hstmtst.cpp*.

The hashing works by combining the current state with the current input symbol into a single *long* integer, which is used as the key. The division method of hashing is used to compute the appropriate bucket. The bucket list is searched looking for an entry with the computed key, and the destination state stored there is returned. The *operator()()* function illustrates this process.

```
unsigned HStm::operator()(unsigned src, unsigned x) const
// Search for the associated edge in the table, returning
// the destination state, or fail_trans if no such edge
// exists.
{
  HStmKey key;
```

```
    key.half.a = src; key.half.b = x;
    unsigned h = unsigned(key.all % num_buckets);
    HStmChain *p = buckets[h];
    while(p) {
      if (p->key.all == key.all) return p->trans_state;
      p = p->next;
    }
    return fail_trans; // Can't find the key.
}
```

The *SetEdge()* function (not shown) uses a similar process when adding an edge to the matrix.

The *HStm* class trades off space for speed. The class may be faster than the both the *LStm* class and *AStm* class under certain conditions. Like the *LStm* class, the hash table uses an array of linked lists, but the array is larger. The idea is to use hashing to randomize where the entries go and thus increase the possibility that the individual lists will be shorter. Shorter lists imply less searching, and the probability of shorter lists increases with more buckets.

In our trivial example, we do have shorter lists, for no list in Figure 16.6 has over one entry, whereas in the *LStm* version, shown in Figure 16.5, one of the list has two entries. Of course, in more realistic examples, the difference is likely to be more significant. To see the benefits of the *HStm* class, you'll need to use more buckets than there are states in the machine, perhaps twice as many. Because of this, the hash table will be a larger structure than that produced by the *LStm* class, but remember that empty buckets simply contain null pointers. The increase in space may not be that dramatic.

As with the *LStm* class, we don't need contiguous handles for the input symbols of a hash-based state transition matrix, so an input symbol table is not required. The inputs can be used directly, assuming they are compatible with the chosen hashing function. In the case of the *HStm* class, this means the inputs need to be unsigned integers, although that's easy to change.

Because no input symbol table is needed for the *HStm* class, it may perform just as well as the *AStm* class. Unless you don't mind sacrificing a lot of space, the array-based version will need an input symbol table. Recognize that most symbol tables are implemented with hashing anyway. With appropriate modifications to the *HStm* class, we could combine the operation of turning the input into a key, with the operation of computing the overall key based on the input and state number. The result is likely to run as fast as, or faster than, the *AStm* class used in conjunction with a symbol table.

The *HStm* class may also yield a smaller table than the *AStm* class, especially under the following conditions: (1) when there are a lot of states, (2) there is a large alphabet size on the inputs, and (3) the matrix is relatively sparse. If the matrix is dense and no input symbol table is used, the *AStm* class will yield a faster and more compact table.

Note that you wouldn't want to use a hash-table for graphs in general. The *HStm* class is biased in finding a destination state given a current state and input symbol. That's due to the way it computes the hashing function. Finding the adjacent edges to a given state would be very expensive. Basically, you would have to search all entries in the table exhaustively. Thus algorithms like finding shortest paths would be miserably slow.

DETERMINISTIC VS. NON-DETERMINISTIC FINITE AUTOMATA

In a state transition diagram, if no two outgoing edges of a state have the same label, then the corresponding machine is called a *deterministic finite automaton* (DFA). In a DFA, there is no ambiguity about what state to go to next, given an input symbol and current state. In a *non-deterministic finite automaton* (NFA), there is such an ambiguity. The same label can appear on two or more edges leaving a state.

Consider the state machine in Figure 16.7, which is a simple model of a vending machine. One of the inputs is *c*, which represents when a coin is inserted into the machine. Another input, *d*, represents the button push made when the customer wants the product dispensed. The *r* input represents when the coin return button is pushed to cancel the transaction.

Notice the edges leaving the "collect coin" state. Two of the edges leaving this state have *c* for their label. As far as the digraph is concerned, it's ambiguous whether to go to the "armed" state or stay in the "collect coins" state. That's the reason for the term *non-deterministic*, and that makes our machine an NFA.

Of course, to actually use an NFA, the ambiguities must be resolved somehow. You can pretend that there is an omnipotent being watching over the machine who knows exactly the correct choice to make when there is an ambiguity. This omnipotent being nudges the machine to make the desired transitions.

In programming terms, the omnipotent being turns out to be some auxiliary code that monitors what's going on, recording information needed to resolve ambiguities. For example, if the NFA is a pattern matcher, a typical

Figure 16.7 A non-deterministic finite automaton.

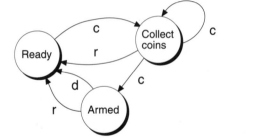

c = coin taken

d = dispense product

r = return coins (cancel)

scenario is for the code to look one or more characters ahead in the input
stream. That way the code can predict what's about to occur and make
decisions accordingly. Many parsers are based on finite automata, and they
often use look-ahead.

Another typical scenario is to keep some type of pushdown stack that
records the inputs seen and perhaps the states traversed as well. The stack is
then used to resolve any ambiguities. In this case, the omnipotent being has
memory and is looking into the past. Such stack-based machines are often
called *pushdown automata,* and they form the basis of many parsers used in
compilers. These types of machines are beyond the scope of what we can
cover here.

In the case of our vending machine, the ambiguity at the "collect coins"
state is resolved by a simple counter that keeps track of how many coins have
been inserted into the machine. Before the proper number has been inserted,
the machine is kept in the "collect coins" state. When the correct number of
coins are received, the machine transitions to the "armed" state, ready to
dispense product.

In all these scenarios, it's important to realize that the code resolving
ambiguities (the omnipotent being, if you will) resides outside the machine
proper. More precisely, the omnipotent being is not a part of the state transition table implementing the machine.

The following *VendorMachine* class shows how we might code the sample
vending machine. Embedded in this machine is a hard-wired NFA. Pay particular attention to case 2 of the *switch* statement inside the *Run()* function. That's
where the non-determinism is resolved.

```cpp
class VendingMachine {
private:
  unsigned state;
  int num_coins_needed, num_coins_obtained;
  void TakeCoin();
  void DispenseProduct() { cout << "Product dispensed\n"; Reset(); }
  void ReturnCoins()     { cout << "Coins returned\n";    Reset(); }
public:
  VendingMachine(int ncn);
  void Reset()           { state = 1; num_coins_obtained = 0; }
  unsigned State()       { return state; }
  void Run();
};

VendingMachine::VendingMachine(int ncn)
: num_coins_needed(ncn)
{
  Reset();
}

void VendingMachine::TakeCoin()
{
```

```cpp
      num_coins_obtained++;
      cout << "Coin taken, thank you. Need "
           << (num_coins_needed - num_coins_obtained) << " more.\n";
}

void VendingMachine::Run()
// NOTE: Use Ctrl-C to get out!
{
   char cmd;

   while(1) {
     switch(state) {
       case 1:
         cout << "Ready for coin (press c): ";
         cin >> cmd;
         if (cmd == 'c') {
            TakeCoin();
            state = 2; // Assuming more than one coin needed
         }
       break;
       case 2:
         cout << "Ready for coin (press c, or r to return coins): ";
         cin >> cmd;
         if (cmd == 'c') {
            TakeCoin();
            if (num_coins_obtained == num_coins_needed) state = 3;
            // Else we stay put in state 2.
         }
         else if (cmd == 'r') ReturnCoins();
       break;
       case 3:
         cout << "Ready to dispense product, "
                 "(press d, or r to return coins): ";
         cin >> cmd;
         if (cmd == 'd') DispenseProduct();
         else if (cmd == 'r') ReturnCoins();
       break;
     }
   }
}

main()
{
   VendingMachine vm(4); // Price of product is four coins.

   cout << "Steal-A-Coin's Vending Machine at your service!\n";
   vm.Run(); // Use Ctrl-C to get out.

   return 0;
}
```

 Code for the *VendorMachine* class is given on disk in the file *vendor.cpp*.

You might wonder why we would use NFAs at all. Aren't DFAs simpler? Actually, NFAs are often easier to construct and are in general more powerful than DFAs. NFAs can also be much smaller, having fewer states than DFAs that handle the same tasks. However, DFAs are usually faster, since there isn't any non-determinism to tackle at runtime. There are methods for turning NFAs into DFAs. (See the famous "dragon book" [Aho 88], for instance.) There are also methods for "compressing" DFAs into fewer states, but these methods are far beyond what we can show here.

Rather than hard-wiring an NFA, we might desire to use a state transition table instead. But could we use the *AStm*, *LStm*, or *HStm* classes? The answer is yes, but modifications are needed. The *AStm* class can't be used as it is given, because there is no way to store the multiple states for a non-deterministic edge. Each cell of the matrix has room for only one state. The class could be modified to handle an NFA by having each cell store a pointer to a list of states.

The *HStm* class suffers from a similar problem because the non-deterministic edges all have the same key. Usually duplicate keys are not allowed in a hash table. We could modify the *HStm* class to handle an NFA by storing, in each linked-list link, not just one destination state but a pointer to another list of states.

In principle, the *LStm* class can handle NFAs just fine, since each edge has its own link. But please note that we have not tested the class in this capacity.

THE AHO-CORASICK MACHINE

While it's okay to hard-wire finite state machines or build them by hand using one of the graph classes, it's more desirable to have the machines built automatically. Some classes of finite state machines lend themselves well to automatic construction, particularly those in the field of compiler parsers. But we don't have room to discuss these machines. However, we will show an example of automatic machine construction with a simple machine that can be used to match multiple patterns in parallel. This machine, developed by A. Aho and M. Corasick, is known as the *Aho-Corasick machine*.

In their seminal paper (see [Aho 75]), Aho and Corasick describe a method for taking a set of (possibly overlapping) patterns to be matched and producing a finite state machine that can be used to match any of the patterns. The machine is made up of three components. (We use their terminology, even though it's somewhat confusing): (1) a *Goto function*, (2) a *Failure function*, and (3) an *Output function*. Suppose our set of patterns is {"bold", "old", and "older"}. Figure 16.8 shows the first two components of an Aho-Corasick machine that can recognize these patterns.

You'll see from the figure that the *Goto* function is nothing more than a state transition diagram (directed graph) that shows the states and transitions necessary to match the set of patterns. In the graph, the edge with the "¿" label

Figure 16.8 The main components of an Aho-Corasick machine.

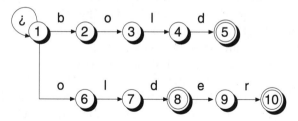

(a) *Goto* function

1	2	3	4	5	6	7	8	9	10
1	1	6	7	8	1	1	1	1	1

(b) *Failure* function

stands for all characters that don't have their own edge. In the real graph, this pseudo-edge is actually composed of many edges, one for each character in the alphabet besides "b" and "o". The states drawn with double circles are accepting states. At these states a match has been found. The accepting states are tied to the *Output* function you'll see later.

The *Goto* function can't handle the pattern matching all on its own. The *Failure* function is used to decide what state to transition to when a failure occurs. Suppose the text being scanned is "bolder". The *Goto* function will fail at state 5, since there isn't any transition from that state using input "e". (In fact, there are no transitions at all.) We don't want the machine to simply give up at this point, because we're actually in the middle of a match for the word "older". The *Failure* function determines that we should transition to state 8. It just happens (not by accident, of course) that state 8 is exactly the state we need to be in to continue the matching.

Consider what happens if the string is "bolt". Now the machine fails at state 4. The *Failure* function tells us to go to state 7, since it thinks we might be matching either "old" or "older". However, we fail at state 7 too, and the *Failure* function tells us to go back to state 1 to start all over. The graph is designed so that we never fail at state 1, so the *Failure* function for that state actually is never used. When a failure occurs, we might spin around using the *Failure* function until we find a state we don't fail at. That may well be the starting state.

In both failure cases of our example, there is an invisible arc going from the state of failure to the state of recovery, all handled by the *Failure* function. In a sense, this arc is non-deterministic. The *Failure* function becomes the "omnipotent being" that magically determines where to go next. In this light, the *Goto* function can be treated as an NFA.

A match is indicated when the machine reaches one of the accept states. In our example, this means states 5, 8, and 10. The *Output* function is used to determine which pattern was matched. The function can be implemented as an array, one entry per state. There can be more than one pattern that matches at a given accept state. This occurs when one pattern is a suffix of another. In our example, the pattern "bold" contains the suffix "old", which is also a pattern. Thus each entry in *Output* is in general a list of matching patterns. Table 16.1 shows the *Output* function for our sample machine.

The ACS Class

Next we'll show a class, *ACS*, that encapsulates the Aho-Corasick machine, and we'll use that as the basis for further discussion. Since this machine is a searching engine, we've modeled it after the search classes of Chapters 11 and 12.

```
class ACS {
protected:
  LStm gotof;              // The goto "function" (digraph)
  RszArr<unsigned> output; // Output symbol codes, one for each state
  unsigned *failure;       // Failure function for each state
  SymTab *symbols;         // Table of patterns to match
  char *text, *t, *et;     // Text being searched
  int n;                   // Length of text
```

Table 16.1 The Output Function for the Machine in Figure 16.8

State	Matching patterns
1	n/a
2	n/a
3	n/a
4	n/a
5	"bold", "old"
6	n/a
7	n/a
8	"old"
9	n/a
10	"older"

```
  unsigned curr_state;       // Current state of state machine
  int shortest_match;        // Controls shortest/longest matching
  const unsigned alphabet_size; // How many unique inputs?
  int AddPattern(unsigned sym);
  int ConstructGoto();
  int ConstructFailure();
public:
  ACS(unsigned init_cap, unsigned alphabet_size_ = 256);
  virtual ~ACS();
  void MatchPreference(int prefer_short);
  void Reset();
  virtual int Reset(SymTab &st);
  int SetPattern(SymTab &st);
  void SetText(char tx[], int n_);
  virtual int Search();
  const char *MatchStr() const;
  unsigned MatchSym() const;
  virtual unsigned NumStates() const;
  virtual unsigned NumEdges() const;
};
```

Complete code for the *ACS* class is given on disk in the files *acs.h* and *acs.cpp*, with test programs in *acstst1.cpp* and *acstst2.cpp*. The files on disk contain error-checking and debugging code that we omit here for the sake of clarity.

The *ACS* class uses a state transition matrix for the *Goto* function, which we arbitrarily use the *LStm* class for. Like other search classes in this book, two functions, *SetText()* and *SetPattern()*, are provided to set up the searching engine. The *SetPattern()* function takes as input a symbol table, which is meant to hold the set of patterns we are matching. *SetPattern()* ultimately causes the two functions *ConstructGoto()* and *ConstructFailure()* to be called. These are responsible for building the corresponding matrix and failure array. Let's look at the *ConstructGoto()* function first.

```
int ACS::ConstructGoto()
// Construct the Goto "function", and partially construct the output
// function. Assumes the Goto function and output array are empty.
// Returns 1 on success, else 0 on memory allocation failure.
{
  // Make room for states 0(null_state) and 1(istate).
  gotof.NewState(); // 1-based
  output.Append(0); output.Append(0); // 0-based

  // Add transitions for each string in the symbol table.
  unsigned n = symbols->NumEntries();
  for (unsigned sym = 1; sym<n; sym++) {
      if (AddPattern(sym) == 0) return 0;
  }

  // If there are no edges from the start state for any
  // given input symbol, add an edge back to the start state.
```

```
    for (unsigned c = 0; c<alphabet_size; c++) {
        if (gotof(istate, c) == null_state)
            gotof.SetEdge(istate, c, istate);
    }

    return 1;
}
```

This function takes each pattern in the symbol table and adds states and edges to the state transition matrix accordingly. The *AddPattern()* function is responsible for most of the work.

```
int ACS::AddPattern(unsigned sym)
// Enter the string referenced by the symbol handle into the state
// transition matrix. ASSUMES room for null_state and istate are
// already allocated. Returns 1 on success, else 0 if allocation failure.
{
    unsigned rv, state, nstate, xstate;

    const char *t = (*symbols)[sym]; // Get the character string.
    state = istate;

    while(1) { // Walk existing prefixes as long as possible.
        xstate = gotof(state, *t);
        if (xstate == null_state) break;
        state = xstate;
        t++;
    }

    while(*t) { // Add a new branch, or extend existing one.
        nstate = gotof.NewState();    // Make room for another state.
        if (nstate == 0) return 0;
        rv = output.Append(0);        // Make room for another output.
        if (rv == 0) return 0;
        if (gotof.SetEdge(state, *t, nstate) == 0) return 0;
        state = nstate;
        t++;
    }

    output[state] = sym; // Record longest matching symbol to this state.

    return 1;
}
```

Incorporating a pattern into the matrix is fairly simple. The basic idea is to try to use any existing prefixes in the graph as long as possible. Branches are added or extended by allocating new states and creating new edges.

Another task of *AddPattern()* is to partially set up the *output* array. Rather than make each entry in the array a list of patterns, we decided to simplify things by recording either the longest pattern or the shortest pattern, based on a preference flag set by the user. The last state added by or encountered by the end of a call to *AddPattern()* is an accepting state, with the pattern just added as one of the possible matches. In fact, the pattern is the longest one possible at that point. So we record its symbol handle in the *output* array. If the user desires to use the shortest matching pattern instead, that is handled later in the *ConstructFailure()* function.

After incorporating all the patterns in the *Goto* transition matrix, the *ConstructGoto()* function performs one additional task. For every input character that doesn't have an edge leaving the initial state (meaning that the character doesn't start a pattern), an edge is added back to the initial state. This prevents a failure from ever occurring on the initial state. If the alphabet is large, a lot of edges may be added at this point. Basically, we're trading space for speed. We could let the machine fail at the initial state and then use the *Failure* function determine what to do next. Of course, that function will turn around and send the machine back to the initial state. We bypass the extra processing by adding the edges. In our example, if you have an alphabet size of 256, the digraph has 263 edges, 254 of which are simply loops around the initial state.

Once the transition matrix is built, the *ConstructFailure()* function is called to build the *failure* array.

```
int ACS::ConstructFailure()
// Construct the failure "function".
// NOTE: ConstructGoto() must be called before this.
// NOTE: ASSUMES failure array is empty coming in.
// Returns 1 if successful, else 0.
{
  LStaq<unsigned> que;
  unsigned c, state, qstate, xstate;

  // Set failure array to have as many states as the goto graph.

  failure = new unsigned[gotof.NumStates()+1]; // emulating 1-based
  if (failure == 0) return 0;
  failure[0] = istate; // Never used, but we'll set anyway.
  failure[1] = istate; // Never used, but we'll set anyway.

  // Queue up all transitions out of the start state that don't
  // loop right back to the start state.

  for(c = 0; c<alphabet_size; c++) {
     state = gotof(istate, c);
     if (state != istate) {
        que.Insert(state);
        failure[state] = istate;
     }
  }

  // Compute the failure function for each state, one level at
  // a time. (A lot like breadth-first traversing)
```

```
while(que.Extract(qstate)) {
   for (c = 0; c<alphabet_size; c++) {
       state = gotof(qstate, c);
       if (state != null_state) {
           // We've found a non-null transition emanating from qstate.
           // Queue up the new state, and then compute its failure
           // function. Basically, we want to find a place where we
           // can continue matching, if at all possible.
           que.Insert(state);
           xstate = failure[qstate];
           while(gotof(xstate, c) == null_state) {
             xstate = failure[xstate];
           }
           failure[state] = gotof(xstate, c);
           // If we'll be doing shortest matching, then
           // keep track of what the shortest symbol for
           // this new state is. More completely, you could
           // append all symbols to a list of outputs for
           // failure[state].
           if (shortest_match) {
               unsigned closest_sym = output[failure[state]];
               if (closest_sym) output[state] = closest_sym;
           }
       }
   }
}

   return 1;
}
```

This functions works by building the *failure* array one level at a time, where the depth of a state is the length of its path from the start state. Note that a queue is used. This suggests that a breadth-first traversal is taking place, which is indeed the case. The queue, in conjunction with the partially built *failure* array, allows us to determine what state to transition to in case of a failure. (It may take awhile before you understand how this function works).

You'll notice that part of *ConstructFailure()*'s task is to finish building the *output* array. This occurs only if the user wants the shortest match returned rather than the longest. It just so happens that if the failure transition state is not null, then that state has for its output the short pattern you're looking for. An example of this can be seen in Figure 16.8. Suppose the state you're interested in is state 5. The longest matching pattern at that point is "bold". The shortest is the suffix "old". But notice how the *Failure* function for state 5 points to state 8, which happens to have a matching pattern of its own, and that pattern, not by accident, is "old".

After the state transition matrix and *failure* array are constructed, the machine is ready for searching. The following *Search()* function uses the text string set by the *SetText()* function as the input string and then runs the state machine for each character of the input string.

```
int ACS::Search()
// Searches for the next match. Returns ending position of
// the match, or -1 if no match.
{
  unsigned new_state;

  while(t < et) {
    while((new_state = gotof(curr_state, *t)) == null_state) {
      curr_state = failure[curr_state];
    }
    curr_state = new_state;
    t++;
    if (output[curr_state]) {
        int endpos = int(t - text) - 1;
        return endpos;
    }
  }
  return -1; // No match
}
```

The searching consists of a nested loop. The inner loop spins around if we have a failure, looking for a state we can transition to that doesn't fail on the given the current input character. Quite often we'll wind up at the initial state, which is guaranteed not to fail. Once we've found a valid state, the *output* array is tested. If it has a non-null symbol, then we have a match, and the ending position of the match is reported. If you need to know what pattern was matched, one of the two following functions can be used.

```
unsigned ACS::MatchSym() const
// Returns the handle to the symbol corresponding
// to the string we just matched. Returns 0 if
// not at a match state.
{
  return output[curr_state];
}

const char *ACS::MatchStr() const
// Returns a pointer to the string that we just matched.
// Returns 0 if not at a match state.
{
  unsigned sym = output[curr_state];
  if (sym) return (*symbols)[sym]; else return 0;
}
```

Building a Deterministic Aho-Corasick Machine

The non-determinism of the Aho-Corasick machine can be seen clearly in the *Search()* function, as many accesses are made to the *failure* array in case of failure, in order to determine what to do next. It's possible to convert the state transition matrix into a DFA. By doing this, the *Search()* function will never make more than one transition for each input character. The *failure* array won't be used at all.

The following *ConstructDFA()* gives the algorithm for creating the *DFA*. The function is part of the *ACSDFA* class, which we derived from the *ACS* class (not shown here but given on disk).

```
int ACSDFA::ConstructDFA()
// Given the goto and failure functions, this constructs
// a deterministic finite automaton.
// NOTE: Must be called only after ConstructFailure(),
// and ASSUMES dfa is empty coming in.
// Returns 1 on success, else 0.
{
  LStaq<unsigned> que;
  unsigned rv, c, qstate, state, xstate;

  // Set it up so that operator()() returns istate on failure.

  dfa.SetFailTrans(istate);

  // The DFA wil have as many states as the goto function.

  unsigned n = gotof.NumStates();
  for (unsigned i = 0; i<n; i++) {
      rv = dfa.NewState();
      if (rv == 0) return 0;
  }

  for (c = 0; c<alphabet_size; c++) {
      xstate = gotof(istate, c);
      if (dfa.SetEdge(istate, c, xstate) == 0) return 0; // Error
      if (xstate != istate) {
        if (que.Insert(xstate) == 0) return 0; // Queue full
      }
  }

  while(que.Extract(qstate)) {
    for (c = 0; c<alphabet_size; c++) {
        state = gotof(qstate, c);
        if (state != null_state) {
           if (que.Insert(state) == 0) return 0;
        }
        else {
           state = dfa(failure[qstate], c);
        }
        // Don't add transitions back to istate, otherwise
        // you'll get hordes of edges. Instead rely on istate
        // being returned on a failed transition.
        if (state != istate) {
            if (dfa.SetEdge(qstate, c, state) == 0) reutrn 0; // Error
        }
    }
  }
```

```
  return 1;
}
```

Complete code for the *ACSDFA* class is given on disk in the files *acsdfa.h* and *acsdfa.cpp*, with test programs in *dfatst1.cpp* and *dfatst2.cpp*.

The logic behind the *ConstructDFA()* function is rather difficult to explain (your author has trouble undertsanding it himself), so we leave the explanation to the original Aho-Corasick paper.

Building the DFA is another example of trading space for speed. Although the DFA will have as many states as the *Goto* digraph, the DFA is likely to have many more edges. In fact, we've seen cases where almost *ten times* as many edges are added. This is due to the fact that the DFA must not have any non-determinism, and it does this by handling every possible transition that could come about by the original *Failure* function.

The DFA would have even more edges if we hadn't made an optimization in *ConstructDFA()*. In the original NFA version of the Aho-Corasick machine, it's likely that the failure transitions will lead to state 1 quite often. In the DFA version, we would have to add a lot edges back to state 1, the number being affected by the alphabet size (the larger the alphabet the worse the number). That's because at any given state, only a few input characters are likely to lead to transitions other than failures. Instead of adding all these edges, we modify the *operator()()* function of the state transition matrix to return the initial state rather than the null state when no appropriate edge is found. This behavior is set up by the call to *SetFailTrans()* in the *ConstructDFA()* function.

The reason to put up with a larger graph for the DFA is that the code for searching is simpler and faster. Here is the *Search()* function of the *ACSDFA* class.

```
int ACSDFA::Search()
// Searches for the next match using the dfa. Returns the
// ending position of the match, or -1 if no match.
{
  while(t < et) {
    curr_state = dfa(curr_state, *t++);
    if (output[curr_state]) {
       int endpos = int(t - text) - 1;
       return endpos;
    }
  }
  return -1; // No match
}
```

In this version, only one transition is made per input character, leading to a faster search.

Applications of the Aho-Corasick Machine

The Aho-Corasick machine can match many patterns in parallel, similar to the way the Shift-Or algorithm in Chapter 12 could. There is a big difference in these two approaches, though. The Shift-Or algorithm is concerned with matching classes of patterns, where the patterns are all related, with slight modifications (such as a different character in a certain spot). In contrast, the Aho-Corasick machine is more concerned with sets of patterns that may be unrelated, although the machine can handle overlapping patterns quite nicely.

The Aho-Corasick machine is at its heart a very simple machine. It's also quite fast for what it does. It's useful in applications such as creating library databases, where the user gives a set of keywords and the database retrieves all articles that contain any of the keywords.

Another intriguing use of the Aho-Corasick machine is to filter "stop words" from a file on the fly, while collecting a set of keywords contained in the file. A *stop word* is a word deemed superfluous and not a keyword proper. For example, it's unlikely that words like "a", "and", "the", and so forth, would be considered keywords.

To filter out stop words, the Aho-Corasick machine could be used as follows: First, use the list of stopwords as the set of patterns to be matched. Then write a program that gathers up all the words seen in a file. Each word must be parsed using some sort of criteria for where the word boundaries are. When scanning each character, run one step of the Aho-Corasick machine. You'll have to construct *Step()* functions, which are the inner guts of the *Search()* functions we've given. Each step is to cycle the machine to the next state. When the end of a word is found by the parser, you can consult the Aho-Corasick machine to see if it's in a matching state. If so, you know you've found a stop word, and you can thus ignore the word.

In the preface of this book, we promised that we would have "no exercises left to the reader." We've waited until the very end of this book to break that promise. You'll have to construct a stopword filter on your own! That being said, we hope you've enjoyed this book!

Bibliography

[Aho 75] "Efficient String Matching: An Aid to Bibliographic Search," A. Aho and M. Corasick, *Communications of the ACM*, vol. 18, no. 6, June 1975, pp. 333-340.

[Aho 83] *Data Structures and Algorithms*, A. Aho, J. Hopcroft, and J. Ullman (Addison-Wesley, 1983), ISBN 0-201-00023-7.

[Aho 88] *Compilers: Principles, Techniques, and Tools*, A. Aho, R. Sethi, and J. Ullman (Addison-Wesley, 1988), ISBN 0-201-10088-6.

[AshWa 85] *Lucid, the Dataflow Programming Language*, W. Wadge and E. Ashcroft (Academic Press, 1985), ISBN 0-12-729650-6.

[Baeza 92] "A New Approach to Text Searching," R. Baeza-Yates and G. Gonnet, *Communications of the ACM*, vol. 35, no. 10, October 1992, pp. 74-91.

[Barr 81] *The Handbook of Artificial Intelligence*, A Barr and E. Feigenbaum (William Kaufmann, Inc., 1981), ISBN 0-86576-005-5.

[Bently 86] *Programming Pearls*, J. Bentley (Addison-Wesley, 1986), ISBN 0-201-10331-1.

[Bently 88] *More Programming Pearls*, J. Bentley (Addison-Wesley, 1988), ISBN 0-201-11889-0.

[BoyerMoore 77] "A Fast String Searching Algorithm," R. Boyer and J. Moore, *Communications of the ACM*, vol. 20, no. 10, October 1977, pp. 762-772.

[Bratley 87] *A Guide to Simulation*, 2nd ed., P. Bratley, B. Fox, and L. Schrage (Springer Verlag, 1987), ISBN 0-387-96467-3.

[Dufrene 92] "An Efficient External Sort Algorithm with No Additional Space," W. Dufrene and F. Lin, *The Computer Journal*, vol. 35, no. 3, 1992, pp. 308-310.

[Flamig 93] *Practical Data Structures in C++*, B. Flamig (John Wiley & Sons, 1993), ISBN 0-471-55863-X.

[Frakes 92] *Information Retrieval*, W. Frakes and R. Baeza-Yates, eds., (Prentice-Hall, 1992), ISBN 0-13-463837-9.

[Gomory 66] "The Travelling Salesman Problem," R. Gomory, *Proceedings of the IBM Scientific Computing Symposium on Combinatorial Problems*, IBM Data Processing Division, New York, pp. 93-121.

[Gonnet 91] *Handbook of Algorithms and Data Structures*, G. Gonnet and R. Baeza-Yates (Addison-Wesley, 1991), ISBN 0-201-41607-7.

[Graham 89] *Concrete Mathematics*, R. Graham, D. Knuth, and O. Patashrik (Addison-Wesley, 1989), ISBN 0-201-14236-8.

[Griswold 83] *The Icon Programming Language*, R. Griswold and M. Griswold (Prentice Hall, 1983), ISBN 0-134-47889-4.

[Harel 87] *Algorithmics*, D. Harel (Addison-Wesley, 1987), ISBN 0-201-19240-3.

[Horowitz 90] *Data Structures in Pascal*, 3rd ed., E. Horowitz and S. Sahni (Computer Science Press, 1990), ISBN 0-7167-8217-0.

[Horspool 80] "Practical Fast Searching in Strings," R. Horspool, *Software Practice and Experience*, vol. 10, 1980, pp. 501-506.

[Knuth 69] *The Art of Computer Programming, Vol. 2: Seminumerical Algorithms*, D. Knuth (Addison-Wesley, 1969), ISBN 0-201-03802-1.

[Knuth 73] *The Art of Computer Programming, Vol. 3: Sorting and Searching*, D. Knuth (Addison-Wesley, 1973), ISBN 0-201-03803-X.

[Knuth 77] "Fast Pattern Matching in Strings," D. Knuth, J. Morris, and V. Pratt, *SIAM Journal on Computing*, vol. 6, 1977, pp. 323-350.

[Koza 92] *Genetic Programming*, J. Koza (MIT Press, 1992), ISBN 0-262-11170-5.

[L'Ecuyer 88] "Efficient and Portable Combined Random Number Generators," P. L'Ecuyer, *Communications of the ACM*, vol. 31, no. 6, June 1988, pp. 742-774.

[Larson 88] "Dynamic Hash Tables," P. Larson, *Communications of the ACM*, vol. 31, no. 4, April 1988, pp. 446-457.

[ParkMill 88] "Random Number Generators Are Hard to Find," S. Park and K. Miller, *Communications of the ACM*, vol. 31, no. 10, October 1988, pp 1192-1201.

[Press 88] *Numerical Recipes in C*, W. Press, B. Flannery, S. Teukolsky, and W. Vetterling (Cambridge Press, 1988), ISBN 0-521-35465-X.

[Rytter 80] "A Correct Preprocessing Algorithm for Boyer-Moore String Searching," W. Rytter, *SIAM Journal on Computing*, vol. 9, 1980, pp. 509-512.

[Sedgewick 77] "Permutation Generator Methods," R. Sedgewick, *Computing Surveys*, vol. 9, no. 2, June 1977, pp. 137-163.

[Sedgewick 92] *Algorithms in C++*, R. Sedgewick (Addison-Wesley, 1992), ISBN 0-201-51059-6.

[Stroustrup 91] *The C++ Programming Language*, 2nd ed., B. Stroustrup (Addison-Wesley, 1991), ISBN 0-201-53992-6.

[Sunday 90] "A Very Fast Substring Search Algorithm," D. Sunday, *Communications of the ACM*, vol. 33, no. 8, August 1990, pp. 132-142.

[Sunday 91] "Fast String Searching," A. Hume and D. Sunday, *Software Practice and Experience*, vol. 21, no. 11, November 1991, pp. 1221-1248.

[Tarjan 72] "Depth-First Search and Linear Graph Algorithms," R. Tarjan, *SIAM Journal on Computing*, vol. 1, no. 2, 1972.

[VanLe 93] *Techniques of Prolog Programming*, T. Van Le (John Wiley & Sons, 1993), ISBN 0-471-57175-X.

[Wu 92] "Fast Text Searching Allowing Errors," S. Wu and U. Manber, *Communications of the ACM*, vol. 35, no. 10, October 1992, pp. 83-91.

Index

A

A* algorithm, 392–402, 403
ABicon class, 360
ACS class, 424–429
ACSDFA class, 430–431
Activity network, 321
Acyclic graphs, 321
AddPattern() function, 426–427
ADFV class, 347
Adjacency lists, 332–336
Adjacency matrices, 323–332
    triangular, 328–332
Adjacent edge iterator class hierarchy and, 338–339
AdjEdges() function, 326, 327, 330, 331, 333, 337–339
Admissability condition, 401
AGraph class, 324–327, 387–388
Aho-Corasick machine, 422–432
    applications of, 432
    deterministic, 429–431
Algorithm analysis (comparing algorithms), 11–21, 23
    asymptotic analysis, 11–12
    big-Oh notation and, 11–12, 18
    combining upper bounds, 14–15
    practical considerations in, 18–21
Algorithmic generators, *see* Generators
Algorithmic solutions, 16–17
Algorithms
    comparing, *see* Algorithm analysis
    complexity functions of, 6–17
    defining steps in, 8–9
    definition of, 1
    effectiveness of, 6
    in pseudo code, 4–5
    as recipes for action, 1–2
All pairs shortest path (APSP) problem, 386
APFV class, 383–385
Approximate pattern matching, 299
ARouteFinder class, 397
Arrange() function in *Heap* class, 194–195
Array-based graphs, priority-first visitor class for, 383
Articulation points, 356
Astm class, 410–412
AStrongcon class, 361
Asymptotic analysis, 11–12

B

Backtracking, 44, 98
Balanced 2-way merge sorting, 248–251
Base cases, 34
Best-cost searching, 393, 400
Biconnected components, 356
Biconnectivity, 355–361
Big-Oh notation, 11–12, 18
Binary search algorithm, 26–29
    recursive, 33
BinarySearch() function, 27–28
Binary trees, complete, 185–186
Binomial coefficients, 101–102
Blind searching, 392
BMLSort() function, 249–251, 254
BMSearch() function, 289–291
Boyer-Moore (BM) algorithm, 282–292
    scan loop skipping and, 283–285
    skipping after the match loop, 285–289
    variations on, 291–296
        Boyer-Moore-Horspool (BMH) algorithm, 292–293
        least-frequent character optimization, 295–296
        quick-search algorithm, 293–295
        Tuned-Boyer-Moore algorithm (TBM), 292
Boyer-Moore-Horspool (BMH) algorithm, 292–293
Breadth-first level, 368
Breadth-first number, 368
Breadth-first spanning forest, 366
Breadth-first spanning tree, 366
Breadth-first traversal, 365–371
    topological sorting and, 369
Break statements, 27–30
Bubble sort algorithm, external, 267–271
BubbleUp() function
    of *Heap* class, 189–190
    of *iHeap* class, 210
BubbleUpMax() function of *Deap* class, 202–203
BubbleUpMin() function of *Deap* class, 203
Bubble-up operation
    in deaps, 200–201
    in heaps, 189–190
Buckets, 150
BucketSort() function, 238–239

C

CCS class, 306–307
Chain class, 156, 157
Character class matching, Shift-OR algorithm and, 305–312
Characteristic vector, 301–302
Character strings, hashing, 153–155
Clear() function, 143
  of *CoalHashTable* class, 176–177
ClosestMatch() function, 316
Clustering
  primary, 163
  secondary, 163
Coalesced chaining strategy of collision resolution, 155, 174–178
CoalHashTable class, 175–178
Collision resolution strategies, 150, 155–178, 180
  coalesced chaining, 155, 174–178
  open addressing, 155, 161–174, 180
  separate chaining, 155–160
Collisions, 150, 151
Combination generator, 102–104
Combinations, 101–104
Combinator class, 104
Combined random number generators, 120–123
Comparing algorithms, *see* Algorithm analysis
Complete binary trees, 185–186
Complexity functions, 6–17
  comparisons of, 15–16
  space, 7–8
  time, 7–11
ConfigureCycler() function, 142
Connected components, 351
  strongly, 354
Connected graphs, 351
ConstructDFA() function, 430, 431
ConstructFailure() function, 427–428
ConstructGoto() function, 425–426
Constructors, gotos and, 30–31
Co-recursive recursion, 53
CRan31 class, 120–123
Cross edge, 354
Curr() function, 58, 67, 350

D

Data flow, generators and, 77–78
Dataflow programming, 78
Deap class, 199–200
DeapPartner() function of *Deap* class, 200–201, 202
Debugging, generators as aids in, 67–70

Delete() function
  of *Deap* class, 208
  of *SepHashTable* class, 160
Deletions
  in deaps, 207–208
  in heaps, 193–194
  open addressing and, 173–174
Depth-first iterator, 347–351
Depth-first level, 343
Depth-first number, 343
Depth-first spanning forest, 352
Depth-first spanning tree, 342
Depth-first traversal, 341–351
  algorithm for, 343–347
  biconnectivity and, 355–361
  connectivity of digraphs and, 354–355
  connectivity of graphs and, 351–354
  depth-first iterator and, 347–351
  transitive closure and, 363–365
Design of algorithms, practical issues in, 21–23
Destructors, gotos and, 31–32
Deterministic algorithms, 17
Deterministic finite automaton (DFA), 419–422
Digraphs, 320
  connectivity of, 354–355
Dijstrka's algorithm, 385–386
Diminishing increment sort, 217
Directed acyclic graphs (dags), 321–322, 369
Direct recursion, 52
Dissolving images, 136, 145–147
Distribute() function, 252–254, 256, 259
Divide-and-conquer technique
  binary search algorithm and, 33–36
  quick sort algorithm and, 222
Division method of hashing, 152
Don't care symbol, 306
Double-ended heaps (deaps), 197–208
  extracting the maximum item from, 205–207
  extracting the minimum item from, 203–205
  inserting items, 200–203
  invalid, 198–199
  replacing and deleting items in, 207–208
  subscripting for, 199–200
Double-ended priority queues, 197
Double hash probing, 164–166
Do-while loops, 25

E

EBSort() function, 269–270
Edges, 319
EdgeWt() function, 334

80-20 percent rule, 22
ElfHash() function, 154–155
EQSort() function, 265–266
Euclid's Algorithm, 109
Exact-sum knapsack problem, 43–49
Exiting loops, 25–26
    from the middle, 26–30
    while in switch statements, 30
Exponentation, integer, 34–36
Exponential algorithms, 51
Extendible hashing, 179–180
External merge sorting, 254–272
    bubble sorting, 267–271
    comparison of algorithms, 271–272
    quick sorting, 263–267, 271
    random-access sorting, 262
    replacement selection and, 258–260
Extract() function, 139
    of *Heap* class, 190–192
Extracting
    the maximum item from a deap, 205–207
    the minimum item from a deap, 203–205
Extractions, in heaps, 190–192
ExtractMax() function of *Deap* class, 205–206
ExtractMin() function of *Deap* class, 204

F

Factor generators, 90–95
    performance issues, 95
    testing for primes and, 94–95
Fibber generator, 67
Fibonacci numbers, 51–52
    generators of, 65–67
FiHeap class, 377
File-based hashing, 178–179
FilePak class, 256–257
Filter() function, 139
Finite state machines (automata), 321, 405–432
    Aho-Corasick machine, 422–432
    deterministic vs. non-deterministic, 419–422
    hard-wired, 408–410
    state transition diagrams and, 405–407
    state transition matrices, 410–419
        hash-based, 416–419
        list-based, 415–416
Floyd's algorithm, 195, 386–389
For loops, 25
FRan31 class, 118
FstarMST() function, 379
FstarSSSP() function, 381
FstarUSR() function, 393–394

Full-cycle generator class, 108–112
Full-cycle linear congruential generators (LCGs), 106–108
FullCycler class, 110–112

G

Gcd() function, 109
Generators, 55–83, 82–83
    algorithm inheritance and, 78–79, 81–82
    canonical form for, 63–65
    combination, 102–104
    creating, 57–58
    data flow and, 77–78
    as debugging aids, 67–70
    factor, 90–95
    instruction, 70–77
    iterators and, 77
    knapsack, 57–63
    linear congruential, 105–125
        combined random number generators, 120–123
        full-cycle, 106–108
        full-cycle generator class for, 108–112
        local vs. global randomness and, 117–118
        Minimal Standard Generator, 114–118
        overflow problem and, 114
        random bit sequences and, 123–125
        random number sequences and, 112–114
        spectral test and, 119
    number, 65–67
    performance issues and, 82
    permutation, 96–101
    prime number, 85–90
    random permutation, 141–145
Genetic algorithms, 403
Global randomness, 117–118
Goto statements, 30–32
Graph data structures, 319–339
    adjacency lists, 332–336
    adjacency matrices, 323–332
    adjacent edge iterator class hierarchy and, 338–339
    packed graphs, 336–338
Graph minimization algorithms
    Floyd's algorithm, 386–389
    intractable problems, 402–403
    minimum spanning trees, 379–381
    Prim's algorithm and Dijstrka's algorithm, 385–386
    priority-first traversal, 373–379
    single source shortest path (SSSP) algorithm,

381–383
state-space searching, 389–402
  with the A^* algorithm, 392–402
  heuristic, 392
Graphs
  connectivity of, 351–354
  directed acyclic (dags), 321–322, 369
  packed, 336–338
  reduced, 371
  terminology used with, 320–322
Graph traversals
  breadth-first, 365–371
  depth-first, 341–365
  priority-first, 373–379
Greatest common divisor (GCD), 109
Greedy algorithms, 45
Grep program, 297
Gstar() function, 394

H

Hard-wired finite automata, 408–410
Hash-based transition matrices, 416–419
Hash() function of *SepHashTable* class,
    157–159
Hash functions, 150–155
  for character strings, 153–155
  for numerical keys, 151–153
  perfect, 151
  uniform hash functions, 150
Hashing (hashing algorithms), 79, 149–181
  comparison of techniques, 180–181
  division method of, 152
  extendible, 179–180
  file-based, 178–179
  rehashing, 179–180
Hash tables, 79, 149, 150
Heap class, 186–188
HeapDistribute() function, 259–260
Heap property, 185, 186
Heaps, 183–211
  as complete binary trees, 185–186
  delayed construction of elements in,
      208–209
  double-ended, *see* Double-ended heaps
      (deaps)
  indirect, 209–211
  operations used on, 188–197
    building a heap bottom up, 194–195
    deleting items, 193–194
    extracting an item, 190–191
    inserting an item (bubble up process),
        189–190

replacing items, 192–193
sorting data, 195–197
Heap sort algorithms, 183, 196–197, 220–222,
    233–234
HeapSort() function of *Heap* class, 220–221
Heuristic searching, 392
Higher() function of *Heap* class, 187
Hstar() function, 394, 397
HStm class, 416–419

I

Image dissolving, 136, 145–147
In-degree of a vertex, 324, 369
Indirect heaps, 209–211
Indirect recursion, 52–53
Indirect sorting, 234–238
Induction, 34
IndxToSmallestKey() function, 135
Inheritance, 78–82
  by breaking up algorithms laterally, 80–81
  by breaking up algorithms serially, 81–82
InPlaceMerge() function, 244–245
Insert() function, 139
  of *CoalHashTable* class, 177–178
  of *Deap* class, 200–201
  of *Heap* class, 189
  of *iHeap* class, 210
  of *OpenHashTable* class, 172–173
  of *SepHashTable* class, 158–159
Insertions
  in deaps, 200–203
  in heaps, 189–190
Insertion sort algorithm, 10, 19–21, 215–216,
    233–234
InsertionSort() function, 229
Instruction generators, 70–77
Integer exponentation, 34–36
Internal sorting, 254
Intractable problems, 17
I/O buffers, controlling the size of, 261
IS-A relationship, 79
Iteration, 25–26
  converting recursion into, 38–50
    exact-sum knapsack problem, 43–49
    with stack-based iteration, 39–43
    with tail recursion optimization, 38–39
    when not to leave things recursive,
        51–52
    when to leave things recursive, 50
  queue-based, 52–53
  stack-based, 39–43
Iterators, generators and, 77

K

Key() function of, 157
KnapSack() function, 55–56
Knapsack generators, 57–63
Knapsack problem, exact-sum version of, 43–49
Knuth-Morris-Pratt algorithm (KMP), 296

L

Labeled graphs, 320
LAstar class, 393–397
LAstar() constructor, 394–395
Lateral breakup of algorithms, 80–81
Lazy evaluation, 136
LBFV class, 367–368
LBicon class, 357–360
LDFI class, 347–354
LDFV class, 344–347
Least-frequent character optimization, 295–296
Level ordering, 185
Lexicographic ordering, 96–100
LGraph class, 332–336
Linear congruential generators (LCGs), 105–125
    combined random number generators, 120–123
    full-cycle, 106–108
    full-cycle generator class for, 108–112
    local vs. global randomness and, 117–118
    Minimal Standard Generator, 114–118
    overflow problem and, 114
    random bit sequences and, 123–125
    random number sequences and, 112–114
    spectral test and, 119
Linear congruential sequences, 105
Linear probing, 161–163
Linked lists, merge sort algorithm for, 246–248
List-based graphs, priority-first visitor class for, 375–379
List-based transition matrices, 415–416
Local randomness, 117–118
Logical pattern, 306
Loop unrolling, 285
Lower() function of *Heap* class, 187
LPFI class, 390–391
LPFV class, 375–379
LRouteFinder class, 397–399
LStm class, 415–416
LStrongcon class, 361

M

Machine alphabet, 406
Maps, 209
    indirect sorting and, 234–238
MapSort() function, 237
MatchFSA class, 412–413
Match loop, 276
    in Boyer-Moore (BM) algorithm, 285–289
Match() of *TYPE* parameter, 157
Max-heaps, 185, 186
Median-of-three partitioning, 226–227
Memory cells, space complexity and, 8
Merge() function, 243, 247
MergeSort() function, 246, 251
Merge sorting, 241–260, 271
    balanced 2-way, 248–251
    external, 254–272
        bubble sorting, 267–271
        comparison of algorithms, 271–272
        quick sorting, 263–267, 271
        random-access sorting, 262
        replacement selection and, 258–260
    linked lists, 246–248
    multiway, 260
    natural merge, 251–254
    polyphase, 260
    in pseudo-code, 241–242
    replacement selection and, 258–260
    variable-length records, 261–262
Min-heaps, 185, 186
Minimal Standard Generator, 114–118
Minimum spanning trees, 379–381
Mixed congruential generators, 107
Move() function, 71–74, 76
Multifunction algorithms, 65
Multigraphs, 321, 407
Multiway merging, 260
MyFSA class, 408–410

N

Natural merge sorting, 251–254
Nested loops, complexity and, 9–11
NewPeriod() function, 110
NewVix() function, 325
Next() function
    of combined random number generator, 121
    of *fRan31* class, 118
    of Minimal Standard Generator, 115–117
NextRandomNumber() function, 115
NextSeq() function, 110–111

NMFSort() function, 254–259, 260
  sorting variable-length records and,
    261–262
NMLSort() function, 252–254
Non-deterministic finite automaton (NFA),
    419–422
NP-complete problems, 17
Number generators, 65–67

O

On-line algorithms, 136
Open addressing strategy of collision resolu-
    tion, 155, 161–174, 180
  deletions with, 173–174
  double hash probing and, 164–166
  linear probing and, 161–163
  quadratic probing and, 163–164
  random probing and, 166–167
  uniform probing and, 164
OpenHashTable class, 167–174
  buckets array in, 168–169
  Insert() and *Search()* functions of, 172–
    173
  inuse array in, 168
  probe sequences in, 169–171
Open vertices, 346
Operator()() function, 326–327, 329–330,
    334, 335, 365
  of *AStm* class, 412
  of *HStm* class, 417–418
  of *LStm* class, 415–416
Optimization, 22
Ordered-state space searching, 391
Out-degree of a vertex, 324
Out-degrees, 336
Overflow, linear congruential generators (LCGs)
    and, 114

P

Packed graphs, 336–338
Parameter lists, reducing, 37–38
Parent() function of *LPFV* class, 382
Partition() function, 223–224
Partitioning, quick sort algorithm and, 222–226
Pascal's Triangle, 102
Pass-by-reference, 37
Perfect hash functions, 151
Period, 106
Permutation generators, 96–101
Permutations. *See also* Random permutations
Permutator template class, 98–100

Permute() function, 97–98
PGraph class, 337–338
Pile class, 138–140
Piles, 138–141
Pivot, quick sort algorithm and, 222, 225–227
Placement new operator, 173
Polyphase merging, 260
PostProcess() function, 346
Power() function, 35–36
  tail recursion optimization and, 38–39
Preprocess() function, 346
  of Boyer-Moore (BM) algorithm, 289–291
Primary clustering, 163
PrimDijk() function, 385–386
Prime number generator, 85–90
Prime numbers, testing for
  with *Primes* generator, 90
  with *UniqueFacators* generator, 94–95
Primitive polynomials modulo 2, 123
Prim's algorithm, 385–386
PrintPath() function, 388
Priority-first spanning tree, 379
Priority-first traversal, 373–379
Priority queues, 183–184
  contiguous, 138–141
  double-ended, 197
Probe sequences, 180
  double hash, 164–166
  linear, 161–163
  in *OpenHashTable* class, 169–171
  quadratic, 163–164
  random, 166–167
  uniform, 164
Procedures, inheriting from, 80
ProcessClass() function, 309, 310
Pseudo code, 4–5
Pseudo-random number sequences, 112
Pure multiplicative congruential generators,
    107, 113, 114
Pushdown automata, 420

Q

Qsort() function, 230
Quadratic probing, 163–164
Queue-based iteration, 52–53
Queues, priority, *see* Priority queues
Quick-search algorithm, 293–295
QuickSearch() function, 294–295
Quick sort algorithm, 222–232, 233–234
  built-in *qsort()* routine and, 230
  external, 263–267, 271
  final optimized version of, 231–232

optimizing, 227–229
partitioning and, 222–226
sort merging and, 242
QuickSort() function, 230–231

R

Radix sorting, 239–240
Ran32 class, 115–116
Rand() function, 125
Random-access sorting, external, 262
Random bit sequences, 123–125
Random combinations, 127–136
    reservoir sampling technique for obtaining,
        132–136
    selection sampling technique for obtaining,
        129–132
    shuffle sampling technique for obtaining,
        127–128
Random draw without replacement, 116
Random number generators, *see also* Linear
    congruential generators (LCGs)
    combined, 120–123
    spectral test and, 119
Random numbers, *see also* Random combinations
    non-uniform deviates, 119
    scaled, 116–117
    uniform deviates, 118–119
Random number sequences, 112–114
Random permutation generators, 141–145
Random permutations, 128, 136–145
RandomPermutator class, 139, 141–145
RandomPermutor class, 167
Random priority queues, contiguous, 138–141
Random probing, 166–167
Random sampling, 127
    hashing and, 151
Realloc() function, 345
Recurrence relation, 51
Recursion, 32–53
    complexity and, 11
    converting into iteration, 38–50
        exact-sum knapsack problem, 43–49
        with stack-based iteration, 39–43
        with tail recursion optimization, 38–39
        when not to leave things recursive,
          51–52
        when to leave things recursive, 50
    direct, 52
    divide-and-conquer algorithms and, 33–36
    Fibonacci numbers and, 51–52
    indirect, 52–53
    tail recursive optimization, 38–39
    well-defined, 33–34

Reduced graphs, 371
Regular expression searching, 317–318
Rehashing, 179–180
Replace() function
    of *Deap* class, 207–208
    in *Heap* class, 193–194
Replacement selection, 258–260
Replacing items
    in deaps, 207
    in heaps, 192–193
ReservoirSampler class, 132–136
Reservoir sampling, 132–136
Reset() function, 76, 103, 104, 115–116
    of combined random number generator, 122
    of *Factors* generator, 91–92
    of *LDFV* class, 345
    of *Primes* generator, 87
    of *Random Permutor* class, 143
    of *ReservoirSampler* class, 135
RmvEdge() function, 326, 327
Roadway problem, solving the, 397–402
Run() 67
RunAll() function, 362
    of *LBicon* class, 358–360
Run() function, 64, 70, 82
    of *LAstar* class, 396–397
    of *LBicon* class, 358–360
    of *LStrongcon* class, 362
    in *Primes* generator, 90
    of *TransitiveCloser* class, 364

S

Scaled random numbers, 116–117
Scan loop
    Boyer-Moore (BM) algorithm and, 283–285
    straightforward search (SFS) algorithm and,
        276–277
Search() function
    of *ACS* class, 428–430
    of *ACSDFA* class, 431
    of *CCS* class, 311–312
    of *CoalHashTable* class, 177, 178
    of *OpenHashTable* class, 172–173
    of *TYPE* parameter, 159
    of *WCS* class, 313–317
Searching
    best-cost, 393, 400
    blind, 392
    complexity of, 149–150
    heuristic, 392
    state-space, *see* State-space searching
    strings of text, *see* Text searching
Search state vector, 300–301

Secondaring clustering, 163
SeinSort(), 54
SelectionSampler class, 129–132, 136
Selection sampling, 129–132
Selection sort algorithm, 2–6, 213–215, 233–234
   in C++, 5–6
   in pseudo code, 4–5
SelnSort() function, 69
Sentinels, 276–280
Separate chaining strategy of collision resolution, 155–160
SepHashTable class, 156–160
Sequential algorithms, 136
Serially breaking up algorithms, 81–82
SetClassFlags() function, 309–310
SetCV() function, 309
SetDirEdge() function, 333–334
SetEdge() function, 326, 327, 333–334
   of *AStm* class, 411–412
SetFstar() function, 377–378
SetMom() function, 379
SetPattern() function, 308–309
Setvbuf() function, 261
SFS class, 280–282
SFSearch() function, 274
SFSearch2() function, 276–277
SFSEarch4() function, 279–280
Shell sort algorithm, 10, 216–220, 233–234
Shift-AND algorithm, 303
Shift-OR algorithm, 299–318
   character class matching and, 305–312
   exact matching and, 300–303
   optimizing, 304–305
   search state vector and, 300–301
   wild card matching and, 312–317
Shortest path ancestor matrix, 387
Shuffle arrays, 122
ShuffleSampler class, 136
ShuffleSampler() function, 128
Shuffle sampling, 127–128
Sieve of Eratosthenes, 85–87
Simulated annealing, 403
Single source shortest path (SSSP) algorithm, 381–383
Snode class, 251
Sort() function, 80
   in heap sort algorithm, 197
Sorting
   heaps used for, 195–196
   indirect, 234–238
   radix, 239–240
Sorting algorithms, 8–9, 213–272
   comparison of, 232–234

heap sort, 220–222
insertion sort algorithm, 10, 19–21, 215–216, 233–234
merge, *see* Merge sorting
with *Permutator* generator, 100–101
quick sort algorithms, 222–232, 233–234
   built-in *qsort()* routine and, 230
   external, 263–267, 271
   final optimized version of, 231–232
   optimizing, 227–229
   partitioning and, 222–226
   sort merging and, 242
selection sort algorithm, 2–6, 213–215, 233–234
   in C++, 5–6
   in pseudo code, 4–5
shell sort, 216–220
specialty, 238–240
speedup techniques for, 260–261
topological sorting, 369–371
SOSearch1() function, 302–303
SOSearch2() function, 304–305
Space complexity of algorithms, 7–8
Spectral test, 119
Splay trees, 183, 262
Split() function, 246–247
Stacks, recursive functions and, 36–37
State-space searching, 389–402
   with the A* algorithm, 392–402
   heuristic, 392
State transition diagrams, 405–407
State transition matrices, 410–419
   hash-based, 416–419
   list-based, 415–416
State vector, 300
StepAll() function, 349
   of *LDFI* class, 352–353
Step() function, 57, 65, 67, 69, 80–81, 82, 103–104
   of *Factors* generator, 91, 92
   of *LAstar* class, 395–396
   of *LDFI* class, 349, 350
   in *Primes* generator, 87–88
   of *Random Permutor* class, 143
Straightforward search (SFS) algorithm, 274–282
   creating a search class for, 280–282
   optimizing, 275–277
   sentinels and, 276–280
StrHash() function, 153–154
Strongly connected components, 354, 361–363
Switch statements, exiting loops while in, 30
Symbol table, 323

T

Templates, 53–54
Temporary files, 261
Text searching, 273–318
    with Boyer-Moore (BM) algorithm, *see*
        Boyer-Moore (BM) algorithm
    in files, 296–297
    regular expression, 317–318
    Shift-OR, 299–318
        character class matching and, 305–312
        exact matching and, 300–303
        optimizing, 304–305
        search state vector and, 300–301
        wild card matching and, 312–317
    size of the alphabet and, 273
    with straightforward search (SFS) algorithm,
        274–282
        creating a search class for, 280–282
        optimizing, 275–277
        sentinels and, 276–280
Time complexities, 7–11
    finding, 9–11
Time-doubling rule, 22–23
Topological sorting, 369–371
TopSort() function, 370–371
Towers of Hanoi algorithm, 70–77
TowerSolver generator, 74–76
TransitiveCloser class, 363–365
Transitive closure, 363–365
TransitiveClosure() function, 365
Traveling salesman problem, 402–403
Traversals of graphs, *see* Breadth-first traversal;
        Depth-first traversal
TraverseAll() function, 349, 351–352
    of *LPFV* class, 380–381
Traverse() function, 345–346
    of *APFV* class, 383–384
    of *LBFV* class, 368
    of *LPFV* class, 376–378
    of *TransitiveCloser* class, 364
Triangular adjacency matrices, 328–332

TrickleDown() function of *Heap* class, 190–192
TrickleDownMax() function of *Deap* class,
    205–206
TrickleDownMin() function of *Deap* class,
    204–205
Trickle-down process in heaps, 190–192
Tuned-Boyer-Moore algorithm (TBM), 292

U

UagPtr class, 330–331
UagPtr constructor, 331
UAGraph class, 328–332
Uniform deviates, 118–119
Uniform hash functions, 150
Uniform probing, 164
UniqueFactors class, 93–94
Unrolling loop, 285
Update() function of *SepHashTable* class, 160
Upper bounds, algorithm analysis and, 11–15
USES-A relationship, 79

V

ValidVtx() function, 326
Variable-length records, sorting, 261–262
VendorMachine class, 420–422
Vertices, 319
    open, 346
    as unsigned integers, 328
Visitor class, depth-first, 343–347

W

Warshall's algorithm, 365
Weakly connected digraph, 354
Weighted graphs, 320
While loops, 25
Wild cards, 299
    Shift-OR algorithm and, 312–317